AF606483

126627

Principles of Experimental Design and Analysis

INTERACTIVE MICROCOMPUTER SOFTWARE

The *Interactive Microcomputer Software for the Statistical Analysis of Experiments* is intended to significantly enhance the learning environment in a course using the textbook *Principles of Experimental Design and Analysis* by A. Garcia-Diaz and D. T. Phillips. The software, coded in Turbo BASIC, executes on PCs and includes several attractive data input capabilities. Input errors are checked and guidelines are given on how to provide experimental data to conduct any of twelve possible analyses. In addition to the files containing the microcomputer programs, the code includes one sample data file for each procedure. Microcomputer code requirements are an MS–DOS operating system and at least 4 Mb of RAM. A color monitor will enhance code usage.

A brief description of code capabilities

The microcomputer program executes the following experimental data analysis procedures:

1. *One-way ANOVA*: analysis of variance tests for fixed-effect and random-effect one-factor experiments.
2. *After-ANOVA procedures*: orthogonal contrasts, Scheffe's method, and range test.
3. *Randomized block design*: ANOVA tests for blocks and treatments in complete randomized block designs.
4. *Balanced incomplete block design*: ANOVA procedures to test the significance of treatment and block effects when a balanced incomplete block design is used.
5. *Latin square design*: analysis of variance tests for an unreplicated Latin square design.
6. *Replicated Latin square design*: analysis of variance tests in a Latin square design which can be replicated in either of the following three ways: (a) holding both blocks (rows) and positions (columns) unchanged; (b) replicating either blocks (rows) or positions (columns); (c) replicating both blocks (rows) and positions (columns).
7. *Graeco-Latin square design*: analysis of variance tests for an unreplicated Graeco-Latin square design.
8. *Youden square-design*: analysis of variance tests for a Youden square design.
9. *General factorial design*: procedures to perform a significance test on main effects and interaction effects in a factorial design with identical number of observations for each experimental condition. In the case of two-level designs, the program also finds confidence intervals for the effects.
10. *Fractional factorial design*: procedures to determine the significance of effects, effect estimates, and alias structures in the case of a two-level fractional factorial design. If desired, the program allows the use of an additional fraction (the 'mirror image' design) to clear main effects.
11. *Probability plots*: procedure for the analysis of effects in the case of an unreplicated two-level factorial experiment. The procedure is based on normal probability plots for effect estimates and residuals.
12. *Response surface methodology*: procedure to analyze first-order and second-order effects in two-level factorial experiments (full or fractional) according to the methods of steepest ascent or descent and canonical analysis.

Availability of the code

If you are interested in a copy of the code, please write to the authors at the following address:

Department of Industrial Engineering,
Zachry Engineering Center,
Texas A&M University,
College Station,
Texas 77843.

Principles of Experimental Design and Analysis

Alberto Garcia-Diaz

and

Don T. Phillips

Department of Industrial Engineering
College of Engineering
Texas A&M University
Texas, USA

CHAPMAN & HALL

London · Glasgow · Weinheim · New York · Tokyo · Melbourne · Madras

Published by Chapman & Hall, 2–6 Boundary Row, London SE1 8HN, UK

Chapman & Hall, 2–6 Boundary Row, London SE1 8HN, UK

Blackie Academic & Professional, Wester Cleddens Road, Bishopbriggs, Glasgow G64 2NZ, UK

Chapman & Hall GmbH, Pappelallee 3, 69469 Weinheim, Germany

Chapman & Hall USA, One Penn Plaza, 41st Floor, New York NY 10119, USA

Chapman & Hall Japan, ITP-Japan, Kyowa Building, 3F, 2-2-1 Hirakawacho, Chiyoda-ku, Tokyo 102, Japan

Chapman & Hall Australia, Thomas Nelson Australia, 102 Dodds Street, South Melbourne, Victoria 3205, Australia

Chapman & Hall India, R. Seshadri, 32 Second Main Road, CIT East, Madras 600 035, India

First edition 1995

Typeset in 10/12 pt Times by Thomson Press (India) Ltd, New Delhi

Printed in Great Britain by T.J. Press, Padstow, Cornwall

ISBN 0 412 60570 8

A catalogue record for this book is available from the British Library

Library of Congress Catalog Card Number: 94-72010

∞ Printed on permanent acid-free text paper, manufactured in accordance with ANSI/NISO Z39.48-1992 and ANSI/NISO Z39.48-1984 (Permanence of Paper).

This book is dedicated to Gabriel Alberto Garcia, another small statistic when he arrived on September 3, 1991.

I am very sorry, Pyrophilus, that to the many (elsewhere enumerated) difficulties which you may meet with, and must therefore surmount, in the serious and effectual prosecution of experimental philosophy I add one discouragement more, which will perhaps as much surprise as dishearten you; and it is, that besides that you will find (as we elsewhere mention) many of the experiments published by authors, or related to you by the persons you converse with, false and unsuccessful (besides this, I say) you will meet with several observations and experiments which, though communicated for true by candid authors or undistrusted eyewitnesses, or perhaps recommended by your own experience may, upon further trial, disappoint your expectation, either not at all succeeding constantly or at least varying much from what you expected.

Robert Boyle, 1673,
Concerning the Unsuccessfulness of Experiments.

Contents

Preface

This textbook is the result of 15 years of teaching an introductory graduate course on the statistical design and analysis of industrial experiments in the Department of Industrial Engineering at Texas A&M University. Our major motivation in writing this text has been the desire to present fundamental concepts, theory and procedures used in the analysis of experimental data in a clear and concise fashion, without allowing the mathematical developments to become unnecessarily burdensome. At the same time, we have tried to present important concepts without sacrificing the need to rigidly justify the validity and generality of known procedures. This is not just another book on design of experiments. It is an introductory text written for engineering students which includes a well-balanced treatment of theory and applications.

The material included in this book assumes that the student is familiar with elementary statistics, calculus and introductory linear algebra. The entire book can be covered in a single semester at the rate of two or three lectures per week. Although the emphasis is placed on engineering applications, this textbook could also be used in other academic departments, such as the social sciences, statistics, agriculture and business administration.

In an effort to support the generality of the procedures developed in the text, we have included the following meaningful case studies: wear test analyses of advanced technology drill bits; a study of factors affecting weld strength of rail steel joints; experimental design for typical solar water-heating systems; investigation of the factors affecting the population growth of pine bark beetle; a study of air conditioner noise levels; factorial analysis of a textile operation; an application of response surface methodology to pavement design; and an application of Taguchi methods in the manufacture of plastic valves.

The content of this book can be conceptually divided into three major areas:

(a) basic concepts of statistical design and analysis;
(b) factor-screening procedures to determine which factors significantly affect the response of an experiment;
(c) model formulation methodologies to investigate how model factors influence the experimental response.

The textbook consists of 16 chapters. Chapters 1 through 6 contain the analysis of variance (ANOVA) concepts which serve as a foundation for all further developments. Chapters 7 through 12 include several methodologies to estimate the effects of main factors and interactions among those factors; procedures are also given to test the significance of effects. Chapter 13 considers the development of mathematical relationships that link the factors to the response, and the use of these relationships to optimize an experimental response. Chapter 14 provides an introduction to the Taguchi methodologies as related to modern experimental design. Chapter 15 presents a comprehensive analysis of nested-factor and split-plot experimental designs. Finally, Chapter 16 reviews the fundamentals of fractional factorial selection procedures and summarizes a few methods for computer-aided experimental design. Some special topics such as the mathematical development of Scheffe's method and the three-level experimental design methodology can be omitted. The review of statistical concepts given in Chapter 2 can be omitted or only partially covered if the students already possess a good working knowledge of applied statistics.

One of the attractive features of this textbook is a microcomputer program that executes most of the procedures covered in the book. This microcomputer program can be used as a teaching aid, but it is especially suitable for self-teaching and requires no computer programming skills. In order to provide the users with an effective microcomputer package, we focused on the areas of documentation, language, data entry, data editing, data manipulation, procedures, output and decision support.

We would like to express our gratitude to the many students at Texas A&M University for their contributions and comments. Over the past five years, we have gradually modified the content of the manuscript and its organization, as a result of their suggestions. In particular, we want to thank Mr Dipak Patel, Mr Tzong-Ming Chang, Dr Kai-I Huang, Dr Rafael Rivera and Ms Lourdes Rodriguez Gonzalez for programming earlier versions of the microcomputer procedures. Through the years these procedures have been enhanced by a selected group of our graduate students. To all of them we remain very grateful. The support received from faculty and friends in the Department of Industrial Engineering at Texas A&M is most appreciated. We want to especially recognize Dr Robert P. Beals for being a good friend and a source of inspiration.

Finally, we want to express our sincere appreciation to Dr Richard E. Devor and Dr Vernon McBride for introducing us to the field of statistical analysis of experiments when they were our instructors at the University of Illinois and the University of Arkansas, respectively. Their excellent disposition and knowledge have left an indelible mark on our commitment to teach applied probability and statistics.

Alberto Garcia-Diaz
Don T. Phillips
College Station, summer 1994

The statistical approach

1

Statistical principles have been applied in agricultural and biological sciences for more than 65 years. Three of the most important reasons for the acceptance and implementation of these principles and related procedures in these fields are:

(a) the heterogeneity of the experimental material;
(b) the presence of a number of data variability sources;
(c) the difficulty of reproducing the experimental conditions.

Although these three issues also exist in engineering and physical sciences, in these fields the experimental material is usually more homogeneous, the measurements more precise and the experimental conditions more reproducible. As a result, in many cases researchers thought that a nonstatistical (i.e. mechanistic) approach would be more appropriate. However, during the last 45 years, the need for a statistical approach in engineering and physical sciences has been widely recognized. In particular, during the 1970s and 1980s the growth of statistical analysis in engineering applications has been steady and strong.

Roughly speaking, statistics can be defined as the science as well as the art of the collection, analysis and interpretation of data. It is a science since the results are only valid when some basic assumptions are satisfied by the mathematical analyses involved in the process. It is also an art because frequently it is very difficult or impossible to satisfy some mathematical assumptions in the context of a number of applications. In actual experimentation practitioners, therefore, need to develop strategies that would guarantee that the statistical tools used best fit the actual conditions of specific experiments.

The fundamental purpose of this book is to study in detail the development and application of statistical models and procedures to draw inferences about the effect of a set of factors on a specified experimental response. This study is important in the context of the following two essential and complementary situations:

(a) how to screen the factors affecting the yield of a given response on the basis of a sample taken at random (perhaps under some restrictions on randomization);

(b) how to collect data (when, what type, how many observations) in order to implement a proposed factor-screening methodology.

In both cases the validity of the approach taken depends on the specific assumptions of the corresponding statistical analysis. Applied statistical methods have been and are widely used in a diversified range of experimental areas, such as agriculture, chemical industry, manufacturing, human factors and engineering design.

A large number of engineers and other types of technical professionals graduate without an adequate background in applied statistics. As a result, when they become involved in the analysis of a problem, the tendency exists always to solve the problem using a nonstatistical approach. In some other cases, statistical analyses are conducted without verifying if the assumptions for the valid use of the techniques are satisfied. This tendency is unfortunately enhanced by the wide availability of computerized packages that perform a collection of statistical techniques, since users without an adequate statistical background may not verify if the available data were collected under only those specific restrictions on randomization allowed by the procedure to be used.

Usually, the nonstatistical approach yields causal functional relationships that explain how a particular response varies as the intensity of the factors involved changes. Such relationships, however, can be developed only when the factors being studied are relatively few and/or when their interactions with one another are not very important. In general, when selected factors are studied on the basis of their effect on a particular experimental response, the only sound and logical alternative is a statistical approach for comparing their systematic effects to the effect of a random source of variation, assumed to be negligible.

In order to present a few fundamental definitions, we will consider a situation where it is possible to measure an 'experimental value' to be compared to a 'true value' which is assumed to be unknown. In this context, **accuracy** is defined as the expected difference between these two values. The ability of the experimental method to yield similar results under identical replications is known as the **reproducibility** of the method. Furthermore, any deviation from the true value due to causes inherent in the experimental method is referred to as the **error**. Errors can be of two types, either random or systematic. A random error is individually unpredictable but in the long run it tends to be equal to zero. We say it can be 'averaged out' by randomizing the data collection process. On the other hand, a systematic error persists after a series of similar measurements. It cannot be eliminated by any randomization procedure.

In a broad way, the purpose of statistical experimental analysis is to investigate the significance of systematic effects. A statistical project consists of five fundamental steps:

(a) problem definition
(b) system identification
(c) statistical model formulation
(d) data collection
(e) statistical analysis and results.

In some traditional textbooks on experimental design and analysis, it is customary to group the above tasks according to three categories known as the experiment, the design and the anaysis [1]. Usually, the problem definition and system identification tasks are referred to as the **experiment**. The phase of formulating a statistical model and determining some fundamental aspects of data collection, such as the sample size, the order of experimentation and the type of data to be collected, is known as the **design**. Additionally, the actual data collection and processing, along with the computation of statistics to perform significance tests and interpreting the results, are usually referred to as the **analysis**. In the remaining sections of this chapter a brief description is given for each fundamental task of any statistical project.

1.1 PROBLEM DEFINITION

The basic purpose of this phase is to specify the scope of the problem to be investigated. This can be done by stating the objective of the study and the major assumptions made concerning the particular conditions that define the environment in which a set of experiments will be conducted to perform the corresponding data analysis. The importance of having a complete definition of a problem cannot be overemphasized; indeed, often a correct solution cannot be found for a problem as a result of an incomplete definition.

1.2 SYSTEM IDENTIFICATION

A system can be defined as a collection of elements acting and interacting in order to perform a common function. The specific purpose of this phase is to identify the variables that are believed to be significant in the study. The variables can be grouped into two major classifications: (1) independent variables or factors, (2) dependent variables or responses.

Once the basic factors affecting a specified experimental response are identified, a set of possible values or levels is identified for each factor. The experimental conditions of the system can now be obtained by identifying all possible combinations of factor levels. These levels may be quantitative or qualitative. For example, if the response of an experiment is the yield of a chemical process, a suitable factor is generally the temperature; three possible quantitative levels for this factor could be 10, 40 and 100 °F (−12, 4 and

38 °C). On the other hand, if the response is horsepower requirements in a metal cutting operation, a meaningful qualitative factor could be type of tool. Two possible qualitative levels for this factor in this case would be ceramic and glass cutting tools.

1.3 STATISTICAL MODEL FORMULATION

A model is an ideal representation of a system. In the case of statistical experiments, a model is a mathematical relationship that describes the value of a random observation in terms of a collection of components with assignable causes and a random error (which by definition does not have an assignable cause). The statistical model should be capable of describing response values for random observations taken from the populations corresponding to the experimental conditions of the system. Those effects having assignable causes in the statistical model are generally referred to as systematic effects, since they persist after a series of similar measurements.

In the case of a tool life study for a particular manufacturing process, the response of a single experiment may be defined as tool life in minutes. This response is affected by a variety of factors such as spindle speed (revolutions per minute), feed rate (inches per minute), cut depth (inches), differences between machines, type of materials, type of tool used and operator skills. In order to provide more insight into the situation under consideration, Figure 1.1 gives a simplified illustration of a cutting operation in a lathe. This figure shows as major factors spindle speed, feed rate and cut depth.

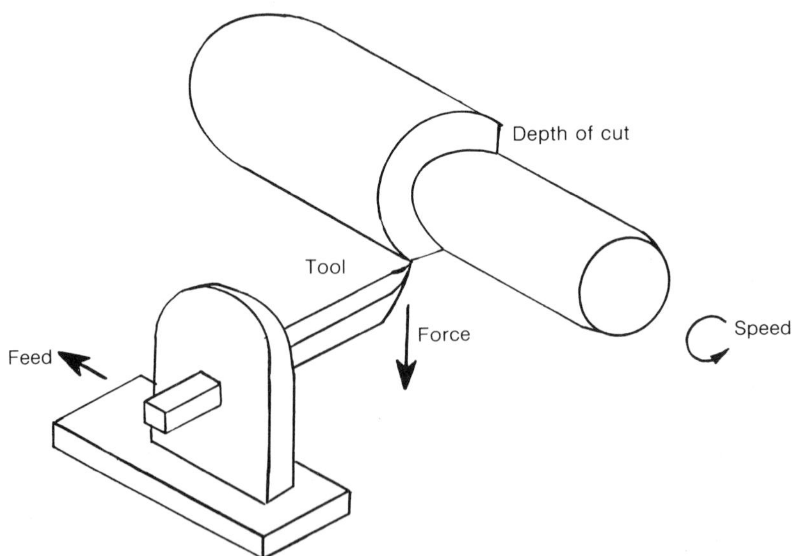

Figure 1.1 Cutting operation in a lathe.

In an effort to simplify the statistical model only major variables should be included. Assuming that these are only the ones shown in Figure 1.1, a suitable mathematical (statistical) model to represent the life of the mth tool tested under an experimental condition defined by selecting the ith level for the feed rate (F_i), the jth level for cut depth (C_j) and the kth level for the spindle speed (S_k) will be

$$Y_{ijkm} = \mu + F_i + C_j + S_k + FC_{ij} + FS_{ik} + CS_{jk} + FCS_{ijk} + \varepsilon_{ijkm} \qquad (1.1)$$

where Y_{ijkm} is the observed tool life, μ is a constant and ε_{ijkm} is the corresponding random error. This error averages out all the effects due to all extraneous factors not included in the model, such as type of lathe, type of tool and operator skills. The terms FC_{ij}, FS_{ik}, CS_{jk} and FCS_{ijk} represent interaction effects that cannot be assigned to individual factors. The model formulated in equation (1.1) and models similar to this one have been extensively studied by a number of manufacturing analysts [2–4].

1.4 DATA COLLECTION

Data are collected for each of the experimental conditions of a system in order to assess the significance of each term having an assignable cause in the statistical model. The assignable cause can be either a factor or an interaction among several factors. The data collection phase includes several basic activities, such as the determination of sample size, the order of experimentation and the randomization method.

For proper use of experimental data, the following three basic principles of the design of experiments must be observed [5]:

(a) randomization
(b) replication
(c) local control.

The purpose of randomization is to allow the random errors in a sample of observations corresponding to the same experimental condition to be considered as if they were stochastically independent.

The discussion of randomness belongs to the foundation of statistical methodology and its applicability to empirical sciences. The vast majority of experimenters are convinced that they know what 'random' is. It is clear to them that a toss of a coin is random; so is a mutation, and so is the emission of an alpha particle. On the other hand, the motion of a planet is not random, and neither is the propagation of sound waves.

Although most people associate randomness with unpredictability, this falls short of satisfactorily explaining what is random. In a very interesting short article on randomness, M. Kac [6] tries to answer this question. He concluded that from the purely operational point of view, the concept of randomness is so elusive as to cease to be viable, but finishes his article with a reassuring note: that whatever our view and beliefs on randomness, if the discipline we practice is sufficiently robust, it contains enough checks and balances to keep us from committing errors that might come from the mistaken belief that we really know what random is.

When a random sample is taken for each of the experimental conditions of a system the experiment is said to be replicated. Usually, all the samples have the same size in order to keep the design balanced. (As will be discussed later in this book, balanced experiments have several attractive features.)

In order for the analysis of an experiment to be valid, the data should not include undesirable systematic effects due to extraneous factors not considered in the statistical model. These systematic deviations can be due to either space or time trends that reduce the homogeneity of the experimental material used to run all experimental conditions of the system. These trends may occur when experimental units adjacent in time or space are used since errors from these experimental units tend to be correlated. Randomization causes the correlation between any two treatments to cancel out by giving any two treatments an equal chance of being adjacent or not.

The purpose of local control is to group the experimental conditions into homogeneous blocks, so that the comparison between two conditions in the same group is not influenced by systematic changes in the experimental material.

1.5 STATISTICAL ANALYSIS

The random error term of the statistical model cannot be predicted for a single observation, but it approaches zero in the long run. On the basis of a random sample it is possible to estimate the value of each term of the model, the error term included; the aim of the statistical analysis is to compare the estimate of each systematic effect to that of the random error to establish if the term with assignable cause is or is not significant. This test is generally known as a **significance test**.

Replication makes the statistical significance test possible since it provides an estimate of the random error; randomization makes the statistical significance test valid since it allows the use of significance tables constructed under the assumption that the errors are independent; local control increases the ability of the statistical test to recognize differences between the experimental conditions.

1.6 AN APPLICATION

The following application is taken from a research project conducted by Hough *et al.* [2, 3] concerning the analysis of a special class of bolt drill bits used in coal mining. The bit used (polycrystalline diamond compact) is an experimental roof bolt bit designed by Sandia Labs. The purpose of the study was to improve productivity and reduce the cost of mining coal through advanced technology. Mancos shale was selected as the medium to be drilled because it is fairly representative of coal measure rock. Shale consists of mud, clay and silt for consolidation. Tests were conducted at the Drilling Research Laboratory, Salt Lake City, Utah, at atmospheric conditions, using air as the flushing medium.

After preliminary conditions and discussions of tentative factors and experimental responses, the dependent and independent variables listed below were selected for further analysis.

Independent variables

(a) Back rake angle: angle between the face of the cutter and a vertical plane in degrees.
(b) Rotary speed: revolutions of the bit per minute.
(c) Thrust: push on the bit in pounds.

Dependent variables

(a) Penetration rate: rate of penetration of the drill bit into the medium measured in inches per minute. This value can be measured directly.
(b) Specific energy: horsepower requirements per cubic inch of material removed per minute, measured in pounds per inch. This value can be computed.
(c) Torque: twisting moment in inch-pounds. This value can be measured directly.

Sketches of the drilling machine and the drill bit to be considered in this sample application are given in Figure 1.2. In the past, drills have been made with hydraulic motors with an auxiliary percussive drill bit which is fitted when difficult or hard strata are encountered. The motor is capable of rotary or rotary percussive action. Since the advent of STRATAPAX blanks, however, the rotary percussion drilling is no longer needed in most underground mines. A bit design such as the one shown in Figure 1.2 performs well in the medium-hard, abrasive sandstone formations [7].

Levels of 7 and 25 degrees were assigned to rake angle, 500 and 750 rev min^{-1} to speed and 100 and 200 lb (45 and 90 kg) to thrust. Figure 1.3 shows a geometrical representation of the $2 \times 2 \times 2$ factorial design used in the experiment. A minimum of $2 \times 2 \times 2 = 8$ test runs are required which form a parallelepiped in the three-dimensional operability region for the variables. Other factors (independent variables) like temperature, depth of hole drilled, drilling medium and interlayer variation, drilling operators' experience and competence,

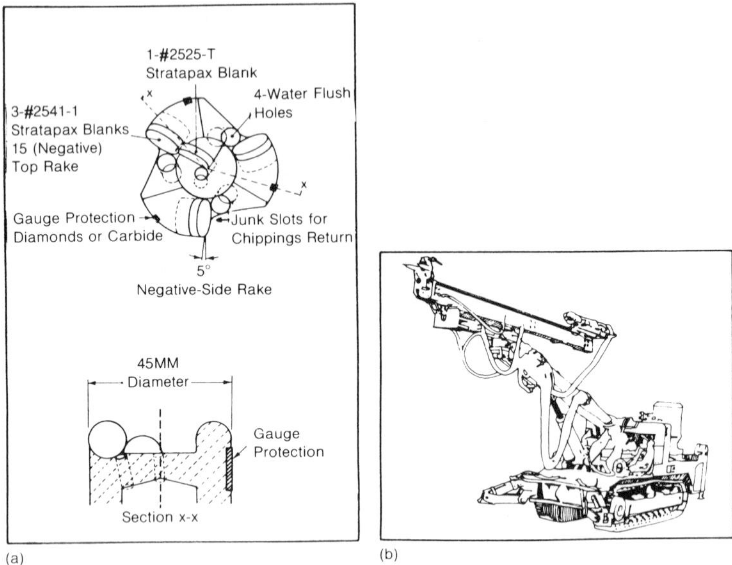

Figure 1.2 (a) Bit design and (b) drilling machine [7].

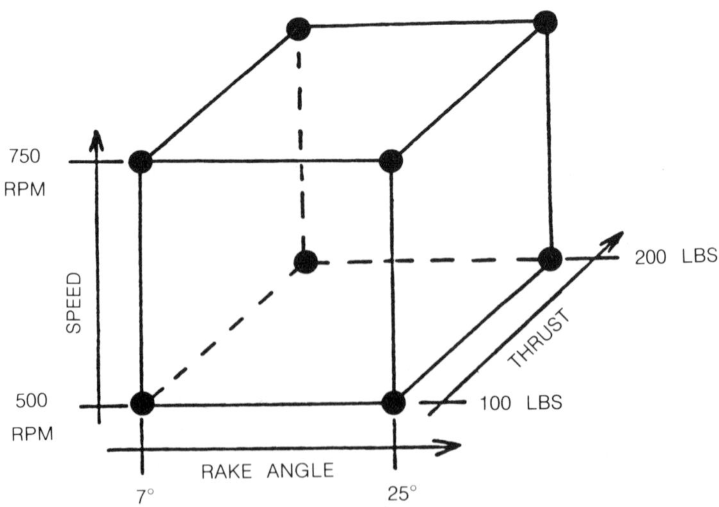

Figure 1.3 Geometrical representation of the sample design.

and bit wear affect performance, but their consideration was deemed beyond the scope of this study.

The mathematical formulation used in describing the random observations collected from each experimental condition is the linear statistical model given in equation (1.2):

$$Y_{ijkm} = \mu + R_i + S_j + L_k + RS_{ij} + RL_{ik} + SL_{jk} + RSL_{ijk} + \varepsilon_{ijkm} \tag{1.2}$$

where Y_{ijkm} is the value of the mth observation corresponding to an experimental condition defined by the combination of ith level of rake angle, jth level of speed and kth level of thrust. In equation (1.2), μ is a parameter common to all test settings called the overall mean, R_i, S_j, L_k are main effects due to rake angle, speed and thrust respectively. R_i, S_j, L_k represent main effects and RS_{ij}, RL_{ik}, SL_{jk}, RSL_{ijk} represent interaction effects. ε_{ijkm} is the random error component in the mth observation taken from the (i, j, k)th experimental condition; here $m = 1, 2$ since each experimental condition was replicated twice.

The data for the eight experimental conditions shown in Figure 1.3 were collected at random. The corresponding 16 observations are summarized in Table 1.1. In Table 1.1, the two levels of each of the factors rake angle, rotary speed and thrust are represented as $-$ and $+$. The lower value of each factor, i.e. 7 degrees in rake angle, 500 rev min^{-1} in speed, and 100 lb (45 kg) in thrust, is assigned to $-$, while the higher value of each factor is assigned to $+$.

Whether an effect is significant or not depends on the amount of change in the response of the experiment as the factor is switched from one level to another. The larger the change, the more significant is the effect. For example, to check whether the main effect due to rake angle is significant, the average response at 7 degrees can be compared to the average response at 25 degrees. The average response at 7 degrees can be calculated by adding the responses

Table 1.1 Experimental results for drill bit analysis

Experimental condition	*Rake angle* (*R*)	*Rotary speed* (*S*)	*Thrust* (*L*)	*Penetration rate* (in min^{-1})		*Torque* (in-lb)		*Specific energy* (lb in^{-2})	
				1st	2nd	1st	2nd	1st	2nd
1	−	−	−	4.00	1.09	13	12	6 896.2	23 200.9
2	+	−	−	4.80	3.43	25	7	11 011.4	4 355.3
3	−	+	−	12.54	5.00	20	3	5 094.0	1 958.1
4	+	+	−	5.62	5.14	20	13	11 284.0	8 038.9
5	−	−	+	6.75	8.18	62	45	19 435.0	11 693.7
6	+	−	+	16.67	22.17	63	110	8 075.2	10 559.9
7	−	+	+	6.25	20.00	25	65	12 741.8	10 377.8
8	+	+	+	26.25	20.77	85	62	10 340.3	9 542.8

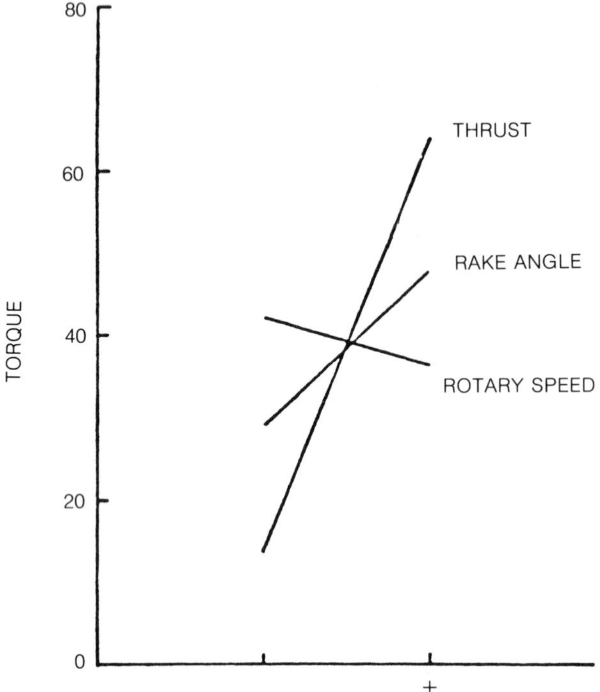

Figure 1.4 Main effects of rake angle, rotary speed, and thrust.

under experimental conditions that correspond to the vertical plane at 7 degrees in Figure 1.3 and then divide by 8; similarly, the average response at 25 degrees can be calculated by adding the responses under experimental conditions that correspond to the vertical plane at 25 degrees and then divide by 8. Therefore, main effects are related to parallel planes. The average value of torque was calculated for each of the three main effects, and the corresponding results are shown in Figure 1.4. The main effect due to thrust seems to be significant since the change in its response is most drastic.

Interaction effects may be similarly analyzed; the interaction effect due to rake angle and thrust, RL_{ik}, is analyzed as an example. Since there are two levels for each of the two factors, there are four possible combinations, i.e. rake angle of 7 degrees with thrust of 100 lb (45 kg), rake angle of 7 degrees with thrust of 200 lb (90 kg), rake angle of 25 degrees with thrust of 100 lb and rake angle of 25 degrees with thrust of 200 lb, corresponding to the four vertical lines in Figure 1.3. The average value of penetration rate for each combination was calculated and corresponding results are shown in Figure 1.5. Figure 1.5 shows that response at thrust of 100 lb decreases as rake angle is changed from 7 to 25 degrees, while at thrust of 200 lb it increases; therefore,

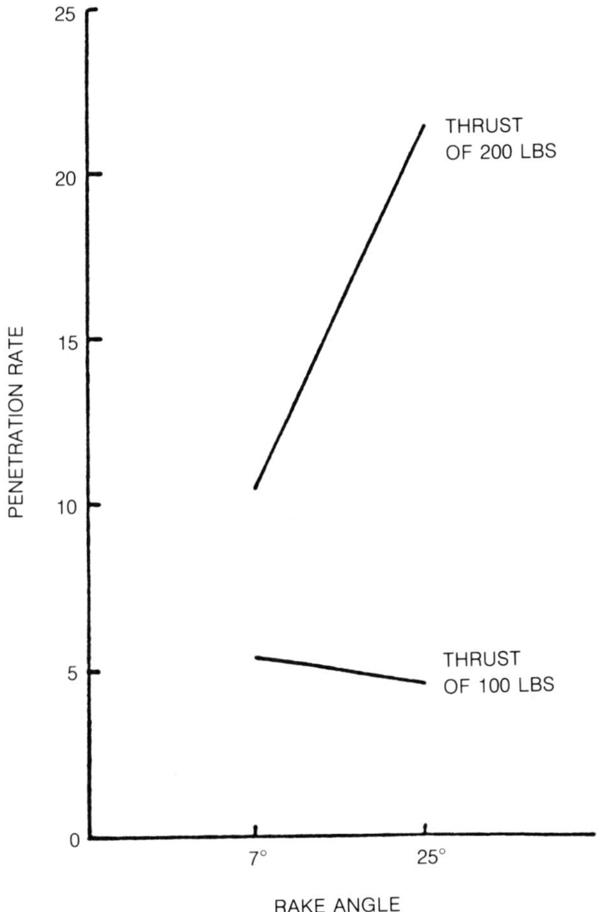

Figure 1.5 Interaction effect due to rake angle and thrust.

the interaction effect of rake angle and thrust is significant. A similar analysis is done for the interaction effect due to rake angle and speed, RS_{ij}, with the corresponding results shown in Figure 1.6. The response increases at both 500 and 750 rev min^{-1} of speed when rake angle is changed from 7 to 25 degrees; therefore, the interaction effect seems to be negligible.

Most of this book is devoted to the statistical technique known as **Analysis of variance** (ANOVA), introduced by Fisher in 1939. Briefly stated, the ANOVA methodology allows us to verify the hypotheses that the effects under consideration are not significant in terms of F-tests. In the example being considered, the ANOVA methodology can be used to test the hypotheses that there is no rake angle effect, no speed effect, no thrust effect, and no

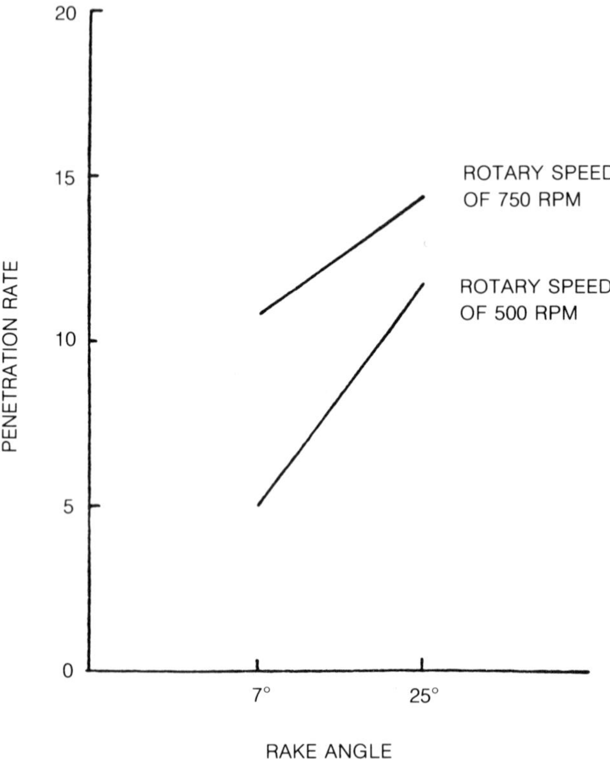

Figure 1.6 Interaction effect due to rake angle and speed.

interaction effects on the response variables penetration rate, specific energy and torque:

$$H_0\colon R_i = 0 \quad \text{for all } i$$
$$H_0\colon S_j = 0 \quad \text{for all } j$$
$$H_0\colon L_k = 0 \quad \text{for all } k$$
$$H_0\colon RS_{ij} = 0 \quad \text{for all } ij$$
$$H_0\colon RL_{ik} = 0 \quad \text{for all } ik$$
$$H_0\colon SL_{jk} = 0 \quad \text{for all } jk$$
$$H_0\colon RSL_{ijk} = 0 \quad \text{for all } ijk$$

A significance level of 5% was chosen as the probability of rejecting a hypothesis when it is actually true. In other words, any conclusion drawn would have a 95% confidence level. The ANOVA results are given in Tables 1.2–1.4 for penetration rate, torque and specific energy, respectively. For each systematic source of variation and the random error, these tables show the appropriate sum of squares (*SS*), their degree of freedom (*df*), mean

Table 1.2 ANOVA table for penetration rate

Source	*df*	*SS*	*MS*	*F*	*Significant*
R	1	105.27	105.27	5.28	
S	1	74.30	74.30	3.73	
$R \times S$	1	10.66	10.66	0.53	
L	1	456.04	456.04	22.89	
$R \times L$	1	145.93	145.93	7.32	*
$S \times L$	1	1.28	1.28	0.06	*
$R \times S \times L$	1	2.87	2.87	0.14	
Error	8	159.41	19.93		
Total	15	955.75			

Table 1.3 ANOVA table for torque

Source	*df*	*SS*	*MS*	*F*	*Significant*
R	1	1 225.00	1 225.00	3.71	
S	1	121.00	121.00	0.37	
$R \times S$	1	2.25	2.25	0.01	
L	1	10 201.00	10 201.00	30.85	*
$R \times L$	1	702.25	702.25	2.12	
$S \times L$	1	110.25	110.25	0.33	
$R \times S \times L$	1	9.00	9.00	0.03	
Error	8	2 645.00	330.63		
Total	15	15 015.75			

Table 1.4 ANOVA table for specific energy

Source	*df*	*SS*	*MS*	*F*	*Significant*
R	1	20 679 074.13	20 679 074.13	0.82	
S	1	41 763 583.13	41 763 583.13	1.66	
$R \times S$	1	82 166 519.93	82 166 519.93	3.26	
L	1	27 373 039.21	27 373 039.21	1.09	
$R \times L$	1	11 006 635.64	11 006 635.64	0.44	
$S \times L$	1	9 498 261.71	9 498 261.71	0.38	
$R \times S \times L$	1	19 678 317.80	19 678 317.80	0.78	
Error	8	201 418 706.70	25 177 338.34		
Total	15	413 584 138.25			

squares (MS) and observed F-values. The procedure is based on comparisons between the F-values computed from the observed results and the critical F-value 5.32 obtained from a significance table. In Tables 1.2–1.4 the symbol * indicates that the observed value of the F-statistic exceeds the critical F-value given by the significance tables and consequently the associated hypothesis cannot be accepted. Note that these results correspond to those obtained from the graphical analysis, i.e. the main effect due to thrust for torque and the interaction effect due to rake angle and thrust for penetration rate are significant. The ANOVA methodology for one-factor and n-factor experiments will be developed in Chapters 3 and 7, respectively. In the latter chapter in particular, the problem being studied here will again be considered to illustrate the ANOVA methodology, and to show how to obtain the numerical results given in Table 1.2.

We would like to close this chapter with the reiteration of the main goal of the book, i.e. to provide both engineering students and professional practitioners with the appropriate statistical methodologies for efficiently designing and conducting planned experiments, and for analysing the corresponding results using different types of ANOVA procedures. All fundamental results will be proved, and each technique will be developed and illustrated on the basis of a meaningful application.

EXERCISES

1. Define statistics.
2. Define (a) experiment, (b) design, (c) analysis.
3. Define the following terms: (a) accuracy, (b) reproducibility, (c) error, (d) mistake, (e) random error, systematic error, (f) random variable. Give examples to illustrate your definitions.
4. State and discuss the basic principles of experimental design.
5. Discuss the steps involved in each of the following phases of a statistical project: (a) the experiment, (b) the design, (c) the analysis.
6. Give a hypothetical example of a statistical project, identifying the following elements: (a) response, (b) factors, (c) levels, (d) experimental conditions.
7. Explain how to use the operation research (scientific method) approach to the problem of developing a statistical project.
8. Assuming that only two factors, each at two levels, are affecting the response of a given experiment, indicate how you would check if main effects and the interaction effect are important.
9. Explain how trends in time and/or space can affect the homogeneity of experimental materials used in the statistical analysis of factor effects. Give examples in each case.
10. A completely randomized experiment with two temperatures (factor A), two lengths of washing time (factor B), and five replications per treatment gave the following results:

(a) average for all experimental conditions having both factors at their low levels: 6.2;
(b) average for all experimental conditions having factor *A* at its high level and factor *B* at its low level: 14.1;
(c) average for all experimental conditions having factor *B* at its high level and factor *A* at its low level: 5.4;
(d) average for all experimental conditions having both factors at their high levels: 7.8.

The variable of interest was the amount of impurities (in grams) remaining in the material washed. Using a graphical approach, estimate main and two-factor interaction effects. Interpret your results.

REFERENCES

1. Hicks, C. R. (1982) *Fundamental Concepts in the Design of Experiments*, 3rd edn, Holt, Rinehart and Wintson, Inc. New York.
2. Hough, C. L. (1986) The effect of back rake angle on the performance of small-diameter polycrystalline diamond rock bits: ANOVA tests. *ASME Journal of Energy Resources Technology*, **108**, 305–9.
3. Hough, C. L., Obioha, O. M. and Garcia-Diaz, A. (1983) Performance, rake angle and wear test analysis of advanced technology coal mine roof bolt drill bits. *Contractor Report SAND83-7006*, Sandia Laboratories. Albuquerque, New Mexico.
4. Wang, K. K. and Devries, M. F. (1969) Investigation of manufacturing processed by statistical experimental design techniques. *University of Wis., Engr. Expt. Sta., Reprint. No. 1396.*
5. Chew, V. (1958) *Experimental Designs in Industry*, John Wiley & Sons, Inc., New York.
6. Kac, M. (1983) What is random? *American Scientist*, **71**, 405–6.
7. Atkins, B. C. (1982) Drilling application successes using STRATAPAX blank bits in mining and construction. Paper presented at the Australian Drilling Association Symposium, January 18–20, 1982.

2 Basic statistical concepts

The purpose of this chapter is to review fundamental concepts of statistical inference. This branch of statistics consists of two major areas known as **parameter estimation** and **hypothesis testing**. Some of the concepts related to these two areas will be directly used in the development of statistical techniques for studying experimental designs and the corresponding experimental results. Other concepts, although not directly used in the developments presented in this book are, however, important prerequisites for a better understanding of the role and scope of statistical inference in the analysis of experiments.

The approach used in this chapter to review fundamental topics in estimation and hypothesis testing is based on brief definitions and/or derivations followed by illustrative examples. A more rigorous treatment of these topics can be found in a good number of textbooks [1–6].

2.1 INTRODUCTION

Statistics can be defined as the art as well as the science of data collection, analysis and interpretation. Theoretical statistics is an exact science based upon rigorous mathematical proofs. However, in real world applications the underlying statistical assumptions can rarely be totally satisfied. The art of statistics comes in knowing which statistical method to employ in order to produce valid conclusions and yield the necessary information at the lowest total cost and expenditure of time.

Ideally a statistician would like to be able to take a complete census of a universe or population. From the census information the statistician would be able to determine the parameters of the population. Due to the time and cost involved in undertaking a complete census of the population, this is usually not a viable option. Instead the statistician will collect a random sample of data and from the information contained in this sample he or she will estimate the parameters that effectively characterize the population sample. The most common parameters are the mean (a measure of central tendency) and the variance (a measure of dispersion). Based upon assumptions concerning the population under investigation and the statistical values calculated from the sample, the statistician will make inferences about the population as an aid in decision making.

2.2 BASIC STATISTICS AND THEIR DISTRIBUTIONS

The statistics that can be computed from the observations included in a random sample are random variables. These random variables are distributed according to certain characteristic probability density or mass functions. Some of the more common statistics and their distributions will now be discussed.

2.2.1 The *Z*-statistic

The Z-statistic is defined as

$$Z = (\bar{X} - \mu)\left(\frac{\sigma}{n^{1/2}}\right)^{-1} \tag{2.1}$$

where $\bar{X} = \sum_{i=1}^{n} x_i/n$ and $x_1, x_2, \ldots, x_n$ are a random sample from a normal population distributed with mean μ and variance σ^2. The Z-statistic follows a standardized normal distribution, that is, normal distribution with mean equal to zero and variance equal to one. This is usually represented by the notation $Z \sim N(0, 1)$.

The density function of the Z-statistic is

$$f(Z) = \frac{1}{(2\pi)^{1/2}} e^{(-1/2)Z^2} \tag{2.2}$$

where $-\infty < Z < +\infty$.

2.2.2 The χ^2-statistic

The χ^2-statistic is

$$\chi^2 = \sum_{i=1}^{n} Z_i^2 \tag{2.3}$$

where all values of Z_i are statistically independent. The number of independent squares in the right-hand side of equation (2.3) is known as the degrees of freedom. The density function of the χ^2-statistic with n degrees of freedom is defined as

$$f(x^2) = x^{n/2-1} e^{-x/2}\left[2^n \Gamma\left(\frac{n}{2}\right)\right]^{-1} \tag{2.4}$$

where $0 < x^2 < \infty$ and $\Gamma(n) = \int_0^\infty e^{-x} x^{(n-1)}\, dx$. Additionally, the expected value of χ^2 is n and the variance of χ^2 is $2n$. This is usually represented by the notation $E[\chi^2] = n$ and $V[\chi^2] = 2n$.

2.2.3 The *t*-statistic

The t-statistic is defined as

$$t = Z\left(\frac{\chi^2}{n}\right)^{-1/2} \tag{2.5}$$

where Z and χ^2 are statistically independent. In equation (2.5), both the t-statistic and the χ^2-statistic have the same number of degrees of freedom, n. The density function of the t-statistic with n degrees of freedom is given by

$$f(t) = \Gamma\left(\frac{n+1}{2}\right)\left\{(n\pi)^{(1/2)}\Gamma\left(\frac{n}{2}\right)\Big/\left[\left(\frac{1+t^2}{n}\right)^{(1/2)(n+1)}\right]\right\}^{-1} \tag{2.6}$$

where $-\infty < t < \infty$. Additionally, $E[t] = 0$ and $V[t] = n(n-2)^{-1}$, where $n > 2$.

2.2.4 The *F*-statistic

The F-statistic is defined as

$$F = \left(\frac{\chi_1{}^2}{n_1}\right)\left(\frac{\chi_2{}^2}{n_2}\right)^{-1} \tag{2.7}$$

where $\chi_1{}^2$ and $\chi_2{}^2$ are two statistically independent χ^2 random variables with degrees of freedom n_1 and n_2 respectively. This is equivalent to saying that the F-statistic has n_1 degrees of freedom in the numerator and n_2 degrees of freedom in the denominator. The density function of the F-statistic is defined as

$$f(F) = \Gamma\left(\frac{n_1+n_2}{2}\right)\left(\frac{n_1}{n_2}\right)^{(n_1/2)} F^{(n_1-2)/2}\left[\Gamma\left(\frac{n_1}{2}\right)\Gamma\left(\frac{n_2}{2}\right)\left(1+\frac{n_1}{n_2}F\right)^{(n_1+n_2)/2}\right]^{-1} \tag{2.8}$$

where $0 < F < \infty$. Additionally, the expected value and variance of F are defined in equations (2.9) and (2.10), respectively.

$$E[F] = \frac{n_2}{n_2 - 2} \quad \text{for } n_2 > 2 \tag{2.9}$$

$$V[F] = \frac{2n_2{}^2(n_1 + n_2 - 2)}{n_1(n_2-2)^2 + (n_2 - 4)} \quad \text{for } n_2 > 4. \tag{2.10}$$

2.3 ESTIMATION

An estimator of a parameter is a function of the observations in a random sample taken from the population with the unknown parameter, such that the distribution of this function is centralized as much as possible about the true value of the parameter. There are two types of estimates:

(a) *Point estimates*: the unknown parameter is estimated by a single numerical value.
(b) *Interval estimates*: a range of numerical values is provided with a given probability that the unknown parameter will be contained in the range.

2.3.1 Point estimates

Point estimates are usually expected to have certain desirable characteristics, such as unbiasedness, consistency and minimum variance. These characteristics are briefly explained as follows:

(a) *Unbiased*: an estimator $\hat{\theta}$ (where the $\hat{}$ symbol denotes that the parameter is being estimated) is called an unbiased estimator of θ if $E[\hat{\theta}]=\theta$. As an example, $E[\bar{X}]=\mu$, therefore $\bar{X}$ is called an unbiased estimator of μ. Additionally, $E[s^2]=\sigma^2$ where $s^2=\sum(x_i-\bar{X})^2/(n-1)$ and the x_i are normally and independently distributed with variance equal to σ^2.
(b) *Consistent*: a statistic $\hat{\theta}$ is called a consistent estimator of θ if as the sample size (from which $\hat{\theta}$ is calculated) becomes larger the value of $\hat{\theta}$ tends toward θ. As an example, as the sample size increases, $\bar{X}$ tends toward μ, therefore $\bar{X}$ is also a consistent estimator of μ.
(c) *Minimum variance*: in general, we prefer an estimator in the class of unbiased estimators that has minimum variance. $\hat{\theta}$ is a minimum variance unbiased estimator if $V[\hat{\theta}]\leqslant V[\hat{\theta}_0]$ for any other unbiased estimator $\hat{\theta}_0$. Thus if $\hat{\theta}_1$ and $\hat{\theta}_2$ are both unbiased estimators of θ and if $V[\hat{\theta}_1]<V[\hat{\theta}_2]$, then we prefer $\hat{\theta}_1$ to $\hat{\theta}_2$.

(a) Mean square error

The mean square error (*MSE*) is one criterion for selecting an estimator. The mean square error is defined as the expected value of the square of the deviation of the estimate from the parameter being estimated, that is, $MSE=E[(\hat{\theta}-\theta)^2]$. It is equal to the variance of the estimate plus the square of the bias (see for instance [2, 3, 6]):

$$MSE_{\hat{\theta}}=V[\hat{\theta}]+(\theta-E[\hat{\theta}])^2. \tag{2.11}$$

EXAMPLE 2.1

Suppose $\hat{\theta}_1$ and $\hat{\theta}_2$ are two estimators of θ such that $E[\hat{\theta}_1]=\theta$, $E[\hat{\theta}_2]=1.2\theta$, $V[\hat{\theta}_1]=25$ and $V[\hat{\theta}_2]=16$. Which estimator would be more desirable under the *MSE* criterion?

Solution

From equation (2.11):

$$MSE_{\hat{\theta}} = V[\hat{\theta}] + (\theta - E[\hat{\theta}])^2.$$

Thus

$$MSE_{\hat{\theta}_1} = V[\hat{\theta}_1] + (\theta - E[\hat{\theta}_1])^2 = 25 + 0 = 25$$

and

$$MSE_{\hat{\theta}_2} = V[\hat{\theta}_2] + (\theta - E[\hat{\theta}_2])^2 = 16 + 0.04\theta^2.$$

Figure 2.1 gives a graphical representation of MSE as a fraction of θ for both $\hat{\theta}_1$ and $\hat{\theta}_2$. As can be seen in Figure 2.1, there is a value of θ at which both $MSE_{\hat{\theta}_1}$ and $MSE_{\hat{\theta}_2}$ are equal. This is obtained by solving the following equation:

$$25 = 16 + 0.04\theta^2$$
$$0.04\theta^2 = 9$$
$$\theta^2 = 225.$$

Therefore

$$\theta = 15.$$

Hence $\hat{\theta}_1$ is preferred when $\theta > 15$ and $\hat{\theta}_2$ is preferred when $\theta < 15$.

(b) Maximum likelihood estimators

One method for finding estimators of unknown parameters is the method of maximum likelihood [1, 4, 6]. Maximum likelihood estimators are consis-

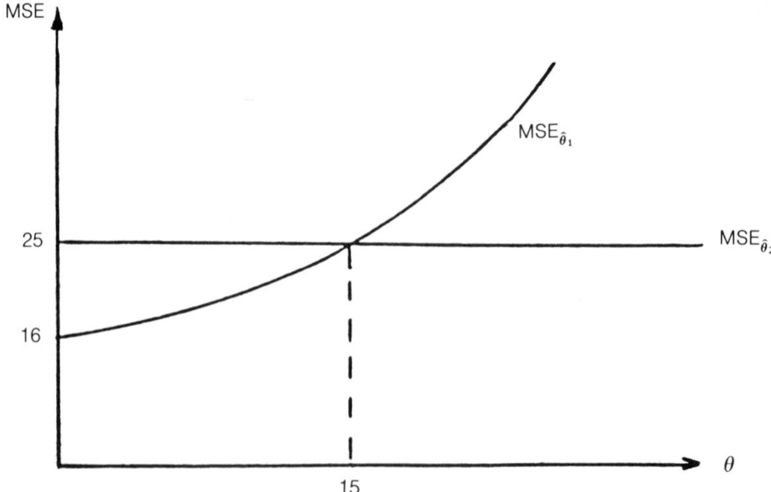

Figure 2.1 Mean square error vs θ in Example 2.1.

tent, and have minimum variance but are not always unbiased. However, the problem of biasedness can, in most cases, be solved by a simple transformation.

Let X be a random variable with density function $f(x, \theta)$. Given a set of independent observations $x_1, x_2, \ldots, x_n$, the likelihood function is defined as the joint probability function of the n random observations.

Let

$$S_i = \left\{ X \mid x_i - \frac{\Delta x_i}{2} \leqslant x \leqslant x_i + \frac{\Delta x_i}{2} \right\} \tag{2.12}$$

where x_i is the ith observed sample value and Δx_i is an infinitesimal quantity. In equation (2.12), S_i is the set of values of x with a probability of occurrence equal to the shaded area indicated in Figure 2.2. As illustrated in Figure 2.2, for $\Delta x_i \rightarrow 0$,

$$P[X \in S_i] = f(x_i; \theta)\Delta x_i \tag{2.13}$$

since the probability of observing a data point falling within the range $x_i \pm \Delta x_i/2$ is equal to the area under the distribution. Let $P[A]$ be defined as

$$P[A] = P[x_1 \in S_1, x_2 \in S_2, \ldots, x_n \in S_n]. \tag{2.14}$$

Simply stated, equation (2.14) indicates that the probability of observing values in the neighborhood of each random observation is equal to the probability that the first data point falls in the range $x_1 \pm \Delta x_1/2$, the second data point falls in the range $x_2 \pm \Delta x_2/2$, ..., and the nth data point falls in the range $x_n \pm \Delta x_n/2$.

Since the random sample consists of independent observations, the probability of observing data points in each of the neighborhoods defined around

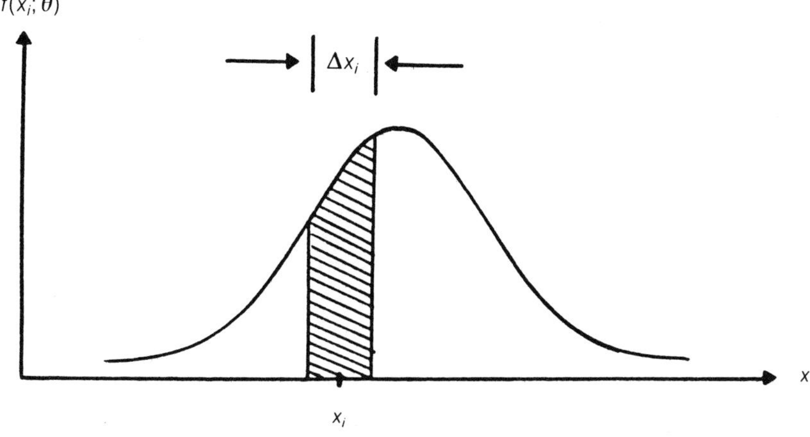

Figure 2.2 Probability associated with values in S_i.

the random observations is equal to

$$P[A] = \{f(x_1;\theta)\Delta x_1\}\{f(x_2;\theta)\Delta x_2\}\cdots\{f(x_n;\theta)\Delta x_n\}. \tag{2.15}$$

Rearranging terms in equation (2.15) yields

$$P[A] = f(x_1;\theta)f(x_2;\theta)\cdots f(x_n;\theta)\Delta x_1 \Delta x_2 \cdots \Delta x_n \tag{2.16}$$

which can also be written as

$$P[A] = \prod_{i=1}^{n} f(x_i;\theta) \prod_{i=1}^{n} \Delta x_i \tag{2.17}$$

where $\prod_{i=1}^{n}$ denotes the product of the first n terms. In order to find the value of θ for which the observed sample would have the highest probability of being extracted, we would maximize $P[A]$, given in equation (2.17), which is equivalent to maximizing

$$L = \prod_{i=1}^{n} f(x_i;\theta) \tag{2.18}$$

since $\prod_{i=1}^{n}\Delta x_i$ does not depend on θ. The function L defined in equation (2.18) is known as the likelihood function.

The maximum likelihood estimator is found by maximizing the likelihood function with respect to the parameter θ being estimated. A well-known fact from elementary calculus is that a necessary condition for the maximum likelihood estimator is

$$\frac{\mathrm{d}L}{\mathrm{d}\theta} = 0. \tag{2.19}$$

In most cases at least one of the solutions to equation (2.19) will yield the maximum likelihood estimator if it exists.

EXAMPLE 2.2

Assume that the random variable X has the exponential distribution

$$f(x;\theta) = \theta\,\mathrm{e}^{-\theta x} \quad \text{if } x > 0, \theta > 0$$

where θ is the parameter of the distribution. Use the method of maximum likelihood to estimate θ based on five random observations $x_1 = 0.9$, $x_2 = 1.7$, $x_3 = 0.4$, $x_4 = 0.3$ and $x_5 = 2.4$.

Solution

We shall first derive a general result for the maximum likelihood estimator for the parameter θ of the exponential distribution. Each variable x_i has the distribution

$$f(x_i;\theta)=\theta\,e^{-\theta x_i}.$$

Since the members of the sample are statistically independent of each other, their joint probability density function is given by equation (2.18):

$$\begin{aligned}L(x_1,\ldots,x_n;\theta)&=\theta\,e^{-\theta x_1}\theta\,e^{-\theta x_2}\cdots\theta\,e^{-\theta x_n}\\&=\theta^n\exp\left(-\theta\sum_{i=1}^{n}x_i\right).\end{aligned}$$

We want to find the value of θ that maximizes the above equation. Note that the value of θ at which $\ln L$ is maximized is identical to the value at which L is maximized. The natural log of the likelihood function is given by

$$\ln\{L(x_1,\ldots,x_n;\theta)\}=n\ln\theta-\theta\sum_{i=1}^{n}x_i.$$

Differentiating $\ln L$ with respect to θ and setting the derivative equal to zero, the following result is obtained:

$$\frac{n}{\theta}-\sum_{i=1}^{n}x_i=0.$$

Solving for θ,

$$\hat{\theta}=\frac{n}{\sum_{i=1}^{n}x_i}.$$

To check if $\hat{\theta}$ is indeed at a maximum, we take the second derivative of $\ln L$ to obtain

$$\frac{d^2}{d\theta^2}\ln L(x_1,\ldots,x_n;\theta)=-\frac{n}{\theta^2}$$

and compute its value at the value $\hat{\theta}=n/\sum_{i=1}^{n}x_i$. Since the second derivative is negative, $\hat{\theta}$ is indeed a maximum likelihood estimator.

Finally, using the given data, $n=5$ and $\sum_{i=1}^{5}x_i=5.7$; therefore

$$\hat{\theta}=\frac{5}{5.7}=0.88.$$

2.3.2 Interval estimates

Point estimates are not of much worth if there is no indication of how precise they are in a particular application. Ideally, a point estimate should be provided with a specified probability of confidence level that the estimate is included in a specified range of likely values for the parameters being estimated. Basically, this is the idea that motivates the use of interval estimates, also referred to as confidence intervals.

This section is divided into two basic parts. The first part will discuss the fundamental procedure for obtaining confidence intervals, assuming that certain special conditions are satisfied. The second part will summarize the general methodology to find confidence intervals.

(a) Special case

Let $\hat{\theta}(x_1, x_2, \ldots, x_n)$ be a statistic defined on the basis of a random sample from a population with density function $f(x;\theta)$, where θ is an unknown parameter. Let also $g(\hat{\theta})$ be the known density function of the statistic $\hat{\theta}$. Therefore, it is possible to find values $\hat{\theta}_{\alpha/2}$ and $\hat{\theta}_{1-\alpha/2}$ such that

$$\int_{-\infty}^{\hat{\theta}_{\alpha/2}} g(\hat{\theta})\mathrm{d}\hat{\theta} = \alpha/2$$

and

$$\int_{\hat{\theta}_{1-\alpha/2}}^{+\infty} g(\hat{\theta})\mathrm{d}\hat{\theta} = \alpha/2.$$

From the two relationships shown above, it follows that

$$P[\hat{\theta}_{\alpha/2} \leqslant \hat{\theta} \leqslant \hat{\theta}_{1-\alpha/2}] = 1 - \alpha. \tag{2.20}$$

Now suppose that there is another statistic $\hat{\mu}(x_1, x_2, \ldots, x_n)$ computed on the basis of the numerical values of the observations $(x_1, x_2, \ldots, x_n)$. Assuming that $\hat{\theta}$ and $\hat{\mu}$ are directly related through a known transformation

$$\hat{\theta} = h(\hat{\mu};\theta) \tag{2.21}$$

and substituting equation (2.21) into equation (2.20), it is concluded that

$$P[\hat{\theta}_{\alpha/2} \leqslant \theta \leqslant \hat{\theta}_{1-\alpha/2}] = 1 - \alpha. \tag{2.22}$$

The range

$$\hat{\theta}_{\alpha/2} \leqslant h(\hat{\mu};\theta) \leqslant \hat{\theta}_{1-\alpha/2}$$

obtained from equation (2.22) is a confidence interval for $h(\hat{\mu};\theta)$; now suppose that an observed value $\hat{\mu}_0$ for the statistic $\hat{\mu}$ is available. Therefore, for the minimum and maximum values of the range previously found for $h(\hat{\mu};\theta)$ it is possible to write

$$\hat{\theta}_{\alpha/2} = h(\hat{\mu}_0;\theta_1) \tag{2.23}$$

and

$$\hat{\theta}_{1-\alpha/2} = h(\hat{\mu}_0;\theta_2) \tag{2.24}$$

where θ_1 and θ_2 are the limits of a $100(1-\alpha)\%$ confidence interval for the parameter θ. In the present discussion it is assumed that equations (2.23) and (2.24) can be directly solved for θ_1 and θ_2.

EXAMPLE 2.3

A random sample of size $n = 21$ was taken from a normal population with unknown mean and standard deviation. Assume that the sample mean and standard deviation are equal to 13.5 and 1.53, respectively. Find a 95% confidence interval for the population mean.

Solution

By definition, $t = Z(\chi^2 n^{-1})^{-1/2}$, where n is the number of degrees of freedom of the χ^2-statistic. In elementary statistics it is proved that

$$\frac{\sum_{i=1}^{n}(X_i - \bar{X})^2}{\sigma^2} \sim \chi^2_{n-1}.$$

This result in turn can be used along with the definition of the t-statistic to prove that

$$(X - \mu)\left(\frac{S}{n^{1/2}}\right)^{-1} \sim t_{n-1}$$

where

$$S = \left[\frac{\sum_{i=1}^{n}(X_i - \bar{X})^2}{n-1}\right]^{1/2}.$$

From a table for the t-distribution, we can obtain the critical value for the 2.5 and 97.5 percentile points of the t-distribution having 20 degrees of freedom. These values are $t_{0.025} = -2.09$ and $t_{0.975} = 2.09$. Therefore, the confidence intervals for μ can be obtained from the following interval for t:

$$-2.09 \leqslant \frac{\bar{X} - \mu}{S/n^{1/2}} \leqslant 2.09$$

which corresponds to $\bar{X} - 2.09Sn^{-1/2} \leqslant \mu \leqslant \bar{X} + 2.09Sn^{-1/2}$. Recalling $n = 21$, $\bar{X} = 13.50$, $S = 1.53$ and $\alpha = 0.05$, we obtain the desired confidence interval:

$$12.802 \leqslant \mu \leqslant 14.198.$$

(b) General case

Consider a given population with density function $f(x; \theta)$. Let $\hat{\theta}(x_1, x_2, \ldots, x_n)$ be an estimator of θ based upon a random sample of n observations; also assume that the density function $g(\hat{\theta}; \theta)$ of $\hat{\theta}$ is known or can be determined.

After specifying a significance level α and choosing an arbitrary value θ' which is substituted for θ in $g(\hat{\theta}; \theta)$, it is possible to find two numbers h_1 and

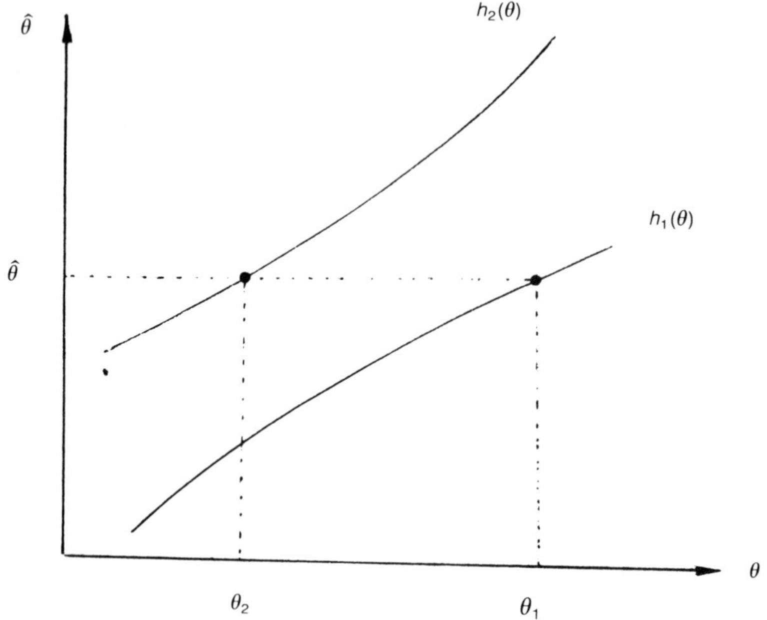

Figure 2.3 h_1 and h_2 as functions of θ.

h_2 such that

$$P[\hat{\theta} < h_1] = \int_{-\infty}^{h_1} g(\hat{\theta}; \theta) \mathrm{d}\hat{\theta} = 1 - (\alpha/2) \tag{2.25}$$

$$P[\hat{\theta} > h_2] = \int_{h_2}^{+\infty} g(\hat{\theta}; \theta) \mathrm{d}\hat{\theta} = 1 - (\alpha/2). \tag{2.26}$$

Since h_1 and h_2 are functions of θ they will be represented more accurately by the symbols $h_1(\theta)$ and $h_2(\theta)$. These functions can be plotted as in Figure 2.3 after considering different values of $\theta = \hat{\theta}'$, and then solving equations (2.25) and (2.26) for $h_1(\theta')$ and $h_2(\theta')$, respectively.

Once the functions of $h_1(\theta)$ and $h_2(\theta)$ have been obtained, the procedure for obtaining a $100(1-\alpha)\%$ confidence interval for θ is as follows. First, for a sample of size n, compute the value of the estimator, say $\hat{\theta}'$. Then draw a horizontal line at that value and project down to the θ-axis where the line intersects the $h_2(\theta)$ and $h_1(\theta)$ curves. The corresponding θ_1 and θ_2 values will be the limits of the confidence interval for θ.

2.4 TESTING OF HYPOTHESES

A goal in many statistical experiments is to arrive at a decision based upon the estimated parameter value(s) from one or more populations. For example, an engineer may wish to determine whether a proposed jet wing configuration is aerodynamically more efficient than the present wing configuration, and would recommend the new design if it appears to be superior. Perhaps an automobile manufacturer wishes to offer a rustproof undercoating as standard equipment on his/her new cars and so would want to test the mean effective life of one type of undercoating against that of another undercoating in order to ensure that 'quality is job 1'. Hypothesis testing allows one to make such comparisons validly.

A test of a statistical hypothesis is a rule which, when the experimental sample values have been obtained, leads to a decision to accept or reject the hypothesis under consideration. Let c be that subset of the sample space which, in accordance with a prescribed test, leads to the rejection of the hypothesis under consideration. Then c is called the **critical region** of the test.

Let H_0 denote a hypothesis which is to be tested against an alternative hypothesis H_1 in accordance with a prescribed test. The rejection of the hypothesis H_0 when it is true is, of course, an incorrect decision or an error. This type of incorrect decision is known as a type I error. The acceptance of H_0 when it is false is known as a type II error. The probability of committing a type I error is equal to the significance level of the test, denoted by α (which can also be thought of as the 'size' of the critical region c). The probability of committing a type II error, denoted by β, is one minus the power of the test when H_1 is true.

The steps involved in hypothesis testing can be summarized as follows:

(a) Set up the null hypothesis and its alternative.
(b) Set the significance level of the test.

Table 2.1 Tests on means of normal distributions – variance known

Hypothesis	*Test statistic*	*Criteria for rejection*
H_0: $\mu = \mu_0$ H_1: $\mu \neq \mu_0$		$\lvert Z_0 \rvert > Z_{1-\alpha/2}$
H_0: $\mu = \mu_0$ H_0: $\mu < \mu_0$	$Z_0 = (\bar{y} - \mu_0)\left(\dfrac{\sigma}{n^{1/2}}\right)^{-1}$	$Z_0 < -Z_{1-\alpha}$
H_0: $\mu = \mu_0$ H_1: $\mu > \mu_0$		$Z_0 > Z_{1-\alpha}$
H_0: $\mu_1 - \mu_2 = \gamma$ H_1: $\mu_1 - \mu_2 \neq \gamma$		$\lvert Z_0 \rvert > Z_{1-\alpha/2}$
H_0: $\mu_1 - \mu_2 = \gamma$ H_1: $\mu_1 - \mu_2 < \gamma$	$Z_0 = [(\bar{y}_1 - \bar{y}_2) - \gamma]\left(\dfrac{\sigma_1^2}{n_1} + \dfrac{\sigma_2^2}{n_2}\right)^{-1/2}$	$Z_0 < -Z_{1-\alpha}$
H_0: $\mu_1 - \mu_2 = \gamma$ H_1: $\mu_1 - \mu_2 > \gamma$		$Z_0 > Z_{1-\alpha}$

Table 2.2 Tests on means of normal distributions – variance unknown

Hypothesis	*Test statistic*	*Criteria for rejection*
$H_0\colon \mu=\mu_0$ $H_1\colon \mu\neq\mu_0$		$\lvert t_0\rvert > t_{1-\alpha/2,n-1}$
$H_0\colon \mu=\mu_0$ $H_1\colon \mu<\mu_0$	$t_0=(\bar{y}-\mu_0)\left(\dfrac{S}{n^{1/2}}\right)^{-1}$	$t_0<-t_{1-\alpha,n-1}$
$H_0\colon \mu=\mu_0$ $H_1\colon \mu>\mu_0$		$t_0>t_{1-\alpha,n-1}$
$H_0\colon \mu_1-\mu_2=\gamma$ $H_1\colon \mu_1-\mu_2\neq\gamma$	$t_0=(\bar{y}_1-\bar{y}_2-\gamma)\left[S_p\left(\dfrac{1}{n_1}+\dfrac{1}{n_2}\right)^{1/2}\right]^{-1}$ $v=n_1+n_2-2$	$\lvert t_0\rvert > t_{1-\alpha/2,v}$
$H_0\colon \mu_1-\mu_2=\gamma$ $H_1\colon \mu_1-\mu_2<\gamma$	$t_0=(\bar{y}_1-\bar{y}-\gamma)\left(\dfrac{S_1^2}{n_1}+\dfrac{S_2^2}{n_2}\right)^{-1/2}$	$t_0<-t_{1-\alpha,v}$
$H_0\colon \mu_1-\mu_2=\gamma$ $H_1\colon \mu_1-\mu_2>\gamma$	$v=\left(\dfrac{S_1^2}{n_1}+\dfrac{S_2^2}{n_2}\right)^2\left[\dfrac{(S_1^2n_1^{-1})^2}{n_1-1}+\dfrac{(S_2^2n_2^{-1})^2}{n_2-1}\right]^{-1}$	$t_0>t_{1-\alpha,v}$

Table 2.3 Tests on variances of normal distributions with unknown mean

Hypothesis	*Test statistic*	*Criteria for rejection*
$H_0\colon \sigma^2=\sigma_0^2$ $H_1\colon \sigma^2\neq\sigma_0^2$		$\chi_0^2>\chi_{1-\alpha/2,n-1}^2$ or $\chi_0^2<\chi_{\alpha/2,n-1}^2$
$H_0\colon \sigma^2=\sigma_0^2$ $H_1\colon \sigma^2<\sigma_0^2$	$\chi_0^2=\dfrac{(n-1)S^2}{\sigma_0^2}$	$\chi_0^2<\chi_{\alpha/2,n-1}^2$
$H_0\colon \sigma^2=\sigma_0^2$ $H_1\colon \sigma^2>\sigma_0^2$		$\chi_0^2>\chi_{1-\alpha,n-1}^2$
$H_0\colon \sigma_1^2=\sigma_2^2$ $H_1\colon \sigma_1^2\neq\sigma_2^2$	$F_0=\dfrac{S_1^2}{S_2^2}$	$F_0>F_{1-\alpha/2,n_1-1,n_2-1}$ $F_0<F_{\alpha/2,n_1-1,n_2-1}$
$H_0\colon \sigma_1^2=\sigma_2^2$ $H_1\colon \sigma_1^2<\sigma_2^2$	$F_0=\dfrac{S_2^2}{S_1^2}$	$F_0>F_{1-\alpha,n_2-1,n_1-1}$
$H_0\colon \sigma_1^2=\sigma_2^2$ $H_1\colon \sigma_1^2>\sigma_2^2$	$F_0=\dfrac{S_1^2}{S_2^2}$	$F_0>F_{1-\alpha,n_1-1,n_2-1}$

(c) Determine the appropriate test statistic to test H_0.
(d) Determine the sample distribution of this test statistic when H_0 is true.
(e) Set up the size of the critical region c equal to the significance level.
(f) Choose a random sample of n observations, compute the test statistic and accept or reject H_0.

2.4.1 Hypothesis tests and appropriate statistics

Tables 2.1–2.3 summarize the test statistics used and the criteria for rejection adopted when verifying typical hypotheses concerning the mean and the variance of normal distributions. Table 2.1 considers tests on means when the variance is known. Tables 2.2 considers the same kind of tests when variance is unknown; in the last two tests of this table the unknown variances are not assumed to be equal and the given procedure is based on an approximation. In this table, for the fourth test, $S_p = [(n_1 - 1)S_1{}^2 + (n_2 - 1)S_2]v^{-1}$. Table 2.3 considers tests on variances assuming that the mean is known.

EXAMPLE 2.4

The product made at a certain industrial plant is mechanically weighed and packed in 50 kg boxes. The average weight of an empty box is 0.5 kg and its variability is practically negligible. On the other hand, the machine introduces a variability of $0.000\,04^{1/2}$ kg around the expected weight according to a normal distribution. In order to check for systematic errors in the weighing process, the quality control department takes regular samples of the product and weighs each loaded box in the sample using a control balance with known standard deviation of $0.000\,01^{1/2}$ kg. When the difference between the actual weight and the expected weight is significant at the 5% level, the weighing and packing machine is adjusted. Otherwise, the machine continues in the process without any adjustment. It is desired to check if any adjustment is needed based on the random sample given in Table 2.4.

Table 2.4

Box	*Weight*	*Box*	*Weight*
1	50.512	9	50.581
2	50.498	10	50.499
3	50.483	11	50.489
4	50.540	12	50.501
5	50.508	13	50.500
6	50.000	14	50.500
7	50.476	15	50.476
8	50.519	16	50.508

Solution

Since we do not want the machines to underweigh or overweigh we use a two-tailed test.

$$H_0: \mu = 50.500$$
$$H_1: \mu \neq 50.500.$$

The following parameters are also given: $\sigma = (0.000\,01 + 0.000\,04)^{1/2} = 0.007\,071\,06$, $n = 16$, $\alpha = 0.05$. The given sample has a mean equal to

$$\bar{X} = 807.590/16 = 50.474.$$

The standard deviation of the sample mean distribution is equal to

$$\sigma/4 = 0.007\,070\,106/4 = 0.002.$$

The critical values for the Z-statistic are obtained from a table for the normal distribution as follows:

$$Z_{0.025} = -1.96$$
$$Z_{0.975} = +1.96.$$

Therefore the critical region for H_0 is defined as the interval containing values

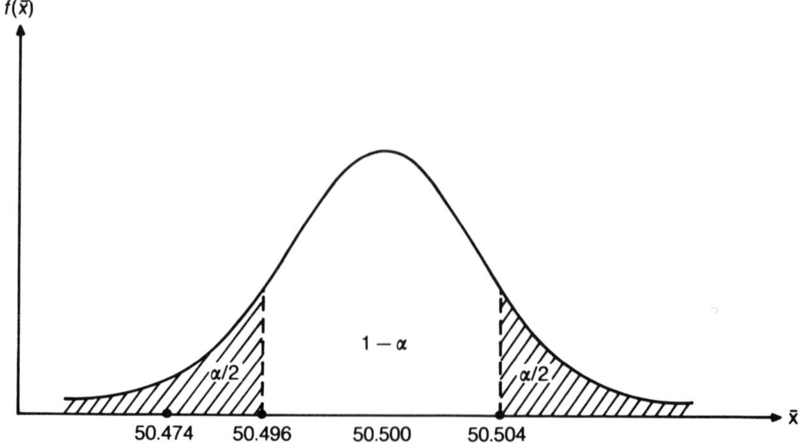

Figure 2.4 Critical region for H_0 in Example 2.6.

of $\bar{X}$ such that

$$\bar{X} > 50.500 + 0.002Z_{0.975} = 50.504$$
$$\bar{X} < 50.500 + 0.002Z_{0.025} = 50.496.$$

Since the observed value $\bar{X} = 50.474$ does not lie in the acceptance region between 50.504 and 50.496 as shown in Figure 2.4, the null hypothesis H_0 is rejected at the 5% level of significance.

EXAMPLE 2.5

Two processes are currently being developed to produce a given type of light bulb. Process 1 yields a bulb with a lifetime being normally distributed with variance ${\sigma_1}^2$. Process 2 also yields a normally distributed lifetime, with variance ${\sigma_2}^2$. Both processes yield the same expected lifetime, but process 1 is slightly less expensive than process 2. In order to verify if the processes have equal variances at the 5% level of significance, two random samples were taken, one from each process. The sample from process 1 had size equal to 8 and sample variance equal to 1560. The sample from the other process had size equal to 10 and sample variance equal to 1000. Perform the appropriate hypothesis test.

Solution

$$H_0 : {\sigma_1}^2 = {\sigma_2}^2.$$

The test statistic is the F-statistic; its observed value is equal to

$$F = {S_1}^2/{S_2}^2 = 1560/1000 = 1.56.$$

The critical region for this test is defined as $F \geqslant F_c$, where F_c is the critical value of the F-statistic for $n_1 - 1 = 7$ degrees of freedom for the numerator, $n_2 - 1 = 9$ degrees of freedom for the denominator, and 5% level of significance. From a table for percentiles of the F distribution, $F_c = 3.29$. Hence, the hypothesis is not rejected.

EXAMPLE 2.6

Given a random sample of size $n = 10$ and the sample observations:

$$x_1 = 108 \qquad x_6 = 138$$
$$x_2 = 124 \qquad x_7 = 163$$
$$x_3 = 124 \qquad x_8 = 159$$
$$x_4 = 106 \qquad x_9 = 134$$
$$x_5 = 115 \qquad x_{10} = 139,$$

it is desired to test the hypothesis that the true variance is equal to 100 against the alternative hypothesis that the true variance does not equal 100. Use a 5% level of significance.

Solution

$$H_0: \sigma^2 = {\sigma_0}^2$$
$$H_1: \sigma^2 \neq {\sigma_0}^2$$

where ${\sigma_0}^2 = 100$. For the given data, $\chi^2 = (\sum_{i=1}^{n}(X_i - \bar{X})^2)({\sigma_0}^2)^{-1} = 34.38$. The rejection criterion for this test is $\chi_0^2 > \chi_{1-\alpha/2,n-1}^2 = 19.0$, or $\chi_0^2 < \chi_{\alpha/2,n-1}^2 = 2.7$. Since $\chi_0^2 = 34.38$ falls in the rejection region, H_0 is rejected at 5% level of significance.

2.4.2 Generalized likelihood ratio test

Let H_0 be a hypothesis stating that the parameters of $f(x; \theta_1 \theta_2, \ldots, \theta_k)$ belong to a subspace ω of the parameter space Ω. Then the generalized likelihood ratio is the quotient defined as [6]:

$$\lambda = \frac{L(\hat{\omega})}{L(\hat{\Omega})} \tag{2.27a}$$

where $L(\hat{\omega})$ is the maximum of the likelihood function in the region ω, and $L(\hat{\Omega})$ is the maximum of the likelihood function in the region Ω.

The critical region for testing H_0 is an interval $0 < \lambda < A$ where A is determined by

$$\int_0^A g(\lambda | H_0) \mathrm{d}\lambda = \alpha \tag{2.27b}$$

where α is the probability of a type I error and $g(\lambda | H_0)$ is the density of λ when H_0 is true.

The case for tests on the mean of a normal population is discussed next. Suppose that we have a sample of n observations, $x_1, x_2, \ldots, x_n$, from a normal population with mean μ and variance σ^2. Define the null and alternative hypotheses as

$$H_0: \mu = \mu_0$$
$$H_1: \mu \neq \mu_0.$$

The parameter space Ω can be graphically thought of as the entire μ-axis and ω as a vertical line intersecting the μ-axis at $\mu = \mu_0$. The likelihood function is

$$L = \left(\frac{1}{(2\pi)^{1/2}\sigma}\right)^n \exp\left(-\frac{1}{2}\right) \sum_{i=1}^{n} [(x_i - \mu)\sigma^{-1}]^2. \tag{2.27c}$$

The maximum value of L in Ω ($L(\hat{\Omega})$) is achieved when $\mu = n^{-1}\sum X_i = \bar{X}$ and $\sigma^2 = n^{-1}\sum(X_i - \bar{X})^2$. Also the maximum value of L in ω ($L(\hat{\omega})$) is obtained when $\mu = \mu_0$ and $\sigma^2 = n^{-1}\sum(X_i - \mu_0)^2$. It can be verified that the likelihood ratio is given by

$$\lambda = \frac{L(\hat{\omega})}{L(\hat{\Omega})} = \left[\frac{\sum(X_i - \bar{X})^2}{\sum(X_i - \mu_0)^2}\right]^{n/2} \tag{2.27d}$$

Since $\sum(X_i - \mu_0)^2 = \sum(X_i - \bar{X})^2 + n(\bar{X} - \mu_0)^2$ equation (2.27d) can be rewritten as

$$\lambda = \left[\frac{1}{1 + [n(\bar{X} - \mu_0)^2(\sum(X_i - \bar{X})^2)^{-1}]}\right]^{(n/2)} \tag{2.27e}$$

The quantity in the denominator in the right-hand side of equation (2.27e) is

$$1 + \frac{t^2}{n-1}$$

where t has the t-distribution with $n - 1$ degrees of freedom when H_0 is true. Therefore the likelihood ratio can be expressed as follows, when H_0 is assumed to be true:

$$\lambda = \left[\frac{1}{1 + (t^2(n-1)^{-1})}\right]^{(n/2)} \tag{2.27f}$$

EXAMPLE 2.7

Solve Example 2.6 using the generalized likelihood ratio test.

Solution

It can be shown [6] that for this test the critical region $0 < \lambda < A$ corresponds to the intervals $0 < \chi^2 < a$ and $\chi^2 > b$ where $\chi^2 = \sum_{i=1}(X_i - \bar{X})^2/\sigma_0^2$ follows a chi-square distribution with $n-1$ degrees of freedom if H_0 is true, and when a and b are such that

$$a/b = e^{(a-b)/n}.$$

For Example 2.6, $a = 2.7$ and $n = 10$. Therefore b can be obtained from the relationship

$$2.7/b = e^{(2.7-b)/10}.$$

It can be verified that the solution of the above equation is the value $b = 24.9$. As a result of the current analysis, H_0 is rejected if $\chi^2 > 24.9$ or $\chi^2 < 2.7$. Since the observed value of χ^2 is $\chi_0^2 = 34.38$ (found in Example 2.6), then H_0

is rejected. Note that the rejection criteria in the generalized likelihood ratio method are not the same as the rejection criteria using the equal tail test (as in Example 2.6).

2.5 POWER CURVES

The power of a test is the probability of rejecting the null hypothesis assuming that a particular choice of the alternative hypothesis is true. The power curve of the test is the graph of the power as a function of the parameter under consideration. Let H_0 denote a hypothesis which is to be tested against an alternative hypothesis H_1 in accordance with a prescribed rule. The 'significance level' of the test (or the 'size') is the maximum value of the power function of the test when H_0 is true. In symbols, the power function can be represented as $\pi(\lambda) = P(\text{rejecting } H|\lambda)$ where λ represents the particular choice of H_1 assumed to be true. Examples 2.8–2.10 illustrate the methodology to find the power function of a test.

EXAMPLE 2.8

In the manufacture of a certain kind of metallic cable, samples of size equal to six are taken at regular intervals for testing the breaking strength of the cable equal to 18 470 kg against the alternative hypothesis that it exceeds this value. Previous experimentation indicates that the variance of breaking strength is equal to 40 kg^2.

(a) Find the power curve of the test assuming that $\alpha = 0.05$ and that the true mean is $\mu = 18\,476$.
(b) Sketch the power curve.

Solution

(a) *Power of the test*:

$$H_0: \mu = 18\,470$$
$$H_1: \mu > 18\,470.$$

If $(\bar{X} - \mu_0)(\sigma n^{-1/2})^{-1} > Z_{0.95}$ we conclude that H_1 is true. Here $\mu_0 = 18\,470$, $\sigma = 6.32$, $n = 6$ and $Z_{0.95} = 1.645$. The power of this test for $\mu > 18\,476$ is, therefore, given by

$$P(\text{rejecting } H_0|18\,476) = P\left(\frac{\bar{X} - 18\,470}{6.32/6^{1/2}} > 1.645|18\,476\right)$$

$$= P\left(\frac{\bar{X} - 18\,476}{6.32/6^{1/2}} + \frac{18\,476 - 18\,470}{6.32/6^{1/2}} > 1.645\right)$$

$$= P(Z > -0.68) = 0.7517)$$

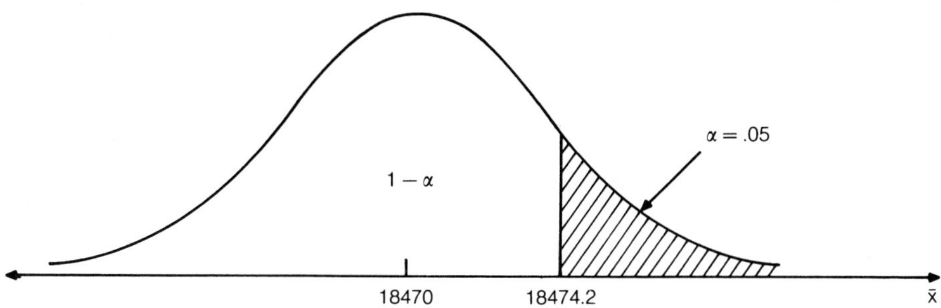

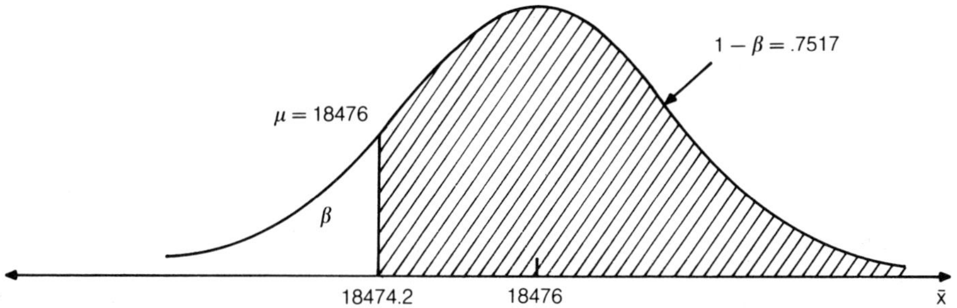

Figure 2.5 Area under the normal curve corresponding to α and β.

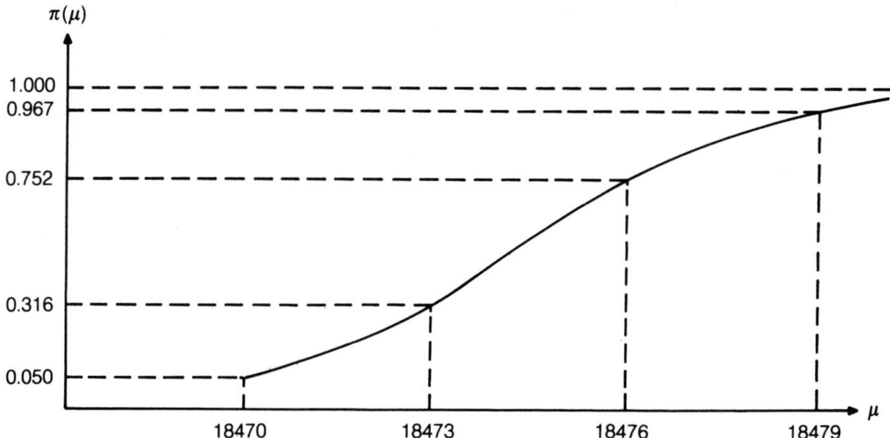

Figure 2.6 Power curve.

Figure 2.5 shows the areas under the curve of a normal density function that correspond to α and β.

(b) *Power curves*: for any value $\mu > 18\,470$ and $\alpha = 0.05$, the power function of this test is given by

$$\pi(\mu) = P\left(Z > 1.645 - \frac{\mu - 18\,470}{2.582} \right).$$

This is an increasing function of μ as can be seen in Figure 2.6.

EXAMPLE 2.9

Find the power function for the test considered in Example 2.5.

Solution

$$H_0: {\sigma_1}^2 = {\sigma_2}^2$$
$$H_1: {\sigma_1}^2 = {\sigma_2}^2.$$

As indicated in Example 2.5, ${S_1}^2/{S_2}^2 = 1560/1000 = 1.56$. If the test hypothesis is true, then ${S_1}^2/{S_2}^2$ is distributed as F_{n_1-1,n_2-1} so that the critical region at the α level of significance becomes

$${S_1}^2/{S_2}^2 > F_{1-\alpha,n_1-1,n_2-1}.$$

Since the probability of this inequality being fulfilled is α for the case when H_0 is true, the power of the test with respect to the alternative hypothesis $\lambda > 1$, where $\lambda = {\sigma_1}^2/{\sigma_2}^2$, is

$$\pi(\lambda) = P[{S_1}^2/{S_2}^2 > F_{1-\alpha,n_1-1,n_2-1} \mid \lambda]$$
$$\pi(\lambda) = P[({S_1}^2/{\sigma_1}^2)/({S_2}^2/{\sigma_2}^2) > (1/\lambda)F_{1-\alpha,n_1-1,n_2-1} \mid \lambda]$$
$$\pi(\lambda) = P[F_{n_1-1,n_2-1} > (1/\lambda)F_{1-\alpha,n_1-1,n_2-2}].$$

It can be verified that this power function is an increasing function of λ.

EXAMPLE 2.10

For Example 2.4, find the probability of a type II error assuming that the true value of the mean is 50.503.

Solution

$$H_0: \mu = 50.500$$
$$H_1: \mu \neq 50.500$$
$$\sigma = 0.007\,071\,06 \qquad n = 16 \qquad \alpha = 0.05.$$

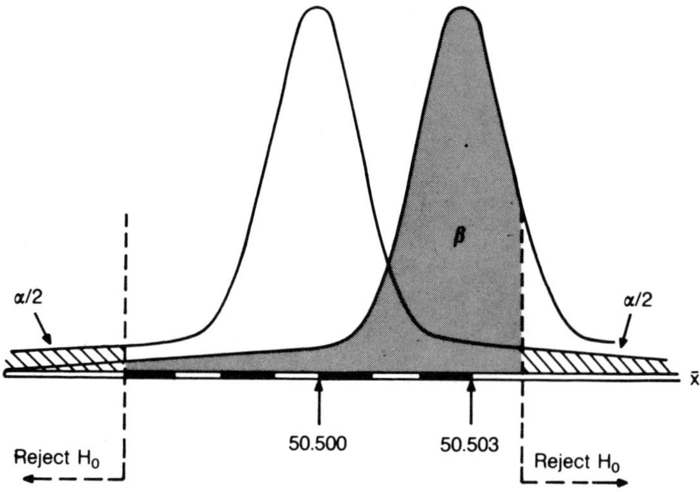

Figure 2.7 Probability of committing a type II error (β) for Example 2.10.

Recall that an error of type II is made when the null hypothesis is accepted given that the alternative hypothesis is true. In this example, the specific alternative being considered is defined as

$$H_1\colon \mu = \mu_1 = 50.503 \quad (\neq 50.500).$$

The distributions corresponding to the null and alternative hypotheses are shown in Figure 2.7. For the specified value of μ_0, the power of the test is:

$$\pi(\mu_1) = P[\text{reject } H_0 \mid \mu_1] = 1 - \beta.$$

We reject H_0 if $\bar{X} < 50.496$ or $\bar{X} > 50.504$. In symbols,

$$\begin{aligned}
\beta &= P[50.496 \leqslant \bar{X} \leqslant 50.504 \mid \mu_1 = 50.503] \\
&= P\left[\frac{(50.496 - 50.503)}{0.002} \leqslant Z \leqslant \frac{(50.504 - 50.503)}{0.002}\right] \\
&= P[-3.50 \leqslant Z \leqslant 0.500] \\
&= P[Z \leqslant 0.500] = 0.6915.
\end{aligned}$$

2.6 SAMPLE SIZE

The sample size generally depends on the following considerations:

(a) the allowable probabilities of making type I and type II errors (α and β);

(b) The desired level of precision, or the magnitude of the difference one wishes to detect;
(c) the variability present in the population.

Methods of calculating the sample are illustrated in Examples 2.11–2.13.

EXAMPLE 2.11

For the case considered in Example 2.8, determine the sample size needed to detect a 10 kg increase in the mean breaking strength of the cable, using $\alpha = 0.05$, $\beta = 0.02$ and $\sigma^2 = 40$.

Solution

$$H_0\colon \mu = 18\,470$$
$$H_1\colon \mu > 18\,470.$$

Alpha equation:

$$\frac{\bar{X}_c - 18\,470}{6.32n^{-1/2}} = 1.645 = Z_{1-\alpha}.$$

Beta equation:

$$\frac{\bar{X}_c - 18\,480}{6.32n^{-1/2}} = -2.055 = Z_{\beta}.$$

Solving for n, $n = 5.5$, that is $n = 6$. Also $\bar{X}_c = 18\,474.2$.

EXAMPLE 2.12

Find the correct sample size for the problem in Example 2.11 by using the power function of the test.

Solution

$$\alpha = 0.05 \qquad \beta = 0.02 \qquad \sigma^2 = 40 \qquad Z_{0.95} = 1.645$$
$$H_0\colon \mu = 18\,470$$
$$H_1\colon \mu > 18\,470$$
$$\pi(\mu_1) = P\left[\frac{\bar{X} - \mu_0}{\sigma/n^{-1/2}} > Z_{0.95} \,|\, \mu_1\right].$$

Recall that one minus the power of the test is equal to the probability of committing a type II error when H_1 is true.

$$\pi(\mu_1) = 1 - \beta = P\left[\frac{\bar{X} - \mu_0 + \mu_1 - \mu_1}{\sigma/n^{-1/2}} > Z_{0.95} | \mu_1\right]$$

$$= 1 - \beta = P\left[\frac{\bar{X} - \mu_1}{\sigma/n^{-1/2}} > Z_{0.95} - \frac{\mu_1 - \mu_0}{\sigma/n^{-1/2}} | \mu_1\right]$$

$$= 0.98 = P\left[\frac{\bar{X} - \mu_1}{6.32/n^{-1/2}} > Z_{0.95} - \frac{\mu_1 - 18\,470}{6.32/n^{-1/2}} | \mu_1\right].$$

Since $\mu_1 = 18\,480$,

$$Z_{0.02} = -2.055 = Z_{0.95} - \frac{18\,480 - 18\,470}{6.32/n^{-1/2}}$$

where $Z_{0.95} = 1.645$, which implies that $n > 5.46$ or $n = 6$. When dealing with sample size calculations involving statistics that have associated degrees of freedom, the solution must be obtained by a 'trial and error' method as in Example 2.13.

EXAMPLE 2.13

Given that the probability of type I error is 0.05 and that the probability of a type II error is 0.20, what is the sample size needed to detect the fact that the true variance of a normal population is actually four times the variance stated in the null hypothesis?

Solution

$$H_0: \sigma^2 = \sigma_2{}^2$$
$$H_1: \sigma^2 > \sigma_0{}^2.$$

For tests on variances of normal distributions with unknown population means, the statistic to be used is the χ^2-statistic. The power of the test is given by

$$P\left[\frac{\Sigma(X_i - \bar{X})^2}{\sigma_0{}^2} > \chi^2_{0.95,n-1} | \sigma^2 = 4\sigma_0{}^2\right] = 0.80.$$

Assuming that the true variance is $4\sigma_0{}^2$ the statistic $(\Sigma(X_i - \bar{X})^2)(\sigma_0{}^2)^{-1}$ is transformed into a χ^2-statistic by dividing it by 4; after doing this the power function is rewritten as follows:

$$P\left[\frac{\Sigma(X_i - \bar{X})^2}{4\sigma_0{}^2} > \frac{\chi^2_{0.95,n-1}}{4} | \sigma^2 = 4\sigma_0{}^2\right] = 0.80$$

Table 2.5

n	$n-1$	$\chi^2_{0.20,n-1}$	$\frac{1}{4}\chi^2_{0.95,n-1}$
10	9	5.38	4.2
9	8	4.59	3.9
8	7	3.82	3.5
7	6	3.07	3.2

or, equivalently,

$$P\left[\chi^2_{n-1} > \frac{\chi^2_{0.95,n-1}}{4}\right] = 0.80.$$

Therefore the sample size required is given by the value n such that

$$\chi^2_{0.20,n-1} = \tfrac{1}{4}\chi^2_{0.95,n-1}.$$

We can determine n by 'trial and error' as indicated in Table 2.5. Since the value of n is between 7 and 8 so choose $n = 8$.

EXERCISES

1. Define the following terms: (a) statistics, (b) statistical inferences, (c) estimation, (d) hypothesis, (e) hypothesis testing.
2. Define: (a) critical region, (b) statistic, (c) level of significance, (d) power.
3. Define the Z, t, χ^2 and F-statistics.
4. Find a confidence interval for the variance of a normal distribution with mean equal to μ.
5. Find the power function for the following cases:
   (a) H_0: $\mu = \mu_0$, H_1: $\mu = \mu_1$, for a normal distribution with unknown mean μ but known standard deviation;
   (b) same as in (a) but assume that the standard deviation is not known.
6. What are the two most important subdivisions of statistical inference? What are the most desirable conditions for point estimates? Define each condition. Provide examples of statistics which satisfy those conditions. Also give examples of statistics which do not satisfy the conditions.
7. What important considerations must be taken into account when choosing the correct sample size for a statistical experiment?
8. Write the alpha and beta equations for the case where it is desired to find the sample size given the standard deviation, the risks and the difference in means to be detected, for (a) one-direction shifts in the mean, and (b) for a shift in any direction. Relate your results to the power function in each case.

9. Draw the power curve for Example 2.11.
10. Let t be the t-statistic distributed with m degrees of freedom. Identify the condition under which the F-statistic is equivalent to t^2.
11. A machine packs a certain product in doses of weight x_1, where x_1 is normally distributed about 25.0 g with standard deviation equal to 0.4 g. The weight of the empty packet, x_2, is also normally distributed with mean of 5.0 g and standard deviation of 0.2 g. Find the range within which 95% of the packets will be contained.
12. A relay is adjusted to be released when a certain action has lasted for μ_1 seconds. The release period, however, is not always exactly μ_1 seconds, but is normally distributed about this value with variance $\sigma_1{}^2$. Another relay, in series with the first, is adjusted to be released after $\mu_2 > \mu_1$ seconds, the release period being normally distributed with mean μ_2 and variance $\sigma_2{}^2$. How large should the difference between the means be in order that the second relay is released before the first one only one out of every 1000 occasions [7]?
13. Suppose that we want to test the hypothesis that the true mean diameter of rivets made under the same conditions is 13.42 mm. Assume that the standard deviation is 0.12 mm, and that the diameters are normally distributed. For $\alpha = 0.01$ and assuming a risk of 0.05 of not rejecting the test hypothesis when $|\mu - \mu_0| = 0.02$ mm, find the corresponding sample size.
14. Develop a relationship to determine the sample size for testing the hypothesis $H_0: \sigma^2 = \sigma_0{}^2 = 14$ against $H_1: \sigma^2 > \sigma_0{}^2$. Assume $\alpha = 0.01$. It is desired to have $\beta = 0.05$ if the true value of the variance is 16. Find the sample size.
15. Let X be a standardized normal random variable. Find the density function for (a) $y = x^2$, and (b) $y = x^{-1}$. Plot both density functions.
16. Let x_1 and x_2 have the bivariate density function

$$f(x_1, x_2) = (\tfrac{1}{2}\pi)\exp[-\tfrac{1}{2}(x_1{}^2 + x_2{}^2)].$$

Find the joint density function for the random variables $y_1 = x_1(x_2)^{-1}$ and $y_2 = x_1$.

REFERENCES

1. Bickel, P. J. and Doksum, K. A. (1977) *Mathematical Statistics*, Holden-Day, Oakland.
2. Fogiel, M. (1978) *The Statistics Problem Solver*, Research and Education Association, New York.
3. Gibra, I. N. (1973) *Probability and Statistical Inference for Scientists and Engineers*, Prentice-Hall, Englewood Cliffs.
4. Hogg, R. V. and Craig, A. T. (1978) *Introduction to Mathematical Statistics*, 4th edn, Macmillan, New York.

5. Hogg, R. V. and Tanis, E. A. (1977) *Probability and Statistical Inference*, Macmillan, New York.
6. Mood, A. M. and Graybill, F. A. (1963) *Introduction to the Theory of Statistics*, 2nd edn, McGraw-Hill, New York.
7. Dano, K. (1944) Tidsaftrapning Ved Oberstromsbeskuttelse, *Ingenioren*, **53**, 13–19.

FURTHER READING

Neter, J., Wasserman, W. and Whitmore, G. A. (1978) *Applied Statistics*, Allyn & Becon, Boston.

Ostle, B. and Mensing, R. W. (1975) *Statistics in Research*, 3rd edn, Iowa State University, Ames.

Statistical analysis of completely randomized one-factor experiments

3

3.1 FUNDAMENTAL ASSUMPTIONS OF THE ANALYSIS OF VARIANCE

In this chapter consideration is given to single-factor experiments with a completely randomized design. The hypothesis to be tested is that all levels (generally referred to as treatments) of the factor have normal populations with equal mean. The statistical analysis conducted for testing this hypothesis is generally known as the **one-way analysis of variance** (ANOVA). The following assumptions must be made concerning the random error in order to use the one-way ANOVA technique:

(a) The errors are statistically independent; that is, the magnitude of the error in one observation is not influenced by the magnitude of the error in another observation.
(b) Each treatment population has the same variance, that is, the random errors have the same variance in experiments designed to compare treatment means.
(c) The errors are normally distributed.

Practical situations occur where one or more basic assumptions of the ANOVA are violated. For example, the assumption of independence is not true when the errors are correlated. The most common causes of positive correlation between errors are time and space trends in experimental materials or equipment. Figure 3.1 illustrates a possible underlying trend that might occur in successive yield results from a batch process. Slight variations in quality of starting material, techniques of different operators, or weather conditions could all contribute to a systematic disturbance of this sort.

Figure 3.2 illustrates one occurrence of trends in space. In this figure, contours of equal thickness of an aluminum sheet are shown as irregular curves across the sheet. The numbered squares, however, designate samples taken without regard for the equal-thickness contours. If the samples are taken as numbered in Figure 3.2, it is easy to see that results of any test of

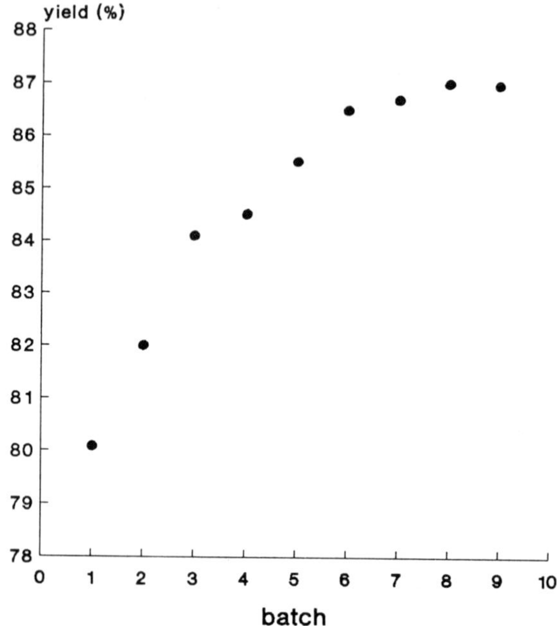

Figure 3.1 Example of positive correlation due to a time trend.

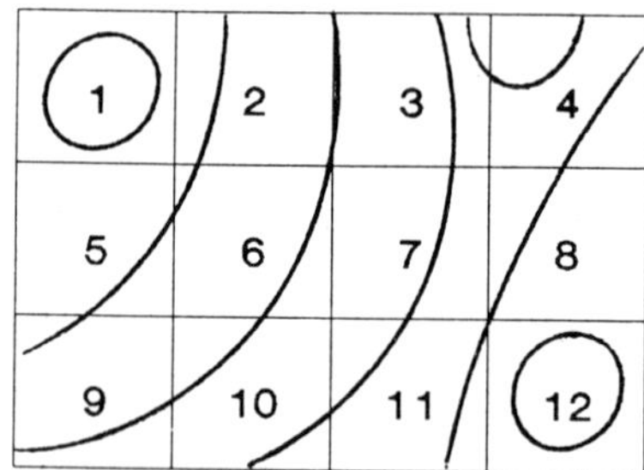

Figure 3.2 Example of correlation due to a spatial trend.

characteristics affected by material thickness will be related to the sample location. Other examples of spatial trends are found in agricultural experiments where such things as water drainage or previous planting patterns will affect the crop yield of a specific plot of ground.

Trends of time and space represent examples of positive correlation. Obviously, negative correlation may also occur. Negative correlation occurs in the case of compensating errors and is seen in data such as those taken for successive batch yields from multi-stage processes [1].

The elimination of trends in experimental errors is best accomplished by removing the cause or source of correlation. While ideal, this may not be a feasible alternative. In such cases, randomization in sampling is used to average out the trends and allow use of the ANOVA techniques.

In many cases the equal-variance assumption does not hold as a result of a correlation between the mean and the standard deviation. Often, this relationship can be eliminated and the variance made uniform by a simple transformation of the data. As an illustration, Table 3.1 prescribes a transformation needed to stabilize the variance in the event that the mean and standard deviation are related by three typical deterministic relationships.

In general, if σ is proportional to $f(\mu)$, the appropriate transformation to stabilize the variance is given by

$$Y^* = \int \frac{\mathrm{d}Y}{f(Y)}. \tag{3.1}$$

The proof of this statement follows. Let the original data Y be distributed with mean μ and variance σ^2. Let Y^* be the transformed data such that $V[Y^*]$ is equal to a constant K. Using the transformation

$$Y^* = g(Y) \tag{3.2}$$

and a Taylor's series, it is possible to expand $g(Y)$ about μ, as shown in equation (3.3):

$$Y^* = g(\mu) + g'(\mu)(Y - \mu). \tag{3.3}$$

Since $V[Y^*] = K$, we can conclude that

$$\begin{aligned} K &= V[g(\mu)] + [g'(\mu)]^2 V[Y - \mu] \\ &= 0 + [g'(\mu)]^2 V[Y] \\ &= [g'(\mu)]^2 \sigma^2. \end{aligned} \tag{3.4}$$

If σ is proportional to $f(\mu)$, then we can write

$$\sigma^2 = C[f(\mu)]^2 \tag{3.5}$$

where C is a proportionality constant. Therefore

$$K = [g'(\mu)]^2 [f(\mu)]^2 C. \tag{3.6}$$

Table 3.1 Transformations for uniform variance

Relationship	*Transformation*
σ proportional to μ^2	Reciprocals of observations
σ proportional to μ	Logarithms of observations
σ proportional to $\mu^{1/2}$	Square root of observations

Solving for $g'(\mu)$,

$$g'(\mu) = \left(\frac{K}{C}\right)^{1/2} \frac{1}{f(\mu)}. \tag{3.7}$$

Setting

$$m = \left(\frac{K}{C}\right)^{1/2}$$

it is possible to write

$$g(\mu) = \int \frac{m\,\mathrm{d}\mu}{f(\mu)} + a \tag{3.8}$$

where a is a constant of integration. Without loss in generality, we can set $m = 1$ and $a = 0$ in order to stabilize the variance of the transformed variable. That is,

$$g(Y) = \int \frac{\mathrm{d}Y}{f(Y)}. \tag{3.9}$$

Some experiments are designed to investigate the frequency of occurrence of a particular event, as illustrated in Example 3.1, which addresses a simple quality control problem in the manufacturing area.

EXAMPLE 3.1

In a test of industrial welding robots samples of size 10 were taken at random. For each sample the proportion of robots operating continuously for over 100 hours was computed. It is believed that the true proportion of the entire population of robots which can operate over 100 hours is 0.97. The results obtained from six random samples are as shown in Table 3.2, where $\bar{p}$ is the observed proportion in each sample.

Using the information in Table 3.2 it is desired to investigate if the variance

Table 3.2

Sample	$\bar{p}$
1	0.95
2	0.96
3	0.94
4	0.93
5	0.98
6	0.97

of $\bar{p}$ is or is not homogeneous; in case that it is not, what is an appropriate transformation for stabilizing the variance?

The expected value of the proportion in n robots is p and the variance of the proportion is given by $p(1-p)n^{-1}$. Thus the variance is proportional to $p(1-p)$. Equivalently, the standard deviation is proportional to $[p(1-p)]^{1/2}$. Then an approximate transformation is obtained by

$$\int \frac{\mathrm{d}p}{[p(1-p)]^{1/2}} = 2 \arcsin p^{1/2} \tag{3.10}$$

Thus, the transformed variable may be defined as $X = \arcsin(\bar{p})^{1/2}$. As an illustration, the random sample $\bar{p}_i = 0.95, 0.96, 0.94, 0.93, 0.98, 0.97$ would be transformed into the random sample $X_i = 1.34, 1.37, 1.32, 1.30, 1.43$ and 1.40, respectively. The value of X_1, for instance, is computed as $X_1 = \arcsin 0.95^{1/2} = 1.34$. The transformation $X = \arcsin(\bar{p})^{1/2}$ can be obtained using the following procedure: let $p = \sin^2\theta$; therefore, $\mathrm{d}p = 2\sin\theta\cos\theta\,\mathrm{d}\theta$ and $p(1-p) = \sin^2\theta\cos^2\theta$. Now it is possible to write

$$\begin{aligned}\int \frac{\mathrm{d}p}{[p(1-p)]^{1/2}} &= \int \frac{2\sin\theta\cos\theta\,\mathrm{d}\theta}{\sin\theta\cos\theta} = 2\theta.\\ &= 2\arcsin p^{1/2}.\end{aligned} \tag{3.11}$$

According to a Taylor's series,

$$\arcsin(\bar{p})^{1/2} = \arcsin p^{1/2} + \frac{1}{2[p(1-p)]^{1/2}}(\bar{p} - p). \tag{3.12}$$

Hence

$$E[\arcsin(\bar{p})^{1/2}] = \arcsin p^{1/2} \tag{3.13}$$

and

$$V[\arcsin(\bar{p})^{1/2}] = \frac{1}{4}\frac{V[\bar{p}]}{p(1-p)} = \frac{1}{4n} \tag{3.14}$$

since $V[\bar{p}] = p(1-p)n^{-1}$. As can be seen here, the transformed data indeed have a constant variance.

Correlation between σ and μ is frequently found in samples exhibiting nonnormality. If the errors are not normally distributed the true probability of accepting the hypothesis that there is no treatment effect is not equal to the probability given by significance tables. Davies [1] discusses the seriousness of nonnormality in two situations:

(a) In experiments using internal controls and comparing the means, normal significance tables allow a sufficient degree of accuracy even when there is a rather large deviation from normality. This assumption will hold for the majority of planned experiments.

(b) Experiments to compare variances of independent groups of observations will show more serious problems with deviations from the assumption of normality.

The following comments given in this paragraph are extracted from Lentner and Bishop [2]. In most real world experiments it is generally agreed that one or more assumptions are violated to some degree. The violation in a given research setting affects the validity of inferences under some statistical procedures more severely than under others. We are interested primarily in validity of inferences under ANOVA as this is the basic tool of analyzing designed experiments. Thus, our concern is whether or not the underlying assumptions of the ANOVA methodology are violated to the extent of appreciably changing the error rates. The validity of inferences from ANOVA has been discussed in three classical papers by Bartlett [3], Cochran [4] and Eisenhart [5]. In summary, these reported results conclude that:

(a) The classical ANOVA F-test is very robust with respect to nonnormality.
(b) When there are equal replications, mild departure from homogeneous variances have little impact on the validity of the ANOVA test.
(c) Unequal variances have a greater impact on estimates of treatments effects and their variances than on tests.

Variances stabilizing transformation are, therefore, more important for estimation than testing purposes. In cases of severe inequalities, they are important for both estimation and testing, of course.

There are typically three distinctive characteristics that affect the normality of a distribution: asymmetry, peakedness, and trends between the mean and the standard deviation. The occurrence of correlation patterns between the mean and standard deviation, as previously discussed, can be corrected with the transformation given in equation (3.9). The other two characteristics – asymmetry and peakedness – refer to the shape of the density distribution and can be studied by plotting the data. Asymmetry (skewness) is reflected in distributions similar to those shown in Figure 3.3. Peakedness is graphically

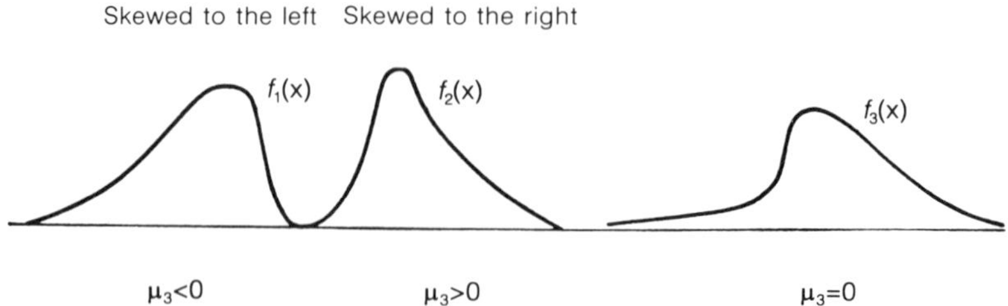

Figure 3.3 Illustration of skewness.

illustrated in Figure 3.4. As can be seen in this figure, there are two typical cases:

(a) distributions which allow less area in a neighborhood of the mean than a normal distribution;
(b) distributions which allow more area in a neighborhood of the mean than a normal distribution.

The first type of distribution is referred to as **platykurtic** and the second type as **leptokurtic**.

The extent of nonnormality can be measured in terms of two coefficients referred to as the **skewness** and **kurtosis** coefficients. The first coefficient is defined as the third central moment of a statistical distribution. The nth central moment of the distribution of a random variable x having mean μ is defined as

$$\mu_n = E(x - \mu)^n.$$

As can be seen, $\mu_1 = E[X - \mu] = E[X] - \mu = \mu - \mu = 0$. Also, by definition, $\mu_2 = E(X - \mu)^2 = \sigma^2$ and, therefore, the second central moment is a measure of dispersion (variance). Moreover, it can be proved that the central moments of the normal distribution are given by the recursive relationship

$$\mu_n = (n-1)\sigma^2 \mu_{n-2}. \tag{3.15}$$

(a)

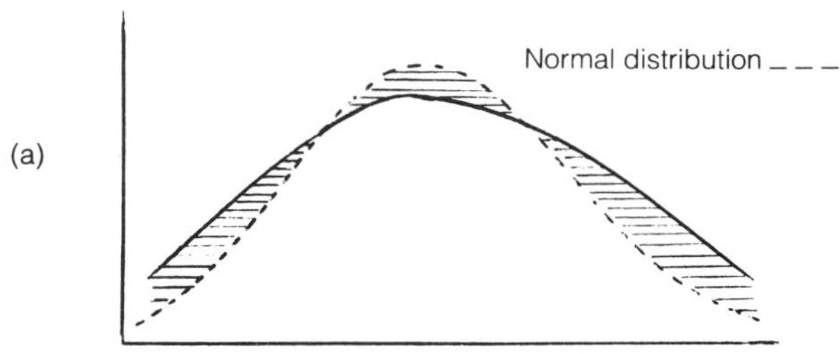

(b)

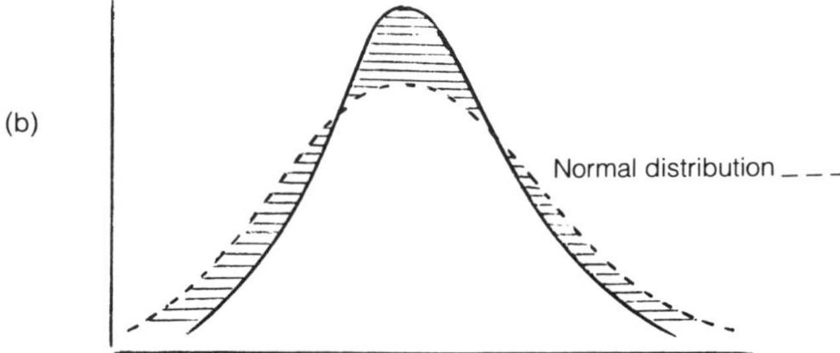

Figure 3.4 Illustration of peakedness: (a) platykurtic, (b) leptokurtic.

For example,

$$\mu_3 = 2\sigma^2\mu_1 = 0$$

and

$$\mu_4 = 3\sigma^2\mu_2 = 3\mu_2{}^2 = 3\sigma^4. \tag{3.16}$$

The quantity

$$K = \frac{\mu_4}{\sigma^4} - 3 \tag{3.17}$$

which is thus zero for the normal distribution is called the kurtosis of a distribution. Distributions with $\mu_3 > 0$ are skewed to the right; if $\mu_3 < 0$ they are skewed to the left. Additionally, platykurtic distributions have $K < 0$ and leptokurtic distributions $K > 0$. Since typically μ_3 and K are not known in advance, then it is possible to estimate them from a sample of random observations. The corresponding estimates, computed on the basis of a sample of size m, are given below:

$$\hat{\mu}_3 = \sum_{i=1}^{m} \frac{(X_i - \bar{X})^3}{m}$$

$$\hat{K} = \sum_{i=1}^{m} \frac{(X_i - \bar{X})^4}{mS^4} - 3$$

where $\bar{X}$ is the sample mean and S, an estimate of the standard deviation of the distribution, is defined as

$$S = \left[\sum_{i=1}^{m} \frac{(X_i - \bar{X})^2}{m} \right]^{1/2}.$$

If $\hat{\mu}_3$ and $\hat{K}$ are significantly different from zero the normality of the sampled population cannot be accepted.

3.2 ANALYSIS OF VARIANCE (ANOVA)

Any random observation used in the analytical methodology presented in this section is assumed to have three additive components:

(a) a common effect
(b) a treatment effect
(c) a random error.

By definition, the common effect is a constant value present in all observations; all observations in the sample taken from a given treatment population have a quantity (fixed or random) that may be different from treatment to treatment; finally, the error is a random quantity that cannot be predicted in advance but whose expected value is equal to zero. The basic purpose of the significance test is to compare an estimate of the treatment effect to an estimate of the random error. Assuming that the random error is negligible, it is possible through this comparison to establish if the treatment effect is or is not negligible itself.

The single factor under consideration may take on k fixed values or a larger number of values which is either very difficult or impossible to incorporate in the analysis. When the number of values or levels is very large, only k levels are chosen at random from the population of levels. Usually, the levels of the factor are also referred to as treatments.

The mathematical model that describes the sources of variation in each of the observations taken at random from the populations corresponding to the k treatments is given by

$$Y_{ij} = \mu + \tau_j + \varepsilon_{ij} \tag{3.18}$$

where Y_{ij} is the ith random observation from jth treatment, μ the common effect, τ_j the effect due to treatment j and ε_{ij} the random error in the ith observation from treatment j.

When all levels are a small set of fixed numerical values or qualitative choices the mathematical model of the experiment is called a **fixed effect model**. In the case where all of the treatments are chosen at random from the population of possible treatments, the model is designated as a **random effect model**. Table 3.3 summarizes the fundamental differences between these models. In this table, $\sigma_e{}^2$ is the variance of the error and $\sigma_\tau{}^2$ the variance of the population of treatments.

In order to clarify the applicability of each model a simple manufacturing problem is considered. Suppose that it is desired to test if all operators in a group can perform a manufacturing process at the same level of performance, measured in terms of percent of defective items produced. It has been established that six batches are appropriate for operators to be efficiently compared to one another. Each operator, therefore, produces six batches and the number of defective items in each batch is counted. If the number of operators is 5 the total number of batches to be manufactured is equal to 30. These 30 batches should be assigned to the operators at random in

Table 3.3 Comparison of random and fixed models

Fixed-effect model	*Random-effect model*
τ_j is a fixed constant	τ_j is a random variable
$\sum_j n_j \tau_j = 0$	$\tau_j \sim N(0, \sigma_\tau^2)$
$H_0: \tau_j = 0, j = 1, 2, \ldots, k$	$H_0: \sigma_3^2 = 0$
$E[Y_{ij}] = \mu + \tau_j$	$E[Y_{ij}] = \mu$
$V[Y_{ij}] = \sigma_e^2 = \sigma^2$	$V[Y_{ij}] = \sigma_\tau^2 + \sigma_e^2$
One-way ANOVA	One-way ANOVA

order to eliminate possible trends. In this case a fixed effect model can be used to investigate the operator effect; the number of treatments (levels) used is equal to the number of treatments available. However, if the number of operators available were equal to 25 a total number of 150 batches would be needed. Assuming that only 30 batches are allowed, then only 5 operators can be used in the experiment. In this case a random effect model would be suitable, with 5 treatments chosen at random from the entire population of 25 operators.

3.2.1 The fixed-effect ANOVA model

Consider a normal distribution with mean equal to μ and variance σ^2. Furthermore, suppose that this population can be divided into k groups or treatments; the jth treatment corresponds to a population which is normally distributed (normality assumption) with mean μ_j and variance σ^2 (equal-variance assumption). As an illustration, assume that there is a factor for which three treatments can be considered. A graphical representation of the three treatment populations is given in Figure 3.5; in this figure, Y_{i1}, Y_{i2}, Y_{i3} are the ith observations from treatment populations $j = 1, 2, 3$. It is possible to express Y_{ij} as the sum of a constant plus a random variable, that is,

$$Y_{ij} = \mu_{.j} + \varepsilon_{ij} \tag{3.18a}$$

where ε_{ij} is the random error contained in Y_{ij}. In this equation, a dot is placed in front of the subscript of the mean of the jth treatment population to indicate that the experimental data are grouped according to one classification way, i.e. treatments (or columns). This notation will become more valuable in further chapters.

From Figure 3.5 it can be observed that if all three treatment populations were identical then $\mu_{.1} = \mu_{.2} = \mu_{.3} = \mu$. In other words, there would be no treatment effect when $\mu_{.1} - \mu = \mu_{.2} - \mu = \mu_{.3} - \mu = 0$.

Let the effect of the jth treatment be defined as

$$\tau_j = \mu_{.j} - \mu. \tag{3.19}$$

As can be seen in Figure 3.5, $\tau_1 < 0$, $\tau_2 < 0$ and $\tau_3 > 0$. Moreover, $\sum_{j=1}^{k} n_j \tau_j = 0$. Substituting equation (3.19) in equation (3.18a) it is possible to reformulate the fixed-effect ANOVA model as indicated in the following relationship:

$$Y_{ij} = \mu + \tau_j + \varepsilon_{ij} \tag{3.20}$$

where $i = 1, 2, \ldots, n_j$, and $j = 1, 2, \ldots, k$.

Table 3.4 summarizes the notation and data format used to rearrange the basic information needed for the ANOVA methodology. From the basic data it is possible to calculate three additional pieces of information used in the analytical development of ANOVA:

(a) the grand total $T_{..} = \sum_{j=1}^{k} T_{.j}$;

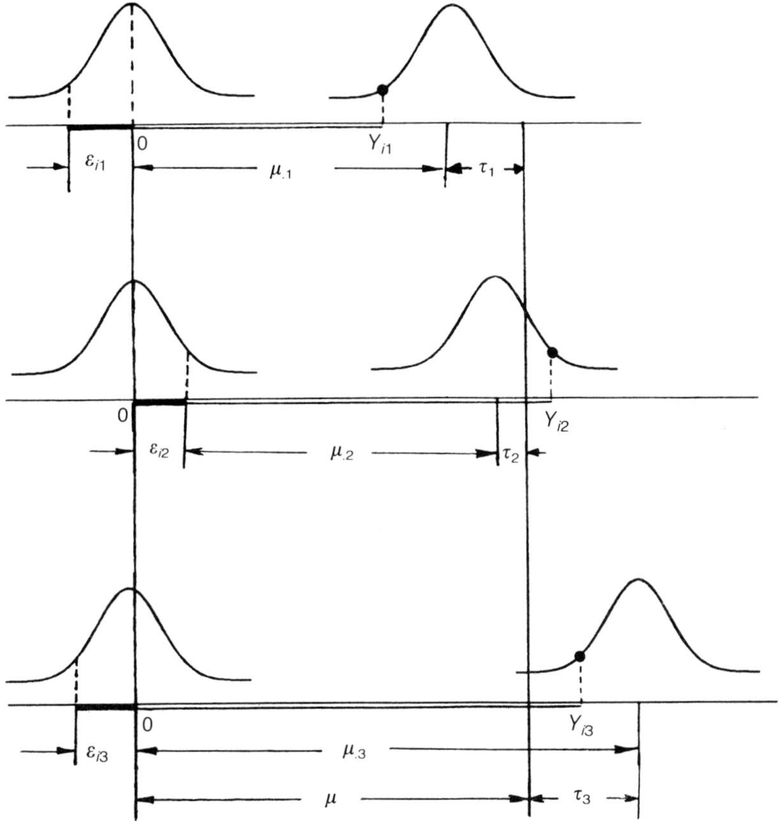

Figure 3.5 The fixed model.

Table 3.4 Basic notation and data format

	Treatments					
	$j=1$	$j=2$	$j=3\cdots$	j	$\cdots j=k-1$	$j=k$
	Y_{11} Y_{21} $\vdots$ $Y_{n_1,1}$	Y_{12} Y_{22} $\vdots$ $Y_{n_2,2}$	Y_{13} Y_{23} $\vdots$ $Y_{n_3,3}$	Y_{1j} Y_{2j} $\vdots$ $Y_{n_j,j}$	$Y_{1,k-1}$ $Y_{2,k-1}$ $\vdots$ $Y_{n_{k-1},k-1}$	Y_{1k} Y_{2k} $\vdots$ $Y_{n_k,k}$
Total	$T_{\cdot 1}$	$T_{\cdot 2}$	$T_{\cdot 3}$	$T_{\cdot j}$	$T_{\cdot k-j}$	$T_{\cdot k}$
Sample size	n_1	n_2	n_3	n_j	n_{k-1}	n_k
Sample mean	$\bar{Y}_{\cdot 1}$	$\bar{Y}_{\cdot 2}$	$\bar{Y}_{\cdot 3}$	$\bar{Y}_{\cdot j}$	$\bar{Y}_{\cdot k-1}$	$\bar{Y}_{\cdot k}$

(b) the total number of observations $N = \sum_{j=1}^{k} n_j$;
(c) the grand average $\overline{Y}_{..} = T_{..}/N$.

The total deviation $Y_{ij} - \mu$ can be broken down into two components:

$$Y_{ij} - \mu = (\mu_{.j} - \mu) + (Y_{ij} - \mu_{.j}). \tag{3.21}$$

The first component is the deviation of the jth treatment mean from the population mean, that is, τ_j. The second component is the deviation of the ith observation in the jth population from the corresponding population mean, that is, ε_{ij}. Since the parameters are not known, the true variations cannot be established. However, a sample equation based on the above equation can be formulated to estimate the total variation using statistics that estimate the parameters:

$$Y_{ij} - \overline{Y_{..}} = (\overline{Y_{.j}} - \overline{Y_{..}}) + (Y_{ij} - \overline{Y_{.j}}).$$

In order to take into consideration the magnitude of the deviations regardless of algebraic signs, the squares of the deviations can be computed:

$$(Y_{ij} - \overline{Y_{..}})^2 = (\overline{Y_{.j}} - \overline{Y_{..}})^2 + (Y_{ij} - \overline{Y_{.j}})^2 + 2(\overline{Y_{.j}} - \overline{Y_{..}})(Y_{ij} - \overline{Y_{.j}}) \tag{3.22}$$

After taking double sum with respect to i and j, equation (3.22) yields

$$\sum_j \sum_i (Y_{ij} - \overline{Y_{..}})^2 = \sum_j \sum_i (\overline{Y_{.j}} - \overline{Y_{..}})^2 + \sum_j \sum_i (Y_{ij} - \overline{Y_{.j}})^2 + \sum_j \sum_i 2(\overline{Y_{.j}} - Y_{..})(Y_{ij} - \overline{Y_{.j}}). \tag{3.23}$$

The last term of equation (3.23) is equal to zero, as can be seen below:

$$\begin{aligned}\sum_j \sum_i (\overline{Y_{.j}} - \overline{Y_{..}})(Y_{ij} - \overline{Y_{.j}}) &= \sum_j (\overline{Y_{.j}} - \overline{Y_{..}}) \sum_i (Y_{ij} - \overline{Y_{.j}}) \\ &= \sum_j (\overline{Y_{.j}} - \overline{Y_{..}}) \sum_i (Y_{ij} - \overline{Y_{.j}}) \\ &= \sum_j (\overline{Y_{.j}} - \overline{Y_{..}})(n_j \overline{Y_{.j}} - n_j \overline{Y_{.j}}) = 0.\end{aligned}$$

Now define the following sums of squares:

$$SS_{\text{total}} = \sum_j \sum_i (Y_{ij} - \overline{Y_{..}})^2 \tag{3.24}$$

$$SS_{\text{treatment}} = \sum_j \sum_i (\overline{Y_{.j}} - \overline{Y_{..}})^2 \tag{3.25}$$

$$SS_{\text{error}} = \sum_j \sum_i (Y_{ij} - \overline{Y_{.j}})^2. \tag{3.26}$$

Since the sum of all cross products is equal to zero, as has been previously shown, then the total sum of squares, SS_{total}, can be written as

$$SS_{\text{total}} = SS_{\text{treatment}} + SS_{\text{error}}. \tag{3.27}$$

Equation (3.27) is the **fundamental equation of the analysis of variance.**

Once the sums of squares for the total, treatments and error are computed, three additional statistics, usually referred to as *mean squares*, can be defined as follows by dividing each sum of squares by the corresponding number of degrees of freedom (the number of degrees of freedom is actually the number of independent squares in the sum):

$$MS_{\text{total}} = \frac{SS_{\text{total}}}{N-1} \tag{3.28}$$

$$MS_{\text{treatment}} = \frac{SS_{\text{treatment}}}{k-1} \tag{3.29}$$

$$MS_{\text{error}} = \frac{SS_{\text{error}}}{N-k}. \tag{3.30}$$

The expected value of the treatment mean square can be computed as follows:

$$E[MS_{\text{treatment}}] = E\left[\frac{SS_{\text{treatment}}}{k-1}\right] = \frac{E[SS_{\text{treatment}}]}{k-1}. \tag{3.31}$$

Additionally,

$$E[SS_{\text{treatment}}] = E\left[\sum_j \sum_i (\overline{Y}_{.j} - \overline{Y}_{..})^2\right]. \tag{3.32}$$

The procedure to find this expected value will be illustrated assuming that all n_j' are equal to n. In this case,

$$E[SS_{\text{treatment}}] = n\sum_j E[(\overline{Y}_{.j} - \overline{Y}_{..})^2]. \tag{3.33}$$

By definition,

$$\overline{Y}_{..} = \frac{1}{k}(\overline{Y}_{.1} + \overline{Y}_{.2} + \cdots + \overline{Y}_{.k}).$$

Therefore

$$\overline{Y}_{.j} - \overline{Y}_{..} = \frac{k-1}{k}\overline{Y}_{.j} - \frac{1}{k}\sum_{m \neq j} \overline{Y}_{.m}. \tag{3.34}$$

Recall that

$$E[X^2] = V[X] + (E[X])^2. \tag{3.35}$$

Therefore

$$E[SS_{\text{treatment}}] = n\sum_j V[\overline{Y}_{.j} - \overline{Y}_{..}] + n\sum_j (E[\overline{Y}_{.j} - \overline{Y}_{..}])^2. \tag{3.36}$$

Equation (3.36) can be rewritten as follows, taking into consideration the definition of treatment effects:

$$E[SS_{\text{treatment}}] = n\sum_j V[\overline{Y}_{.j} - \overline{Y}_{..}] - n\sum_j \tau_j^2. \tag{3.37}$$

Using equation (3.34), and assuming that all sample averages have equal variance and are independently distributed, it is possible to conclude that

$$\begin{aligned} V[\overline{Y}_{.j} - \overline{Y}_{..}] &= \frac{(k-1)^2}{k^2} V[\overline{Y}_{.j}] + \frac{(k-1)}{k^2} V[\overline{Y}_{.j}] \\ &= \frac{(k-1)}{k} V[\overline{Y}_{.j}]. \end{aligned} \tag{3.38}$$

But, as can be seen in Table 3.3, $V[Y_{ij}] = \sigma^2$, and $V[\overline{Y}_{.j}] = \sigma^2 n^{-1}$. Therefore

$$V[\overline{Y}_{.j} - \overline{Y}_{..}] = \frac{k-1}{k}\frac{\sigma^2}{n} \tag{3.39}$$

and, consequently,

$$E[SS_{\text{treatment}}] = (k-1)\sigma^2 + n\sum_j \tau_j^2. \tag{3.40}$$

Finally,

$$E[MS_{\text{treatment}}] = \sigma^2 + n\sum_j \frac{\tau_j^2}{k-1}. \tag{3.41}$$

Following a similar procedure, it is possible to prove that

$$E[MS_{\text{error}}] = \sigma^2. \tag{3.42}$$

If $H_0: \tau_j = 0$, for all j, is true, $E[MS_{\text{treatment}}] = \sigma^2$. Therefore, in this case, two independent unbiased estimates of σ^2, namely $MS_{\text{treatment}}$ and MS_{error}, are available. Then, under the assumption that H_0 is true, it is possible to define two independent chi-square statistics, as shown in equations (3.43) and (3.44) below, where the subscripts of the chi-square terms indicate the corresponding number of degrees of freedom:

$$\frac{SS_{\text{treatment}}}{\sigma^2} \sim \chi^2_{k-1} \tag{3.43}$$

$$\frac{SS_{\text{error}}}{\sigma^2} \sim \chi^2_{N-k}. \tag{3.44}$$

The above statistics are independent chi-square random variables as demonstrated by Cochran [4]. Cochran proved that if $Z_1, Z_2, \ldots, Z_v$ are normally and independently distributed with mean of 0 and standard deviation of 1, and

$$\sum_{i=1}^{v} Z_i^2 = \chi^2_{v_1} + \chi^2_{v_2} + \cdots + \chi^2_{v_s} \tag{3.45}$$

then $\chi^2_{v_1}, \chi^2_{v_2}, \ldots, \chi^2_{v_s}$ are independent if and only if $v_1 + v_2 + \cdots + v_s = v$. Since the sum of degrees of freedom for $SS_{\text{treatment}}$ and SS_{error} is equal to

$$(K-1)+(N-k)=N-1$$

then $SS_{\text{treatment}}/\sigma^2$ and $SS_{\text{error}}/\sigma^2$ are independently distributed chi-square random variables. Recalling the definition of the F-statistic, it is concluded that

$$F_{k-1,N-k} = \frac{SS_{\text{treatment}}}{\sigma^2(k-1)}\left[\frac{SS_{\text{error}}}{\sigma^2(N-k)}\right]^{-1} \tag{3.46}$$

that is,

$$F_{k-1,N-k} = \frac{MS_{\text{treatment}}}{MS_{\text{error}}} \tag{3.47}$$

under the assumption that the null hypothesis is true. It is left as an exercise for the reader to prove that the sums of squares can be expressed in terms of the observed totals $T_{..}$ and $T_{.j}$ as indicated below:

$$SS_{\text{treatment}} = \sum_j \frac{T_{.j}^2}{n_j} - \frac{T_{..}^2}{N} \tag{3.48}$$

$$SS_{\text{total}} = \sum_i \sum_j Y_{ij}^2 - \frac{T_{..}^2}{N}. \tag{3.49}$$

Once the sums of squares for the treatments and the total are computed, SS_{error} can be found by subtraction, as indicated by equation (3.27). That is,

$$SS_{\text{error}} = SS_{\text{total}} - SS_{\text{treatment}}. \tag{3.50}$$

The general format for the ANOVA table, used to summarize the results of the analysis of variance, is given in Table 3.5.

Table 3.5 The general analysis of variance table

Source of variation	*df*	*SS*	*MS*	*F*
Between treatments	$k-1$	$SS_{\text{treatment}}$	$\frac{SS_{\text{treatment}}}{(k-1)}$	$\frac{MS_{\text{treatment}}}{MS_{\text{error}}}$
Within treatments	$N-k$	SS_{error}	$\frac{SS_{\text{error}}}{(N-k)}$	
Total variation	$N-1$	SS_{total}		

EXAMPLE 3.2

Three computer-controlled lathes are being tested with interest in the consistency of performance of the machines with the same program. A specific spindle diameter is to be measured for all three machines. Since only three machines are used, the experiment has only one factor at three fixed levels. Also, no numerical values can be assigned to the machines so the levels are qualitative. Six observations per machine are considered adequate, and the order for testing can be completely randomized. The mathematical model is

$$Y_{ij} = \mu + \tau_j + \varepsilon_{ij}$$

where $i = 1, 2, 3, 4, 5, 6$ and $j = 1, 2, 3$. There are three treatments (machines) and six observations per machine. Data are as shown in Table 3.6.

The original data were coded by subtracting 0.25 from each observation and then multiplying by 10^5. Coded results for the analysis of variance are shown in Table 3.7. From these results, it is possible to calculate the grand total and the sum of squared observations:

$$T_{..} = 295$$

$$\sum_j \sum_i Y_{ij}^2 = 9661.$$

Table 3.6 Original data

I	II	III
0.25040	0.24995	0.24999
0.25035	0.25000	0.25020
0.25042	0.25010	0.25017
0.25037	0.25007	0.25010
0.25030	0.25001	0.25012
0.25041	0.24997	0.25002

Table 3.7 Coded data

	I	II	III
	40	−5	−1
	35	0	20
	42	10	17
	37	7	10
	30	1	12
	41	−3	2
$T_{.j}$	225	10	60
n_j	6	6	6
$\sum_i Y_{ij}^2$	8539	184	938

Table 3.8 ANOVA table

Source	*df*	*SS*	*MS*	*F*
Treatments	2	4219.7	2109.70	52.10
Error	15	606.8	40.46	
Total	17	4826.5		

Using equations (3.48)–(3.50), the results obtained are summarized in the ANOVA table (Table 3.8). The critical value of the F-statistic with 2 degrees of freedom in the numerator and 15 degrees of freedom in the denominator is equal to 3.68, when the level of significance is set equal to 0.05. As can be seen, the observed value of F of 52.1 is highly significant and H_0 is, therefore, rejected. Thus, it is possible to conclude that a difference in the average diameter of the test part produced by the three machines exists. However, it is not possible to establish which of the three treatments are significantly different at this point.

3.2.2 The random-effect ANOVA model

When the one-factor experiment is conducted after choosing the treatments at random, a meaningful analysis of the results must include estimates of the variation due to differences in treatments, as well as the variation due to the random error. For this reason, the random model is generally referred to as a **components-of-variance model**. A graphical illustration of the meaning of each parameter and each random variable involved in the formulation of the components-of-variance model is given in Figure 3.6. In this figure, the basic components of Y_{ij} are:

(a) the common effect μ;
(b) the random treatment effect τ_j;
(c) the random error.

It is assumed that $\tau_j \sim N(0, \sigma_\tau^2)$, $\varepsilon_{ij} \sim N(0, \sigma_e^2)$ and that τ_j and ε_{ij} are independent random variables. According to the random model

$$Y_{ij} = \mu + \tau_j + \varepsilon_{ij} \tag{3.51}$$

where $i = 1, 2, \ldots, n_j$ and $j = 1, 2, \ldots, k$, the mean and variance of the random variable Y_{ij} are equal to μ and $\sigma_\tau^2 + \sigma_e^2$, respectively. Note that if this were a fixed effect model, the mean and variance would instead be equal to $\mu + \tau_j$ and σ_e^2.

The sum of squares of the treatments is again defined as in the case of the fixed effect model. That is,

$$SS_{\text{treatment}} = \sum_j \sum_i (\overline{Y}_{.j} - \overline{Y}_{..})^2. \tag{3.52}$$

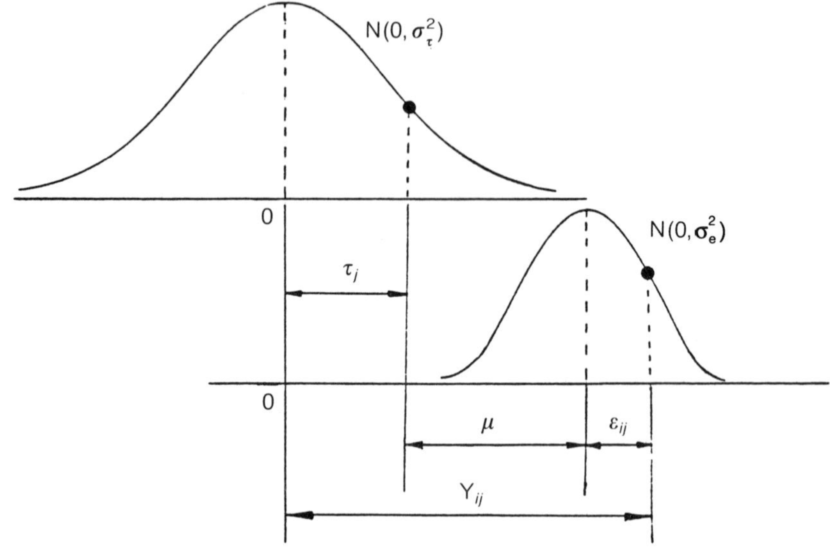

Figure 3.6 The random ANOVA model.

From equation (3.51) it is concluded that

$$\overline{Y_{.j}} = \frac{\sum_i Y_{ij}}{n_j} = \mu + \tau_j + \frac{\sum_i \varepsilon_{ij}}{n_j} \tag{3.53}$$

and

$$\overline{\overline{Y_{..}}} = \frac{\sum_j \sum_i Y_{ij}}{N} = \mu + \frac{\sum_j n_j \tau_j}{N} + \frac{\sum_j \sum_i \varepsilon_{ij}}{N}. \tag{3.54}$$

Therefore

$$\begin{aligned}\overline{Y_{.j}} - \overline{\overline{Y_{..}}} &= \left(\tau_j - \frac{\sum_j n_j \tau_j}{N} \right) + \left(\frac{\sum_i \varepsilon_{ij}}{n_j} - \frac{\sum_j \sum_i \varepsilon_{ij}}{N} \right) \\ &= (\tau_j - \bar{\tau}) + (\overline{\varepsilon_{.j}} - \overline{\overline{\varepsilon_{..}}})\end{aligned} \tag{3.55}$$

where for notational convenience the symbols $\bar{\tau}$, $\overline{\varepsilon_{.j}}$ and $\overline{\overline{\varepsilon_{..}}}$ are used to represent the following averages:

$$\bar{\tau} = \frac{\sum_j n_j \tau_j}{N}$$

$$\overline{\varepsilon_{.j}} = \frac{\sum_i \varepsilon_{ij}}{n_j}$$

$$\overline{\overline{\varepsilon_{..}}} = \frac{\sum_j \sum_i \varepsilon_{ij}}{N}.$$

From equation (3.55),

$$(\overline{Y_{.j}} - \overline{Y_{..}})^2 = (\tau_j - \bar{\tau})^2 + (\overline{\varepsilon_{.j}} - \overline{\overline{\varepsilon_{..}}})^2 + 2(\tau_j - \bar{\tau})(\overline{\varepsilon_{.j}} - \overline{\overline{\varepsilon_{..}}}).$$

Therefore, multiplying each term by n_j,

$$n_j(\overline{Y_{.j}} - \overline{Y_{..}})^2 = n_j(\tau_j - \bar{\tau})^2 + n_j(\overline{\varepsilon_{.j}} - \overline{\overline{\varepsilon_{..}}})^2 + 2n_j(\tau_j - \bar{\tau})(\overline{\varepsilon_{.j}} - \overline{\overline{\varepsilon_{..}}}).$$

Taking sums over all possible treatments,

$$\sum_j n_j(\overline{Y_{.j}} - \overline{Y_{..}})^2 = \sum_j n_j(\tau_j - \bar{\tau})^2 + \sum_j n_j(\overline{\varepsilon_{.j}} - \overline{\overline{\varepsilon_{..}}})^2 + 2\sum_j n_j(\tau_j - \bar{\tau})(\overline{\varepsilon_{.j}} - \overline{\overline{\varepsilon_{..}}}).$$

Dividing each term by the number of degrees of freedom of the treatments,

$$\frac{\sum_j n_j(\overline{Y_{.j}} - \overline{Y_{..}})^2}{k-1} = \frac{\sum_j n_j(\tau_j - \bar{\tau})^2}{k-1} + \frac{\sum_j n_j(\overline{\varepsilon_{.j}} - \overline{\overline{\varepsilon_{..}}})^2}{k-1} + \frac{2\sum_j n_j(\tau_j - \bar{\tau})(\overline{\varepsilon_{.j}} - \overline{\overline{\varepsilon_{..}}})}{k-1}.$$

Finally, taking expected values, $E[MS_{\text{treatment}}]$ can be written as follows:

$$E\left[\frac{\sum_j n_j(\overline{Y_{.j}} - \overline{Y_{..}})^2}{k-1}\right] = E\left[\frac{\sum_j n_j(\tau_j - \bar{\tau})^2}{k-1}\right] + E\left[\frac{\sum_j n_j(\overline{\varepsilon_{.j}} - \overline{\overline{\varepsilon_{..}}})^2}{k-1}\right] + 2E\left[\frac{\sum_j n_j(\tau_j - \bar{\tau})(\overline{\varepsilon_{.j}} - \overline{\overline{\varepsilon_{..}}})}{k-1}\right]. \qquad (3.56)$$

The third term of the right-hand side of equation (3.56) is equal to zero, since by assumption τ_j and $\overline{\varepsilon_{.j}}$ are independent random variables.

Since by assumption $\tau_1, \tau_2, \ldots, \tau_k$ are normally and independently distributed random variables with mean equal to zero and variance σ_τ^2,

$$E[\tau_j^2] = \sigma_\tau^2 \qquad (3.57)$$

and

$$E[\tau_i \tau_j] = 0. \qquad (3.58)$$

The expected value of the first term of the right-hand side of equation (3.56) can be obtained as follows. First, for $\bar{\tau} = \sum n_j \tau_j N^{-1}, \sum_j n_j = N$, it is concluded that $E[\bar{\tau}] = 0$ and that

$$V[\bar{\tau}] = \sum \left(\frac{n_j}{N}\right)^2 \sigma_\tau^2. \qquad (3.59)$$

It is also noted that $E[(\bar{\tau})^2] = V[\bar{\tau}]$. Now let us consider the product

$$\tau_j \bar{\tau} = \tau_j \frac{\sum_j n_j \tau_j}{N} = \sum_{i \neq j} \left(\frac{\sum_i n_i}{N}\right) \tau_i \tau_j + \left(\frac{n_j}{N}\right) \tau_j^2. \qquad (3.60)$$

Using equation (3.58) and (3.60) we concluded that

$$E[\tau_j \bar{\tau}] = \frac{n_j}{N} \sigma_\tau^2. \qquad (3.61)$$

Finally, it is noted that

$$E\left[\sum_j n_j(\tau_j-\bar{\tau})^2\right]=E\left[\sum_j n_j\tau_j^2\right]-2E\left[\sum_j n_j\tau_j\bar{\tau}\right]+E\left[\sum_j n_j(\bar{\tau})^2\right]$$

$$=\sum_j n_j\sigma_j^2-2\sum_j n_j\left(\frac{n_j}{N}\right)\sigma_\tau^2+\sum_j n_j\sum_j\left(\frac{n_j}{N}\right)^2\sigma_\tau^2$$

$$=\left(\sum_j n_j\right)\sigma_\tau^2-\frac{2}{N}\left(\sum_j n_j^2\right)\sigma_\tau^2+\frac{(\sum_j n_j)}{N^2}\sum_j(n_j)^2\sigma_\tau^2$$

$$=N\sigma_\tau^2-\frac{2\sum_j n_j^2}{N}\sigma_\tau^2+\frac{\sum_j n_j^2}{N}\sigma_\tau^2$$

$$=N\sigma_\tau^2-\frac{\sum_j n_j^2}{N}\sigma_\tau^2$$

$$=\left(\frac{N^2-\sum_j n_j^2}{N}\right)\sigma_\tau^2.$$

Similarly, it is possible to show that

$$E\left[\frac{\sum_j n_j(\bar{\varepsilon}_{.j}-\bar{\bar{\varepsilon}}_{..})^2}{k-1}\right]=\sigma_e^2.$$

From the above results it follows that

$$E[MS_{\text{treatment}}]=\frac{N^2-\sum_j n_j^2}{N(k-1)}\sigma_r^2+\sigma_e^2. \tag{3.62}$$

In the particular case that $n_1=n_2=\cdots=n_k=n$, equation (3.62) yields the result

$$E[MS_{\text{treatment}}]=n\sigma_\tau^2+\sigma_e^2.$$

Note that in the case that $\sigma_\tau^2=0$ then equation (3.62) reduces to

$$E[MS_{\text{treatment}}]=\sigma_e^2. \tag{3.63}$$

It is also possible to prove that

$$E[MS_{\text{error}}]=\sigma_e^2. \tag{3.64}$$

Therefore, as in the analysis of the fixed effect model, there are two independent unbiased estimates of σ_e^2 assuming that $\sigma_\tau^2=0$. As a result of this, equation (3.47) also holds in the case of a random effect model. In this case, the hypothesis $H_0:\sigma_\tau^2=0$ can be tested as follows:

(a) Calculate $MS_{\text{treatment}}$.
(b) Calculate MS_{error}.
(c) Compute the ratio $F_0=MS_{\text{treatment}}(MS_{\text{error}})^{-1}$.

(d) From F-tables obtain the critical value $F_c = F_{1-\alpha}(k-1, N-k)$.
(e) If $F_0 > F_c$ reject H_0. Otherwise, accept it.

As in the fixed effect model, it can be proved that the total sum of squares is still equal to the treatment sum of squares plus the error sum of squares, that is

$$SS_{\text{total}} = SS_{\text{treatment}} + SS_{\text{error}}.$$

The above identity states the fact that the total variability in the observations can be partitioned into two components, the first component representing variation between treatments, and the second representing variation within each treatment (random error). A meaningful question in this case is: what fraction of the total variability corresponds to treatment variation and what fraction to random error? This question can be answered proceeding as follows.

Let S_E^2 and S_T^2 be the components of variance due to the random error and treatment variation, respectively. Equations (3.62) and (3.64) suggest the following system of two simultaneous equations in S_E^2 and S_T^2:

$$MS_{\text{treatment}} = \frac{N^2 - \sum_j n_j^2}{N(k-1)} S_T^2 + S_E^2 \tag{3.65}$$

$$MS_{\text{error}} = S_E^2. \tag{3.66}$$

Once S_E^2 and S_T^2 are obtained, the fractions of variability due to error (P_E) and treatments (P_T) can be determined as shown below:

$$P_E = \frac{S_E^2}{(S_E^2 + S_T^2)} \tag{3.67}$$

$$P_T = \frac{S_T^2}{(S_E^2 + S_T^2)}. \tag{3.68}$$

EXAMPLE 3.3

The National Business Machine (NBM) company is putting together a computer system for sale to business. While they manufacture most of their own equipment, they do not wish to take the time at present to develop a letter quality printer (LQP) to go with their system. For this reason, they wish to find the best reasonably priced LQP available to subcontract as a system option. One consideration in the selection is the hours of operation until failure. In order to determine whether there might be some difference between machines, NBM approached three randomly chosen manufacturers, who supplied them with six of the requested models. NBM then tested the three manufacturers' printers, arriving at the results in hours to first failure shown in Table 3.9.

Table 3.9

	Model I	*Model II*	*Model III*
	60	102	121
	45	96	132
	72	105	118
	68	99	128
	71	103	131
	52	95	126
$T_{.j}$	368	600	756
n_j	6	6	6

The analysis of the random-effect model for this particular application is conducted through a sequence of steps as indicated below:

(a) Grand total: $T_{..} = 368 + 600 + 756 = 1724; N = 18$.
(b) Sum of squared observations: $\sum_i \sum_j y_{ij}^2 = 178\,668$.
(c) Total sum of squares, equation (3.49):

$$SS_{\text{total}} = 178\,668 - \frac{(1724)^2}{18} = 178\,668 - 165\,120.89 = 13\,547.11.$$

(d) Treatment sum of squares, equation (3.48):

$$\begin{aligned} SS_{\text{treatment}} &= \frac{(368)^2 + (600)^2 + (756)^2}{6} - 165\,120.89 \\ &= 177\,826.67 - 165\,120.89 \\ &= 12\,705.78. \end{aligned}$$

(e) Error sum of squares, equation (3.50): $SS_{\text{error}} = 13\,547.11 - 12\,705.78 = 841.33$.
(f) ANOVA table ($\alpha = 0.01$, $F_c = 6.36$) (Table 3.10). The value of $F = 113.26$ is highly significant and, hence, the hypothesis that $\sigma_\tau{}^2 = 0$ is rejected.

Table 3.10

Source	*df*	*SS*	*MS*	*F*
Treatments	2	12 705.78	6352.89	113.26
Error	15	841.33	56.09	
Total variation	17	13 547.11		

(g) Components of variance: from equation (3.65):

$$6S_T^2 + S_E^2 = 6352.89.$$

From equation (3.66):

$$S_E^2 = 56.09$$

Solving these equations:

$$S_E^2 = 56.09 \qquad S_T^2 = 1049.47.$$

From equation (3.67):

$$P_E = \frac{56.09}{1049.47 + 56.09} = 0.05.$$

From equation (3.68):

$$P_T = \frac{1049.47}{1049.47 + 56.09} = 0.95.$$

3.3 LEAST-SQUARES REGRESSION SIGNIFICANCE TEST

The purpose of this section is to introduce a general methodology to test models describing a given observed process. The methodology to be considered can be actually used instead of ANOVA to perform hypothesis testing, but it is more general than the analysis of variance technique.

3.3.1 An example of a nonlinear model

As an illustration [6], consider the relationship

$$P = P_0 - (P_0 - P_f)\,e^{-(\rho/W)^\beta} \tag{3.69}$$

where P_0 is a given constant and P_f, ρ and β are unknown parameters; in the above relationship, W and P are the independent and dependent variables, respectively. In order to estimate the parameters of the given model for a particular choice of P_0 the following procedure, known as the least-squares method, can be used. According to this method, a sample of W_i and P_i values is taken at random and then the parameters are estimated by the choices of P_f, ρ and β that will minimize the error sum of squares, which is defined as

$$\sum_{i=1}^{m} \varepsilon_i^2 = \sum_{i=1}^{m} [P_i - P_0 + (P_0 - P_f)\,e^{-(\rho/W_i)^\beta}]^2 \tag{3.70}$$

where m is the sample size. Instead of minimizing the function defined in equation (3.70) an alternative procedure will be used.

The original relationship defined in equation (3.69) can be expressed as

$$P_0 - P = \alpha\, e^{-(\rho/W)^\beta} \tag{3.71}$$

where

$$\alpha = P_0 - P_f.$$

Taking the natural logarithm of equation (3.71) yields

$$\ln(P_0 - P) = \ln(\alpha) - \left(\frac{\rho}{W}\right)^\beta$$

which can also be written as

$$\ln(P_0 - P) = \ln(\alpha) - \rho^\beta\left(\frac{1}{W}\right)^\beta. \tag{3.72}$$

Using the transformation $e^\tau = W^{-1}$, equation (3.72) becomes

$$\ln(P_0 - P) = \ln(\alpha) - \rho^\beta (e^\beta)^\tau$$

which is equivalent to

$$z = a - bc^\tau$$

where variables of substitution are

$$z = \ln(P_0 - P) \tag{3.73}$$

$$a = \ln(\alpha) \tag{3.74}$$

$$b = \rho^\beta \tag{3.75}$$

$$c = e^\beta \tag{3.76}$$

Therefore, the observed values of P_i and W_i are transformed into values of z_i and τ_i, respectively. The statistical model to be used is defined as

$$z_i = a - bc^{\tau_i} + \varepsilon_i \tag{3.77}$$

where ε_i is the random error corresponding to the value z_i associated with τ_i.

The basic procedure to estimate the parameters a, b and c is the well-known 'least-squares' method. This method computes a, b and c in such a way that the quantity $\sum_{i=1}^{m} \varepsilon_i^2$ is minimized. This quantity can be obtained from equation (3.77) as

$$\sum_{i=1}^{m} \varepsilon_i^2 = \sum_{i=1}^{m} (z_i - a + bc^{\tau_i})^2.$$

The necessary (and in this case sufficient) conditions for a minimum are given by the so-called 'normal equations':

$$\frac{\partial(\sum_{i=1}^{m}\varepsilon_i^{2})}{\partial a}=0$$

$$\frac{\partial(\sum_{i=1}^{m}\varepsilon_i^{2})}{\partial b}=0$$

$$\frac{\partial(\sum_{i=1}^{m}\varepsilon_i^{2})}{\partial c}=0.$$

The above conditions can be shown to be equivalent to

$$\sum_{i=1}^{m}(z_i-a+bc^{\tau_i})=0 \tag{3.78}$$

$$\sum_{i=1}^{m}(z_i-a+bc^{\tau_i})c^{\tau_i}=0 \tag{3.79}$$

$$\sum_{i=1}^{m}(z_i-a+bc^{\tau_i})\tau_i c^{\tau_i-1}=0. \tag{3.80}$$

It is noted that equations (3.78) and (3.79) are linear in a and b; therefore, both parameters can be obtained in terms of z_i, τ_i and c. The corresponding results are

$$a=\frac{(\sum_{i=1}^{m}c^{2\tau_i})(\sum_{i=1}^{m}z_i)-(\sum_{i=1}^{m}c^{\tau_i})(\sum_{i=1}^{m}z_ic^{\tau_i})}{m(\sum_{i=1}^{m}c^{2\tau_i})-(\sum_{i=1}^{m}c^{\tau_i})(\sum_{i=1}^{m}c^{\tau_i})} \tag{3.81}$$

$$b=\frac{-m(\sum_{i=1}^{m}z_ic^{\tau_i})+(\sum_{i=1}^{m}c^{\tau_i})(\sum_{i=1}^{m}z_i)}{(\sum_{i=1}^{m}c^{2\tau_i})-(\sum_{i=1}^{m}c^{\tau_i})(\sum_{i=1}^{m}c^{\tau_i})}. \tag{3.82}$$

The values of a and b given by equations (3.81) and (3.82) can be substituted into equation (3.80) to obtain the following final result:

$$\sum_{i=1}^{m}z_i\tau_ic^{\tau_i}-\frac{(\sum_{i=1}^{m}c^{2\tau_i})(\sum_{i=1}^{m}z_i)-(\sum_{i=1}^{m}c^{\tau_i})(\sum_{i=1}^{m}z_ic^{\tau_i})}{m(\sum_{i=1}^{m}c^{2\tau_i})-(\sum_{i=1}^{m}c^{\tau_i})(\sum_{i=1}^{m}c^{\tau_i})}\left(\sum_{i=1}^{m}\tau_ic^{\tau_i}\right)$$
$$+\frac{-m(\sum_{i=1}^{m}z_ic^{\tau_i})+(\sum_{i=1}^{m}c^{\tau_i})(\sum_{i=1}^{m}z_i)}{(\sum_{i=1}^{m}c^{2\tau_i})-(\sum_{i=1}^{m}c^{\tau_i})(\sum_{i=1}^{m}c^{\tau_i})}\left(\sum_{i=1}^{m}\tau_ic^{2\tau_i}\right)=0. \tag{3.83}$$

Equation (3.83) can be solved for c using a trial-and-error procedure or a simple numerical analysis method. Once c is determined, a and b can be computed from equations (3.81) and (3.82), respectively, and their correspon-ding values can be used to estimate α (and thus P_f), ρ and β from equations (3.74)–(3.76).

EXAMPLE 3.4

Given the model

$$P = 5 - (5 - P_f)\mathrm{e}^{-(\rho/W)^\beta}$$

find the estimates of the parameters ρ, β and P_f using the data in Table 3.11.

Solution

Table 3.12 summarizes the basic computations $z_i = \ln(5 - P_i)$ and $\tau_i = \ln W_i^{-1}$. The results given in Table 3.13 are obtained by the bisection method for the

Table 3.11

W_i	P_i
0.90	4.8
0.95	4.7
1.10	4.0
1.50	3.1
1.60	3.0

Table 3.12

i	W_i	P_i	z_i	τ_i
1	0.90	4.8	−1.609	0.1053
2	0.95	4.7	−1.204	0.0513
3	1.10	4.0	0	−0.0953
4	1.50	3.1	0.6418	−0.4054
5	1.60	3.0	0.6931	−0.4700

Table 3.13

c	$f(c)$
150	3.6×10^{-4}
125	-1.4×10^{-3}
137.5	-4.9×10^{-4}
143.7	-6.4×10^{-5}
146.875	1.4×10^{-5}
145.313	4.2×10^{-5}
144.531	-1.1×10^{-5}
144.922	1.5×10^{-5}
144.727	2.05×10^{-6}
144.629	-4.5×10^{-6}

solution of the equation

$$f(c) = 0$$

where $f(c)$ is the left-hand term of equation (3.83). By plotting the points given in Table 3.12, it is possible to obtain an estimate of the value of c at which $f(c) = 0$, as shown in Figure 3.7. The value of c equals 144.7 approximately and the corresponding parameters are: $\beta = 4.97$, $\rho = 1.08$ and $P_f = 2.66$.

In order to illustrate the basic concept behind the general least-squares significance test, suppose that it is desired to test that $a = 0$ in the model formulated in equation (3.77). Under the assumption that this hypothesis is true, this model reduces to

$$z_i = -bc^{\tau_i} + \varepsilon_i. \tag{3.84}$$

The model of equation (3.84) is usually referred to as a **reduced model**. The normal equations of the reduced model can be obtained following the same procedure as that for the original model; it can be verified that the

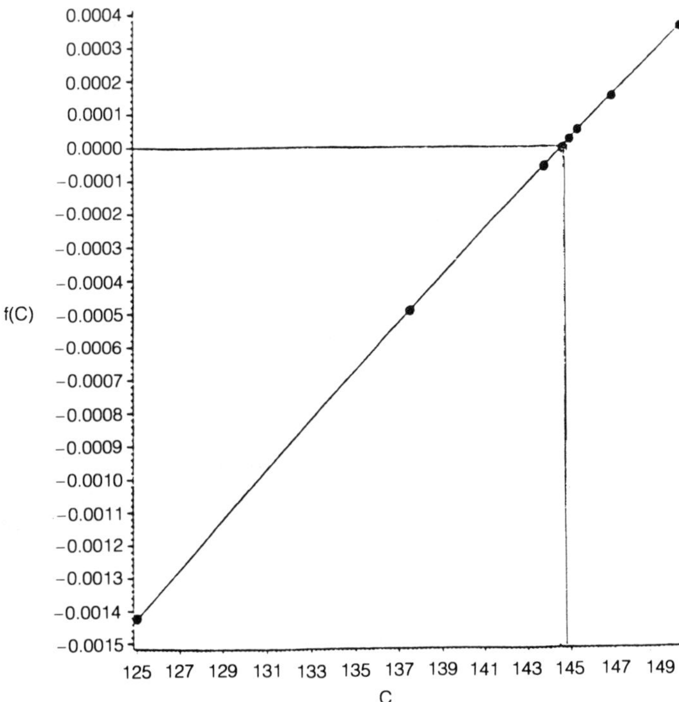

Figure 3.7 $f(c) = 0$ vs c.

corresponding normal equations are given by

$$\sum_{i=1}^{m} (z_i + bc^{\tau_i})c^{\tau_i} = 0 \tag{3.85}$$

$$\sum_{i=1}^{m} (z_i + bc^{\tau_i})\tau_i c^{\tau_i - 1} = 0. \tag{3.86}$$

Eliminating b, the estimate of c can be obtained as the solution of the non-linear equation

$$\frac{\sum_i c^{\tau_i}}{\sum_i c^{2\tau_i}} = \frac{\sum_i z_i \tau_i c^{\tau_i}}{\sum_i \tau_i c^{2\tau_i}}. \tag{3.87}$$

Once c is estimated, b can be estimated using the relationship

$$b = \frac{\sum_i c^{\tau_i}}{\sum_i c^{2\tau_i}} \tag{3.88}$$

which can be derived from equation (3.85).

In order to test if the hypothesis $H_0{:}a = 0$ can be accepted on the basis of a random sample of data (τ_i, z_i), $i = 1, 2, \ldots, m$, a comparison must be made between the sum of squares attributed to the parameter a and the sum of squares attributed to the experimental error in the original model.

Let $SS_{\text{error}}(a, b, c)$ be the sum of squares due to the random error according to the original model; also, let $SS_{\text{error}}(b', c')$ be the sum of squares due to the random error according to the reduced model. In symbols,

$$SS_{\text{error}}(a, b, c) = \sum_{i=1}^{m} (z_i - a + bc^{i})^2$$

$$SS_{\text{error}}(b', c') = \sum_{i=1}^{m} (z_i + b'(c')^{\tau_i})^2.$$

Once the above sums of squares are computed, the sum of squares due to the parameter a is estimated as

$$SS_a = SS_{\text{error}}(b', c') - SS_{\text{error}}(a, b, c).$$

Finally, the value of SS_a can be compared to the sum of squares $SS_{\text{error}}(a, b, c)$ in order to decide if indeed the hypothesis that $a = 0$ can be accepted. The sums of squares associated with the original and reduced models are defined by the following relationships where each SS_{reg} term is a regression sum of squares:

$$\sum_i z_i^2 = SS_{\text{reg}}(a, b, c) + SS_{\text{error}}(a, b, c)$$

and

$$\sum_i z_i^2 = SS_{\text{reg}}(b', c') + SS_{\text{error}}(b', c')$$

Using the above relationships, SS_a can alternatively by rewritten as

$$SS_a = SS_{\mathbf{reg}}(a, b, c) - SS_{\mathbf{reg}}(b', c').$$

It should be noted that the number of degrees of freedom can be estimated by subtracting the number of parameters from the number of observations. Once the mean square associated with SS_a is computed, it can be compared to the error mean square to decide if the hypothesis that $a = 0$ should be accepted or not.

3.3.2 Linear models (ANOVA)

After a nonlinear case has been illustrated, the simpler linear case considered in the ANOVA problems will be studied. As a result of the discussion presented in this section the reader will be able to see that the ANOVA test is only a special case of the more general regression significance test. Consider the linear model

$$Y_{ij} = \mu + \tau_j + \varepsilon_{ij} \tag{3.89}$$

where $i = 1, 2, \ldots, n_j$ and $j = 1, 2, \ldots, k$. Let m be the estimate of μ and let t_j be the estimate of τ_j. The corresponding model for the least-squares analysis can be written as

$$Y_{ij} = m + t_j + e_{ij}. \tag{3.90}$$

The procedure is to minimize the quantity

$$\sum_j \sum_i e_{ij}^2 = \sum_j \sum_i (Y_{ij} - m - t_j)^2. \tag{3.91}$$

By setting the derivative of $\sum_j \sum_i e_{ij}^2$ with respect to m equal to zero, we get

$$\sum_j \sum_i Y_{ij} = \sum_j \sum_i m + \sum_j \sum_i t_j$$

or

$$T_{..} = Nm + \sum_j n_j t_j. \tag{3.92}$$

Similarly, taking the derivative with respect to t_j and setting the results equal to zero, we obtain

$$\sum_i Y_{ij} = \sum_i m + \sum_i t_j$$

or

$$T_{.j} = n_j m + n_j t_j \quad \text{for } j = 1, 2, \ldots, k. \tag{3.93}$$

For the special case where all $n_j = n$, it may be easily shown that the normal equations (3.92) and (3.93) reduce to

$$T_{..} = Nm + n \sum_j t_j \tag{3.94}$$

and

$$T_{.j} = nm + nt_j \quad \text{for } j = 1, 2, \ldots, k. \tag{3.95}$$

Recalling that the null hypothesis to be tested establishes that $\sum_j \tau_j = 0$, it is therefore possible to impose the condition that $\sum_j t_j = 0$ in this case, and equations (3.94) and (3.95) become

$$T_{..} = Nm \tag{3.96}$$

and

$$T_{.j} = nm + nt_j \quad \text{for } j = 1, 2, \ldots, k. \tag{3.97}$$

In the general case, where $n_j \neq n$, if $\sum_{j=1}^{k} n_j t_j = 0$, then the normal equations (3.92) and (3.93) can be written as follows:

$$T_{..} = Nm \tag{3.96a}$$

$$T_{.j} = n_j m + n_j t_j \quad \text{for } j = 1, 2, \ldots, k. \tag{3.97a}$$

Therefore

$$m = \frac{T_{..}}{N} \tag{3.96b}$$

$$t_j = \frac{T_{.j}}{n_j} - m = \frac{T_{.j}}{n_j} - \frac{T_{..}}{N} \quad \text{for } j = 1, 2, \ldots, k. \tag{3.97b}$$

The sum of squares of the regression due to the estimates m and t_j is defined as

$$SS_{\text{reg}} = \sum_j \sum_i Y_{ij}^2 - SS_{\text{error}}(m, t_j) \tag{3.98}$$

where the sum of squares of the error $SS_{\text{error}}(m, t_j)$ can be obtained as follows:

$$\begin{aligned}
SS_{\text{error}}(m, t_j) &= \sum_j \sum_i e_{ij}^2 = \sum_j \sum_i (Y_{ij} - m - t_j)^2 \\
&= \sum_j \sum_i Y_{ij}^2 + \sum_j \sum_i (m + t_j)^2 - 2\sum_j \sum_i Y_{ij}(m + t_j) \\
&= \sum_j \sum_i Y_{ij}^2 - 2m\sum_j \sum_i Y_{ij} - 2\sum_j \sum_i Y_{ij} t_j \\
&\quad + \left[m^2 N + 2m\sum_j \sum_i t_i + \sum_j \sum_i t_j^2 \right] \\
&= \sum_j \sum_i Y_{ij}^2 - 2m\sum_j \sum_i Y_{ij} - 2\sum_j \sum_i Y_{ij} t_j \\
&\quad + Nmm + m\sum_j n_j t_j + m\sum_j n_j t_j + \sum_j n_j t_j t_j \\
&= \sum_j \sum_i Y_{ij}^2 - 2m\sum_j \sum_i Y_{ij} - 2\sum_j \sum_i Y_{ij} t_j \\
&\quad + m(Nm + \sum_j n_j t_j) + \sum_j t_j (n_j m + n_j t_j).
\end{aligned}$$

Using the results obtained in equations (3.92) and (3.93) the above expression can be rewritten as

$$SS_{\text{error}}(m, t_j) = \sum_j \sum_i Y_{ij}^2 - 2mT_{..} - 2\sum_j T_{.j}t_j + mT_{..} + \sum_j T_{.j}t_j.$$

Simplifying the above expression,

$$SS_{\text{error}}(m, t_j) = \sum_j \sum_i Y_{ij}^2 - mT_{..} - \sum_j T_{.j}t_j. \tag{3.99}$$

Now it is possible to rewrite the expression for the regression sum of squares, as follows:

$$SS_{\text{reg}}(m, t_j) = \sum_j \sum_i Y_{ij}^2 - SS_{\text{error}}(m, t_j),$$

that is,

$$SS_{\text{reg}}(m.t_j) = mT_{..} + \sum_j T_{.j}t_j. \tag{3.100}$$

Under the condition $\sum_j n_j t_j = 0$,

$$\begin{aligned} SS_{\text{reg}}(m.t_j) &= mT_{..} + \sum_j T_{.j}t_j \\ &= T_{..}^2/N + \sum_j \frac{T_{.j}^2}{n_j} - \frac{T_{..}^2}{N} = \sum_j \frac{T_{.j}^2}{n_j} \end{aligned}$$

with k degrees of freedom. At this point, it is convenient to summarize the ideas underlying the current procedure:

(a) The response of the experiment is approximated by the liner model

$$Y_{ij} = m + t_j + e_{ij},$$

since use cannot be made of the 'exact' model

$$Y_{ij} = \mu + \tau_j + \varepsilon_{ij}.$$

(b) It is desired to test the hypothesis that the effect of the treatment is not significant. That is, it is desired to test the hypothesis H_0: $\tau = 0$ for all j.
(c) Using a sample from the distribution corresponding to each treatment, we can compute $T_{..}$ and $T_{.j}$ for all j.
(d) Using the value of $\sum_j \sum_i Y_{ij}^2$ and the totals $T_{..}$ and $T_{.j}$ found in step (c), it is possible to compute $SS_{\text{error}}(m, t_{ij})$ and $SS_{\text{reg}}(m, t_j)$ by equations (3.99) and (3.100), respectively.

Continuing with the presentation of the general least-squares significance test, assume, for the moment, that H_0 is actually true. Under the assumption, the model formulated in equation (3.90) reduces to

$$Y_{ij} = m' + e'_{ij}. \tag{3.101}$$

The normal equation for the reduced model of equation (3.101) is found to be

$$\sum_j \sum_i (Y_{ij} - m') = 0$$

or, more simply,

$$T_{..} = Nm'. \tag{3.102}$$

The corresponding sum of squares for the regression under the reduced model is given by

$$SS_{\text{reg}}(m') = m'T_{..}. \tag{3.103}$$

If the null hypothesis is actually true, then the regression sum of squares would be approximately the same under the original and the reduced model. Therefore, the sum of squares due to the treatments can be evaluated as

$$SS_{\text{treatment}} = SS_{\text{reg}}(m, t_j) - SS_{\text{reg}}(m'). \tag{3.104}$$

Using equations (3.100) and (3.103), equation (3.104) can be also written as

$$SS_{\text{treatment}} = T_{..}(m - m') + \sum_j T_{.j} t_j. \tag{3.105}$$

For the condition $\sum_j n_j t_j = 0$,

$$SS_{\text{reg}}(m') = m'T_{..} = \frac{T_{..}^2}{N}$$

with one degree of freedom. Finally,

$$\begin{aligned} SS_{\text{treatment}} &= SS_{\text{reg}}(m, t_j) - SS_{\text{reg}}(m') \\ &= \sum_j \frac{T_{.j}^2}{n_j} - \frac{T_{..}^2}{N}. \end{aligned} \tag{3.106}$$

In order to perform a significance test, the observed value of the F-statistic can be computed as

$$F_0 = \left(\frac{SS_{\text{treatment}}}{k-1}\right)\left(\frac{SS_{\text{error}}(m, t_j)}{N-k}\right)^{-1} \tag{3.107}$$

which is the same expression used in the ANOVA table.

Summarizing, the following steps must be taken in order to apply the general least-squares significance test:

(a) Write the normal equations.
(b) Solve the normal equations.
(c) Find the sum of squares of the regression on m and the t'_j.
(d) Find the reduced normal equations.
(e) Solve the reduced normal equation.
(f) Find the regression sum of the squares corresponding to the reduced model.

(g) Find the sum of squares of the treatments.
(h) Find the error sum of squares.
(i) Find the observed value of the F-statistic.

EXAMPLE 3.5

A manufacturing company wants to decide if there is a significant difference in three available methods of making steel cable. A random sample of several pieces of cable made by each process was examined to determine the breaking strength of each piece of wire. The results in Table 3.14 (in 10^3 psi) correspond to $n_1 = 3$, $n_2 = 2$ and $n_3 = 4$.

First this problem will be solved following the ANOVA methodology described in section 3.2. The results given in Table 3.15 are obtained in this case. As can be seen in the ANOVA table (Table 3.15), the null hypothesis H_0: $\tau_j = 0, j = 1, 2, 3$ cannot be accepted. Now this problem is solved by means of the general least-squares significance test. We will consider the following two cases:

(a) $\sum_{j=1}^{3} t_j = 0$. The normal equations for the given data are

$$89 = 9m + 3t_1 + 2t_2 + 4t_3$$
$$14 = 3m + 3t_1$$
$$21 = 2m + 2t_2$$
$$54 = 4m + 4t_3$$
$$0 = t_1 + t_2 + t_3.$$

Table 3.14

Method 1	*Method 2*	*Method 3*
4	10	13
7	11	13
3		14
		14

Table 3.15

Source	*df*	*SS*	*MS*	*F*	$F_{0.95}(2,6)$
Treatments	2	134.72	67.360	39.74	5.4
Error	6	10.17	1.695		
Total	8	144.89			

Solving these equations simultaneously yields the results

$$m = \frac{86}{9} \qquad t_1 = -\frac{88}{18} \qquad t_2 = \frac{17}{18} \qquad t_3 = \frac{71}{18}.$$

From equation (3.100) the value of $SS_{\text{reg}}(m, t_j)$ is computed as

$$SS_{\text{reg}}(m, t_j) = \left(\frac{86}{9}\right)(89) + \left(\frac{-88}{18}\right)(14) + \left(\frac{17}{18}\right)(21) + \left(\frac{71}{18}\right)(54)$$
$$= 1014.83.$$

(b) $\sum_{j=1}^{3} n_j t_j = 0$. To obtain an unbiased estimator for μ, a better choice is to solve the normal equations along with $\sum_j n_j t_j = 0$. The normal equations for the given data are

$$89 = 9m + 3t_1 + 2t_2 + 4t_3$$
$$14 = 3m + 3t_1$$
$$21 = 2m + 2t_2$$
$$54 = 4m + 4t_3$$
$$0 = 3t_1 + 2t_2 + 4t_3.$$

Solving these equations, we obtain the following results (estimators):

$$m = \frac{89}{9} \qquad t_1 = -\frac{94}{18} \qquad t_2 = \frac{11}{18} \qquad t_3 = \frac{65}{18}.$$

From equation (3.100) the value of $SS_{\text{reg}}(m, t_j)$ is computed as

$$SS_{\text{reg}}(m, t_j) = \left(\frac{89}{9}\right)(89) + \left(\frac{-94}{18}\right)(14) + \left(\frac{11}{18}\right)(21) + \left(\frac{65}{18}\right)(54)$$
$$= 1014.83.$$

The normal equation for the reduced model is, according to equation (3.102),

$$89 = 9m'.$$

Therefore $m' = \frac{89}{9}$ and, from equation (3.103), it is possible to compute $SS_{\text{reg}}(m')$ as

$$SS_{\text{reg}}(m') = \left(\frac{89}{9}\right)(89) = \frac{7921}{9}$$
$$= 880.11.$$

The treatment sum of squares is obtained from equation (3.104):

$$SS_{\text{treatment}} = 1014.83 - 880.11 = 134.72.$$

The sum of squares of the error can now be obtained from equation (3.98):

$$SS_{\text{error}}(m, t_j) = \sum_j \sum_i Y_{ij}^2 - SS_{\text{reg}}(m, t_j)$$
$$= 1025 - 1014.83 = 10.17$$

Finally, a significance test can be conducted by comparing the observed value of the F-statistic

$$F_0 = \frac{MS_{\text{treatment}}}{MS_{\text{error}}} = \frac{\frac{1}{2}(134.72)}{\frac{1}{6}(10.17} = 39.74$$

against a critical value, such as $F_{0.95}(2, 6) = 514$. In this case, the null hypothesis that there is no treatment effect must be rejected. It can be verified that both the ANOVA and the general significance test yield the same results.

EXERCISES

1. Consider the data in Table 3.16 ($n = 10, k = 4$) for an experiment with four levels A, B, C and D for a given factor. Investigate the assumptions of normality and homogeneity of variance. Consider the use of an appropriate transformation in case the variance of the original data is not stable. Do the analysis of variance after all assumptions are reasonably satisfied. Assume that the level of significance is equal to 0.05.
2. The data in Table 3.17 correspond to a one-factor experiment with four levels (A, B, C, D), $n = 10$, and no restrictions on the randomization.

Table 3.16

A		*B*		*C*		*D*	
20	42	8	22	49	122	151	179
28	25	12	11	66	70	126	150
36	24	14	10	80	65	178	148
16	31	10	18	105	70	152	165
25	30	7	12	90	85	172	132

Table 3.17

A		*B*		*C*		*D*	
40	42	8	22	49	122	151	139
48	25	12	11	66	70	126	110
56	24	14	10	80	65	178	108
36	31	10	18	105	70	152	125
45	30	7	12	90	85	172	92

(a) Perform a one-way analysis of variance. Consider $\alpha = 0.05$.
(b) Solve this problem by the general least-squares significance test.

3. Prove that $E[MS_{\text{error}}] = \sigma^2$ for the fixed effect model. Repeat for the random effect model.
4. Find the expected value of the treatment mean square for a fixed-effect model assuming different sample sizes for the treatments.
5. Three fuses are made by different manufacturers. The duration, in months, of reliable operation for three random samples of size $n = 8$, are given in Table 3.18. Test the hypothesis of equal lives assuming a totally randomized design; use $\alpha = 0.05$.
6. Prove equations (3.48) and (3.49).
7. Find the transformation which will stabilize the variance in each of the three following cases:
   (a) The standard deviation is directly proportional to the square of the mean.
   (b) The standard deviation is directly proportional to the mean.
   (c) The standard deviation is directly proportional to the square root of the mean.
8. Explain why only one-tailed tests are considered in ANOVA.
9. Solve Exercise 2 after (a) coding the data by subtracting 20 from each random observation; (b) coding by subtracting 5 and multiplying each observation by 2.
10. Assuming that the runs of the experiment given in Exercise 2 correspond to random samples of levels, find variance components for both treatments and random error.
11. Use the general least-squares significance test on the data given in Exercise 5.
12. Solve Exercise 5 using the general least-squares significance test, and assuming that the first observation of treatment A, and the first and the last observations of treatment C were lost.
13. Find the normal equations to generate the least-squares estimates for the following two models, where Y and X are random variables:
    (a) $Y = h + a(X - k)^3$;
    (b) $Y = 1 + (a - 1)e^{-bX}$.

Table 3.18

A		*B*		*C*	
19	21	21	42	12	15
22	21	21	32	15	18
28	16	32	22	17	17
17	33	25	19	20	19

14. Discuss each of the assumptions of the analysis of variance, and indicate which procedures can be followed to test each assumption.
15. Find the transformation required to stabilize the variance of a Poisson-distributed random variable.
16. Develop the normal equations for the case where all n'_j are equal.
17. Verify the results obtained by the bisection method for the data given in Example 3.4.
18. The ANOVA table (Table 3.19) resulted from an experiment designed to investigate the strength of steel having different percentages of carbon. In Table 3.19, EMS stands for 'expected mean square'.
    (a) Give the linear model for this experiment. Explain all terms used.
    (b) Complete the table. Assuming equal replications and fixed-effect treatments.
    (c) Test the equality of mean strengths for the different percentages.
    (d) Give the estimated standard deviation of the difference between any two treatment sample averages.
    (e) If the average of the first sample is 13 and the average of the second sample is 11, set and interpret a 95% confidence interval for the true mean difference.
19. An entomologist performed an experiment to study bollworm control by five hormones. Bollworm larvae for one hybrid strain were raised under controlled laboratory conditions. At 15 days of age, 50 uniform larvae were each injected with a fixed dose of one hormone, randomly selected from the five. After 24 hours, the number of deaths and survivors was recorded. Each hormone was used on 4 groups of 50 larvae. The number of deaths were as given in Table 3.20.

    Choose a valid experimental response and perform an ANOVA test. Use a 5% level of significance.
20. Consider the experimental situation given in Example 3.3. Find 95% confidence intervals for the error variance and treatment population variance. See results given in [2] p. 50.
21. The data in Table 3.21 came from a completely randomized experiment to compare the reflective properties of four kinds of paint. Only 19 test specimens were available, so that one treatment had to be replicated

Table 3.19

Source	*df*	*SS*	*MS*	*EMS*
Percentages		120		
Experimental error	24	144		
Total	29	264		

Table 3.20

Hormone				
1	*2*	*3*	*4*	*5*
9	14	6	13	24
15	16	4	16	22
18	19	10	17	28
8	19	12	21	25

Table 3.21

Paint 1	*Paint 2*	*Paint 3*	*Paint 4*
195	45	195	120
150	40	230	55
205	195	115	50
110	65	235	80
160	145	225	
820	490	1000	305

only four times. The test specimens were alloted to the paints at random, and the experimental response was obtained by using an optical instrument.

Perform a significance test (use a 5% level of significance) using the general least-squares procedure. Write the normal equations, and solve them under a condition that will make the solution a legitimate set of estimators. Indicate which parameters are estimated by the corresponding estimators.

22. A factory manager wishes to buy machines to perform a certain operation in a production process. There are three companies which make such machines, and the manager obtains one on trial from each company with a view to determining which of the three is best suited to the factory's purposes. Each machine can be operated by one worker. The manager intends to have several operators using the machines for a few days to discover which machine produces the most items per day. Six operators are to be used in the experiment, two being assigned at random to each machine, and each operator will work one day on the particular machine assigned to him. Consider the data in Table 3.22.
    (a) Indicate which model you will use to represent the given experimental data. Identify the response and factor considered.
    (b) State which assumptions are made for the analysis of results. Why is each assumption needed?

Table 3.22

1	*2*	*3*
64	41	65
39	48	57

Table 3.23

23.42	23.70	23.49	23.70	23.85	23.99	23.28
23.42	23.77	23.49	23.85	23.92	23.85	23.56
23.42	23.63	23.49	23.85	23.92	23.92	23.42
23.45	23.67	23.48	23.90	23.92	23.99	23.27
23.43	23.71	23.47	23.90	23.92	23.99	23.50

(c) State the hypothesis to be tested and conduct the appropriate ANOVA test.
(d) Write the normal equations including the condition needed for the solution to yield valid estimators.
(e) From the normal equations find the approrpriate estimators and indicate which unknown quantities (parameters) are being estimated.

23. An instrument has been developed for the measurement of the bulk density of precipitated chalk, and it is proposed to use the machine in testing chalk conformity with a specification. It was decided to send a machine to each of seven laboratories and to send samples taken from chalks covering a range of bulk densities to each laboratory for test. The main problem is to test whether different laboratories agreed in their average results. Because of the nature of the test it is expected that the error would be proportional to the bulk density so that the coefficient of variation rather than the standard deviation is expected to be constant for chalks of different bulk densities. Consider the experimental results shown in Table 3.23 corresponding to five tests for each laboratory. Do the corresponding analysis of variance and explain the results. Discuss all the assumptions made.

REFERENCES

1. Davies, Owen L. *et al.* (1956) *The Design and Analysis of Industrial Experiments*, 2nd edn, Imperial Chemical Industries Ltd, London.
2. Lentner M., and Bishop T. (1986) *Experimental Design and Analysis*, Valley Book Company, Blacksburg.
3. Bartlett, M. S. (1947) The use of transformations. *Biometrics*, **3**, 39–52.

4. Cochran, W. G. (1947) Some consequences when the assumptions for the analysis of variances are not satisfied. *Biometrics*, **3**, 22–38.
5. Eisenhart, G. (1947) The assumption underlying the analysis of variance. *Biometrics*, **3**, 1–21.
6. Garcia-Diaz, A. and Riggins, M. (1984) Serviceability and distress methodology for predicting pavement performance. *Transportation Research Record 997*, pp. 56–61.

FURTHER READING

Cochran, W. G. and Cox, G. M. (1957) *Experimental Designs*, 2nd edn, John Wiley & Sons. Inc., New York.

Hicks, Charles R. (1982) *Fundamental Concepts in the Design of Experiments*, 3rd edn, Holt, Rinehart & Winston, New York.

After-ANOVA tests 4

In section 3.2.1 the fixed-effect ANOVA model was presented and the corresponding statistical testing procedure was illustrated in Example 3.2. The hypothesis being tested in that example was that no significant difference exists between the spindle diameters of parts produced by three computer-controlled lathes. Since the null hypothesis was rejected ($F_0 = 52.10$ and $F_c = 3.68$), it is concluded that at least one of the three lathes is significantly different from the others. The purpose of this chapter is to present and discuss procedures for testing the effect associated with proposed combinations of treatment effects. These procedures, for example, can be used to determine which computer-controlled lathes consistently produce spindle parts of a specific diameter in Example 3.2. The statistical techniques which allow further comparisons to be made between treatment means are referred to as **multiple comparison methods**.

Due to the extensive number of multiple comparison methods available (in excess of 150 reported in the literature), only the more commonly used procedures will be considered in this chapter:

(a) comparisons made on individual treatment means, called orthogonal contrasts;
(b) methods for comparing any and all possible contrasts between treatment means (Scheffe's method);
(c) four different methods to compare all pairs of treatment means, namely the least significant difference method, Tukey's range procedure test, Duncan's multiple range test and Newman–Keuls studentized range test;
(d) Dunnett's method to compare treatments against a control.

4.1 ORTHOGONAL CONTRASTS

The method of orthogonal contrasts is an approach to test whether a set of independent linear combinations of sample means significantly differs from zero. A contrast is defined as a linear combination of the sample means $\bar{Y}_{.j}$

$$C = \sum_{j=1}^{k} c_j \bar{Y}_{.j}$$

where

$$\sum_{j=1}^{k} c_j = 0.$$

Assuming that the treatment sample means are statistically independent and normally distributed with means μ_j and variances $\sigma^2 n_j^{-1}, j = 1, 2, \ldots, k$, the random variable C is also normally distributed with parameters

$$\mu_C = \sum_{j=1}^{k} c_j \mu_j$$

and

$$\sigma_C^{\ 2} = \sum_{j=1}^{k} \frac{c_j^{\ 2} \sigma_j^{\ 2}}{n_j}.$$

Suppose that it is desired to test the hypothesis that μ_C is equal to zero. Under this hypothesis and the assumption that all $\sigma_j^{\ 2}$ are equal to σ^2, the random variable C follows a normal distribution with mean equal to zero and variance equal to

$$\sigma^2 \sum_{j=1}^{k} \frac{c_j^{\ 2}}{n_j}.$$

The above result, in turn, implies that

$$C\left[\sigma^2\left(\sum_j \frac{c_j^{\ 2}}{n_j}\right)\right]^{-1/2} \sim N(0, 1).$$

Using the definition of the chi-square statistic, equation (2.3), it is concluded that

$$C^2\left[\sigma^2\left(\sum_j \frac{c_j^{\ 2}}{n_j}\right)\right]^{-1} \sim \chi_1^{\ 2}.$$

This result can now be rewritten as

$$C^2\left(\sum_j \frac{c_j^{\ 2}}{n_j}\right)^{-1} (\sigma^2)^{-1} \sim \chi_1^{\ 2} \tag{4.1}$$

In Chapter 3 it was shown that the treatment sum of squares divided by σ^2 is distributed as χ^2 with $k-1$ degrees of freedom, under the assumption that all treatment populations have the same mean. Therefore, it is possible to write the sum of squares for treatments as the sum of $k-1$ statistically independent sums of squares, each with one degree of freedom.

Two linear functions or contrasts C_1 and C_2 defined as

$$C_1 = \sum_j c_{1j} \bar{Y}_{.j}$$

$$C_2 = \sum_j c_{2j} \bar{Y}_{.j}$$

are said to be orthogonal (statistically independent) if

$$\sum_j \frac{c_{1j}c_{2j}}{n_j} = 0. \tag{4.2}$$

The above condition comes from the requirement of independence, i.e. that the covariance of C_1 and C_2 be equal to zero.

As illustrations of the basic definitions already presented, consider an experiment with three treatments and sample sizes given by $n_1 = 4$, $n_2 = 10$ and $n_3 = 7$. Examples of orthogonal contrasts are

$$C_1 = 2\bar{Y}_{.1} - \bar{Y}_{.2} - \bar{Y}_{.3}$$
$$C_2 = 3\bar{Y}_{.1} - 45\bar{Y}_{.2} + 42\bar{Y}_{.3}$$

since

$$\sum_j \frac{c_{1j}c_{2j}}{n_j} = \frac{(2)(3)}{4} + \frac{(-1)(-45)}{10} + \frac{(-1)(42)}{7} = 0.$$

Examples of nonorthogonal contrasts for the same case considered above are

$$C_1 = 2\bar{Y}_{.1} - \bar{Y}_{.2} - \bar{Y}_{.3}$$
$$C_2 = 4\bar{Y}_{.1} - \bar{Y}_{.2} - 3\bar{Y}_{.3}$$

since, as can be verified, $\sum_j c_{1j}c_{2j}n_j^{-1} = 2.528 \neq 0$.

A simple proof of the orthogonality condition given in equation (4.2) follows. Let C and D be two contrasts with coefficients c_j and d_j, respectively, for $\bar{Y}_{.j}$. That is

$$C = \sum_j c_j \bar{Y}_{.j}$$
$$D = \sum_j d_j \bar{Y}_{.j}.$$

In general,

$$V[C + D] = V[C] + V[D] + 2\,\mathrm{cov}[C, D] \tag{4.3}$$

where cov$([C, D]$ is the covariance of C and D. Using the result given in equation (4.3), it is concluded that

$$2\,\mathrm{cov}[C + D] = V[C + D] - V[C] - V[D]. \tag{4.4}$$

The variance of C plus D is also given by

$$V[C + D] = V\left[\sum_j c_j\bar{Y}_{.j} + \sum_j d_j\bar{Y}_{.j}\right]$$

or, on rewriting,

$$V[C + D] = V\left[\sum_j (c_j + d_j)\bar{Y}_{.j}\right]. \tag{4.5}$$

Under the assumption that the random variables $\bar{Y}_{.j}$ are statistically independent, equation (4.5) yields

$$V[C+D]=\sum_j (c_j+d_j)^2 V[\bar{Y}_{.j}]. \tag{4.6}$$

Additionally, if $\bar{Y}_{.j}$ has variance equal to $\sigma^2 n_j^{-1}$, equation (4.6) can be rewritten as indicated below:

$$\begin{aligned} V[C+D] &= \sum_j \frac{(c_j+d_j)^2\sigma^2}{n_j} \\ &= \sigma^2 \sum_j \frac{c_j^2+d_j^2+2c_jd_j}{n_j}. \end{aligned} \tag{4.7}$$

Using the result developed in equation (4.7), equation (4.4) can be expressed as

$$2\operatorname{cov}[C,D]=\sigma^2\sum_j \frac{c_j^2+d_j^2+2c_jd_j-c_j^2-d_j^2}{n_j}.$$

That is,

$$\operatorname{cov}[C,D]=\sigma^2\sum_j \frac{c_jd_j}{n_j}. \tag{4.8}$$

After setting $\operatorname{cov}[C,D]=0$ since C and D are statistically independent, equation (4.8) yields the final result

$$\sum_j \frac{c_jd_j}{n_j}=0 \tag{4.9}$$

which is the same orthogonality condition as given in equation (4.2).

The contrasts $C_1, C_2, C_3, \ldots, C_{k-1}$, defined on the same set of sample averages, are said to be mutually orthogonal if all possible pairs of contrasts are orthogonal. A fundamental result concerning a set of $k-1$ orthogonal contrasts defined on k independent sample means now follows.

Let $C_1, C_2, \ldots, C_{k-1}$ be a set of $k-1$ mutually orthogonal contrasts defined on the same set of k sample means. Therefore, under the null hypothesis $H_0\colon \mu_j=\mu$ for all j, we can write

$$\frac{SS_{\text{treatment}}}{\sigma^2}=\left(\frac{C_1^2}{\sum_j c_{1j}^2 n_j^{-1}}+\frac{C_2^2}{\sum_j c_{2j}^2 n_j^{-1}}+\cdots+\frac{C_{k-1}^2}{\sum_j c_{k-1,j}^2 n_j^{-1}}\right)\frac{1}{\sigma^2}$$

which implies that

$$SS_{\text{treatment}}=\sum_{i=1}^{k-1}\frac{C_i^2}{\sum_j c_{ij}^2 n_j^{-1}}. \tag{4.10}$$

The ith term of the summation given in equation (4.10) is actually the sum of squares associated with contrast C_i. That is,

$$SS_{C_i} = \frac{C_i^2}{\sum_j c_{ij}^2 n_j^{-1}}. \tag{4.11}$$

Suppose that after computing the F-statistic as

$$F = \frac{MS_{\text{treatment}}}{MS_{\text{error}}}$$

the null hypothesis H_0 is rejected. In order to test if the rejection is due to contrast C_i, we verify the new hypothesis

$$H_{0i}: \mu_{C_i} = 0$$

by computing the F-statistic as

$$F = \frac{SS_{C_i}}{MS_{\text{error}}} = \frac{MS_{C_i}}{MS_{\text{error}}}$$

which is distributed as F with 1 degree of freedom in the numerator and $N-k$ degrees of freedom in the denominator, where, as in Chapter 3, $N = \sum_j n_j$.

In order to preserve the validity of the specified significance level, contrasts should be identified prior to experimentation. The rationale for this is that if comparisons are not preselected, but rather are determined on the basis of observed results, experimenters may construct contrasts to test the largest observed differences in the means, rendering the level of significance less meaningful than when contrasts are chosen before experimentation.

EXAMPLE 4.1

Consider the following orthogonal contrasts:

$$C_1 = \bar{Y}_{.1} - \bar{Y}_{.4}$$
$$C_2 = \bar{Y}_{.2} - \bar{Y}_{.3}$$
$$C_3 = \bar{Y}_{.1} - \bar{Y}_{.2} - \bar{Y}_{.3} + \bar{Y}_{.4}$$

defined on the treatment means of four random samples, each having a sample size equal to five. It is desired to test the following hypotheses:

$$H_0': \mu_1 - \mu_4 = 0$$
$$H_0'': \mu_2 - \mu_3 = 0$$
$$H_0''': \mu_1 - \mu_2 - \mu_3 + \mu_4 = 0$$

using a significance level of 5% and the following results: $T_{.1} = 42$, $T_{.2} = 36$, $T_{.3} = -32$, $T_{.4} = -40$ and $SS_{\text{total}} = 1338.2$.

To verify the orthogonality of the contrasts, the coefficients can be arranged as shown in Table 4.1. As an illustration, it can be seen that C_1 and C_2 are orthogonal since equation (4.2) is satisfied as shown below:

$$\frac{(1)(0)}{5}+\frac{(0)(1)}{5}+\frac{(0)(-1)}{5}+\frac{(-1)(0)}{5}=0.$$

Following the above approach, it is possible to infer that the contrasts C_1, C_2 and C_3 are orthogonal. The sum of squares of a contrast C is defined as follows:

$$SS_C=\left(\sum_j c_j\bar{Y}_{.j}\right)^2\left(\sum_j\frac{c_j^2}{n_j}\right)^{-1}.$$

From equation (4.10), after setting all n_j equal to $n=5$, it is possible to express the contrast sum of squares SS_C as

$$\begin{aligned}SS_C&=\left(\sum_j c_j\bar{Y}_{.j}\right)^2\left(\sum_j\frac{c_j^2}{n_j}\right)^{-1}\\&=\left(\sum_j\frac{c_jT_{.j}}{n}\right)^2\left(\sum_j\frac{c_j^2}{n_j}\right)^{-1}\\&=\frac{(\sum_j c_jT_{.j})^2}{n\sum_j c_j^2}.\end{aligned}\tag{4.12}$$

Using the result given by equation (4.12), the sum of squares of the contrasts under consideration can be computed as follows:

$$SS_{C_1}=\frac{[(1)(42)+(-1)(-40)]^2}{5[(1)^2+(-1)^2]}=672.4$$

$$SS_{C_2}=\frac{[(1)(36)+(-1)(-32)]^2}{5[(1)^2+(-1)^2]}=462.4$$

$$SS_{C_3}=\frac{[(1)(42)+(-1)(36)+(-1)(-32)+(1)(-40)]^2}{5[(1)^2+(-1)^2+(-1)^2+(1)^2]}=0.2.$$

Table 4.1 Matrix of contrast coefficients

	Treatments			
Contrast	*I*	*II*	*III*	*IV*
C_1	+1	0	0	−1
C_2	0	+1	−1	0
C_3	+1	−1	−1	+1

Table 4.2 ANOVA – Example 4.1

Source of variation	*df*	$MS = SS$	$F = \frac{MS}{MS_{\text{error}}}$	*Significant?* $(F_{0.95,1,16} = 4.49)$
C_1	1	672.4	52.900	Yes
C_2	1	462.4	36.400	Yes
C_3	1	0.2	0.016	No

Therefore

$$SS_{\text{treatment}} = \sum_j SS_{C_j} = 1135$$

and

$$MS_{\text{error}} = \frac{(SS_{\text{total}} - SS_{\text{treatment}})}{N - k}$$

$$= \frac{1338.2 - 1135}{16} = 12.7.$$

The above calculations are summarized in Table 4.2. From the results shown in Table 4.2, it is concluded that the hypotheses H'_0 and H''_0 are rejected at the 5% level of significance. Thus, it is concluded that $\mu_1 \neq \mu_4$ and $\mu_2 \neq \mu_3$. Also, hypothesis H'''_0 is not rejected; therefore, no significant difference exists at the 5% level between $\mu_1 + \mu_4$ and $\mu_2 + \mu_3$.

If it is essential to study a set of contrasts which are not orthogonal, then another approach should be used, as will be shown in the following section.

4.2 SCHEFFE'S METHOD

In 1953 Henry Scheffe introduced the S-method which is known in the statistical literature as Scheffe's method. Ideally, the contrasts to be studied should be selected before actual experimentation in order to keep the validity of the level of significance. This restriction often places a heavy burden on the user. In many instances, contrasts can only be identified after the experiment is conducted due to lack of sufficient information concerning the process. However, Scheffe's method allows the definition of the contrasts after experimentation without inflating the probability of a type I error. Basically, the procedure examines all possible contrasts in terms of a set of simultaneous confidence intervals.

The decision (rejection) criterion for Scheffe's method will be developed in the remaining portion of this section. Matrix notation will be used in order to generalize results and simplify the analytical development.

Suppose that it is desired to test the hypothesis $H_0: \boldsymbol{h}'\boldsymbol{CB} = 0$, where $\boldsymbol{CB}$ is a set of contrasts

$$E[C_1] = \sum_{j=1}^{k} c_{1j}\mu_j$$

$$E[C_2] = \sum_{j=1}^{k} c_{2j}\mu_j$$

$$\vdots$$

$$E[C_d] = \sum_{j=1}^{k} c_{dj}\mu_j$$

and, by definition,

$$\sum_{j=1}^{k} c_{ij} = 0 \quad \text{for } i = 1, 2, \ldots, d.$$

In the formulation of the above null hypothesis, $\boldsymbol{C}$ is a matrix defined as $\boldsymbol{C} = [c_{ij}]$, where $i = 1, 2, \ldots, d$ and $j = 1, 2, \ldots, k$, and $\boldsymbol{B}$ is a vector defined as $\boldsymbol{B}' = [\mu_1 \quad \mu_2 \quad \cdots \quad \mu_k]$. The meaning of $\boldsymbol{h}'$ can be explained for the case where there are two contrasts to be tested. If C_1 is considered in H_0, then $\boldsymbol{h}' = [1 \quad 0]$; but if C_2 is considered in H_0, then $\boldsymbol{h}' = [0 \quad 1]$. In general, if the null hypothesis to be tested is

$$H_0: \sum_{j} c_{ij}\mu_j = 0$$

$\boldsymbol{h}'$ is defined as the vector having the ith element equal to one and all other elements being equal to zero.

In order to illustrate the notation that will be used in this section, the fixed-model $Y_{ij} = \mu + \tau_j + \varepsilon_{ij}$ will be considered for the case $k = 3$, $n_1 = 3$, $n_2 = 2$, $n_3 = 4$ and $d = 2$. As shown in Chapter 3, the fixed-effect model can also be formulated as $Y_{ij} = \mu_j + \varepsilon_{ij}$, where μ_j is the mean of the population associated with the jth treatment. According to this model, the nine observations can be expressed as indicated below:

$$\begin{aligned} Y_{11} &= \mu_1 + \varepsilon_{11} \\ Y_{21} &= \mu_1 + \varepsilon_{21} \\ Y_{31} &= \mu_1 + \varepsilon_{31} \\ Y_{12} &= \mu_2 + \varepsilon_{12} \\ &\vdots \\ Y_{43} &= \mu_3 + \varepsilon_{43}. \end{aligned}$$

The above relationships can all be written using matrix notation as in

$Y = XB + \varepsilon$, where

$$Y = \begin{pmatrix} Y_{11} \\ Y_{21} \\ Y_{31} \\ Y_{12} \\ Y_{22} \\ Y_{13} \\ Y_{23} \\ Y_{33} \\ Y_{43} \end{pmatrix} \quad X = \begin{pmatrix} 1 & 0 & 0 \\ 1 & 0 & 0 \\ 1 & 0 & 0 \\ 0 & 1 & 0 \\ 0 & 1 & 0 \\ 0 & 0 & 1 \\ 0 & 0 & 1 \\ 0 & 0 & 1 \\ 0 & 0 & 1 \end{pmatrix} \quad B = \begin{pmatrix} \mu_1 \\ \mu_2 \\ \mu_3 \end{pmatrix} \quad \varepsilon = \begin{pmatrix} \varepsilon_{11} \\ \varepsilon_{21} \\ \varepsilon_{31} \\ \varepsilon_{12} \\ \varepsilon_{22} \\ \varepsilon_{13} \\ \varepsilon_{23} \\ \varepsilon_{33} \\ \varepsilon_{43} \end{pmatrix}.$$

Proceeding with our development, let T be a statistic defined as follows:

$$T = \frac{b'L^{-1}b}{(k-1)S^2}. \tag{4.13}$$

In equation (4.13),

$$L = C(X'X)^{-1}C' \tag{4.14}$$

$$b = C\hat{B} - CB \tag{4.15}$$

where $\hat{B}$ is an estimator of B, and S^2 is the mean square error. Continuing with the example with $d = 2$, the matrix L and the vector b are defined as follows, where $k = 3$:

$$L = \begin{pmatrix} \sum_{j=1}^{k} \frac{c_{1j}^2}{n_j} & \sum_{j=1}^{k} \frac{c_{1j}c_{2j}}{n_j} \\ \sum_{j=1}^{k} \frac{c_{1j}c_{2j}}{n_j} & \sum_{j=1}^{k} \frac{c_{2j}^2}{n_j} \end{pmatrix}$$

$$b = \begin{pmatrix} \sum_j c_{1j}\bar{Y}_{.j} - \sum_j c_{1j}\mu_j \\ \sum_j c_{2j}\bar{Y}_{.j} - \sum_j c_{2j}\mu_j \end{pmatrix}.$$

Note that $\sigma^2 L$ is the variance–covariance matrix of C_1 and C_2.

The derivation of the distribution of the statistic T is outside of the scope of this book; for this purpose, the reader could consult Scheffe [1], Seber [2] or Miller [3]. For example, Seber [2] shows that T follows an $F_{k-1,N-k}$ distribution where k is the total number of treatments and N is the total number of observations in the experiment. Thus, for a given level of significance

$$P[T < F_c] = 1 - \alpha \tag{4.16}$$

or, equivalently, after using equation (4.13):

$$P\left[\frac{\boldsymbol{b}'\boldsymbol{L}^{-1}\boldsymbol{b}}{(k-1)S^2} < F_c\right] = 1 - \alpha \tag{4.17}$$

where $F_c = F_{1-\alpha,k-1,N-k}$. Hence

$$P[\boldsymbol{b}'\boldsymbol{L}^{-1}\boldsymbol{b} < (k-1)S^2F_c] = 1 - \alpha. \tag{4.18}$$

Now, consider the Cauchy–Schwartz inequality [2]

$$|\boldsymbol{u}|^2|\boldsymbol{v}|^2 > (\boldsymbol{u}'\boldsymbol{v})^2 \tag{4.19}$$

where

$$|\boldsymbol{u}|^2 = \boldsymbol{u}'\boldsymbol{u} \quad \text{and} \quad |\boldsymbol{v}|^2 = \boldsymbol{v}'\boldsymbol{v}.$$

From equation (4.19) it follows that the supremum $(\boldsymbol{u}'\boldsymbol{v})$ $(\boldsymbol{u}'\boldsymbol{v})^2$ is $|\boldsymbol{u}|^2|\boldsymbol{v}|^2$ since the value $|\boldsymbol{u}|^2|\boldsymbol{v}|^2$ is an upper bound on $(\boldsymbol{u}'\boldsymbol{v})^2$. Hence

$$\sup_{\boldsymbol{v}:\,\boldsymbol{v}\neq\boldsymbol{0}}\left[\frac{(\boldsymbol{u}'\boldsymbol{v})^2}{|\boldsymbol{v}|^2}\right] = \frac{|\boldsymbol{u}|^2|\boldsymbol{v}|^2}{|\boldsymbol{v}|^2} = |\boldsymbol{u}|^2 = \boldsymbol{u}'\boldsymbol{u}. \tag{4.20}$$

Since $\boldsymbol{L}$ is a positive definite matrix [2] then there exists a nonsingular matrix $\boldsymbol{R}$ such that $\boldsymbol{L} = \boldsymbol{R}'\boldsymbol{R}$. In order to relate equations (4.18) and (4.19) it is required to define

$$\boldsymbol{v} = \boldsymbol{R}'\boldsymbol{h} \quad \text{and} \quad \boldsymbol{u} = \boldsymbol{R}^{-1}\boldsymbol{b}.$$

Thus

$$\boldsymbol{u}'\boldsymbol{u} = \boldsymbol{b}'(\boldsymbol{R}^{-1})'\boldsymbol{R}^{-1}\boldsymbol{b} = \boldsymbol{b}'\boldsymbol{L}^{-1}\boldsymbol{b} \tag{4.21}$$

and

$$\frac{(\boldsymbol{u}'\boldsymbol{v})^2}{|\boldsymbol{v}|^2} = \frac{(\boldsymbol{b}'(\boldsymbol{R}^{-1})'\boldsymbol{R}'\boldsymbol{h})^2}{\boldsymbol{v}'\boldsymbol{v}} = \frac{(\boldsymbol{b}'\boldsymbol{h})^2}{\boldsymbol{h}'\boldsymbol{R}'\boldsymbol{R}\boldsymbol{h}} = \frac{(\boldsymbol{b}'\boldsymbol{h})^2}{\boldsymbol{h}'\boldsymbol{L}\boldsymbol{h}}. \tag{4.22}$$

Substituting equations (4.21) and (4.22) in equation (4.20), the following result is obtained:

$$\sup\left[\frac{(\boldsymbol{b}'\boldsymbol{h})^2}{\boldsymbol{h}'\boldsymbol{L}\boldsymbol{h}}\right] = \boldsymbol{b}'\boldsymbol{L}^{-1}\boldsymbol{b}. \tag{4.23}$$

Equation (4.23) states that an upper bound on $(\boldsymbol{b}'\boldsymbol{h})^2(\boldsymbol{h}'\boldsymbol{L}\boldsymbol{h})^{-1}$ is $\boldsymbol{b}'\boldsymbol{L}^{-1}\boldsymbol{b}$. This implies that

$$\frac{(\boldsymbol{b}'\boldsymbol{h})^2}{\boldsymbol{h}'\boldsymbol{L}\boldsymbol{h}} \leqslant \boldsymbol{b}'\boldsymbol{L}^{-1}\boldsymbol{b}. \tag{4.24}$$

Now, using equations (4.18) and (4.24), equation (4.25) can be obtained for an α significance level:

$$\frac{(\boldsymbol{b}'\boldsymbol{h})^2}{\boldsymbol{h}'\boldsymbol{L}\boldsymbol{h}} \leqslant (k-1)S^2F_c. \tag{4.25}$$

Taking the square root of both sides of equation (4.25), and rearranging, we obtain

$$|\boldsymbol{b}'\boldsymbol{h}| \leqslant [(k-1)S^2F_c\boldsymbol{h}'\boldsymbol{L}\boldsymbol{h}]^{1/2}. \tag{4.26}$$

Since $|\boldsymbol{b}'\boldsymbol{h}| = |\boldsymbol{h}'\boldsymbol{C}\hat{\boldsymbol{B}} - \boldsymbol{h}'\boldsymbol{C}\boldsymbol{B}| = |\boldsymbol{h}'\boldsymbol{C}\boldsymbol{B} - \boldsymbol{h}'\boldsymbol{C}\hat{\boldsymbol{B}}|$, then equation (4.26) can be written as follows:

$$|\boldsymbol{h}'\boldsymbol{C}\boldsymbol{B} - \boldsymbol{h}'\boldsymbol{C}\hat{\boldsymbol{B}}| \leqslant \boldsymbol{r} \tag{4.27}$$

where

$$\boldsymbol{r} = [(k-1)S^2F_c\boldsymbol{h}'\boldsymbol{L}\boldsymbol{h}]^{1/2}. \tag{4.28}$$

Equation (4.27) can also be expressed as

$$-\boldsymbol{r} \leqslant \boldsymbol{h}'\boldsymbol{C}\boldsymbol{B} - \boldsymbol{h}'\boldsymbol{C}\hat{\boldsymbol{B}} \leqslant \boldsymbol{r} \tag{4.29}$$

or, equivalently,

$$\boldsymbol{h}'\boldsymbol{C}\hat{\boldsymbol{B}} - \boldsymbol{r} \leqslant \boldsymbol{h}'\boldsymbol{C}\boldsymbol{B} \leqslant \boldsymbol{r} + \boldsymbol{h}'\boldsymbol{C}\hat{\boldsymbol{B}}. \tag{4.30}$$

Since equation (4.30) is equivalent to equation (4.25), the probability statement defined in equation (4.18) can be expressed as follows:

$$P[\boldsymbol{h}'\boldsymbol{C}\hat{\boldsymbol{B}} - \boldsymbol{r} < \boldsymbol{h}'\boldsymbol{C}\boldsymbol{B} < \boldsymbol{r} + \boldsymbol{h}'\boldsymbol{C}\hat{\boldsymbol{B}}] = 1 - \alpha. \tag{4.31}$$

Note that the probability statement given in equation (4.31) defines a $100(1-\alpha)\%$ confidence interval for any set of linear combinations $\boldsymbol{h}'\boldsymbol{C}\boldsymbol{B}$. Thus, when $H_0{:}\,\boldsymbol{h}'\boldsymbol{C}\boldsymbol{B} = 0$ is true, then equation (4.31) becomes

$$P[\boldsymbol{h}'\boldsymbol{C}\hat{\boldsymbol{B}} - \boldsymbol{r} < 0 < \boldsymbol{r} + \boldsymbol{h}'\boldsymbol{C}\hat{\boldsymbol{B}}] = 1 - \alpha. \tag{4.32}$$

Therefore, the null hypothesis $H_0{:}\,\boldsymbol{h}'\boldsymbol{C}\boldsymbol{B} = 0$ is rejected when zero is outside the interval $\boldsymbol{h}'\boldsymbol{C}\hat{\boldsymbol{B}} \pm \boldsymbol{r}$.

In summary, Scheffe's method rejects $H_0{:}\,\boldsymbol{h}'\boldsymbol{C}\boldsymbol{B} = 0$ if $|\boldsymbol{h}'\boldsymbol{C}\hat{\boldsymbol{B}}| > \boldsymbol{r}$, where $\boldsymbol{r} = S[(k-1)F_c\boldsymbol{h}'\boldsymbol{L}\boldsymbol{h}]^{1/2}$, $S^2 = MS_{\text{error}}$ and $F_c = F_{1-\alpha,k-1,N-k}$. The following steps must be performed to assess the significance of a contrast according to Scheffe's method:

Step 1. Calculate the numerical value of each contrast:

$$\hat{C}_i = \sum_{j=1}^{k} c_{ij}\bar{Y}_{.j} \quad \text{for } i = 1, 2, 3, \ldots, d.$$

Step 2. Specify the critical value F_c:

$$F_c = F_{1-\alpha,k-1,N-k}.$$

where k is the number of treatments and N the total number of observations.

Step 3. Compute r_i (the ith element of $\boldsymbol{r}$):

$$\boldsymbol{r} = S[(k-1)F_c\boldsymbol{h}'\boldsymbol{L}\boldsymbol{h}]^{1/2}$$

$$r_i = \left[(k-1)F_cMS_{\text{error}}\sum_{j=1}^{k}\frac{c_{ij}^2}{n_j}\right]^{1/2} \quad \text{for } i = 1, 2, 3, \ldots, d.$$

Step 4. Reject the hypothesis that $C_i = \sum_{j=1}^{k} c_{ij}\mu_j = 0$ if $|\hat{C}_i| > r_i$ for $i = 1, 2, 3, \ldots, d$; otherwise, accept the hypothesis.

EXAMPLE 4.2

Suppose that it is desired to examine the following contrasts:

$$C_1 = 2\bar{Y}_{.1} - \bar{Y}_{.3} - \bar{Y}_{.4}$$
$$C_2 = 2\bar{Y}_{.1} - 3\bar{Y}_{.2} - \bar{Y}_{.3} + 2\bar{Y}_{.4}.$$

In this example there are four treatments and the sample size is equal to five for every treatment. The corresponding hypotheses

$$H_1: 2\mu_1 - \mu_3 - \mu_4 = 0$$
$$H_2: 2\mu_1 - 3\mu_2 - \mu_3 + 2\mu_4 = 0$$

will be tested using a 5% level of significance and the following numerical results: $\bar{Y}_{.1} = 8.4$, $\bar{Y}_{.2} = 7.2$, $\bar{Y}_{.3} = -6.4$, $\bar{Y}_{.4} = -8.0$, $k = 4$ and $MS_{\text{error}} = 12.7$.

Step 1. Numerical values of contrasts:

$$\hat{C}_1 = (2)(8.4) + (-1)(-6.4) + (-1)(-8.0) = 31.20$$
$$\hat{C}_2 = (2)(8.4) + (-3)(7.2) + (-1)(-6.4) + (2)(-8.0) = -14.40.$$

Step 2. Critical value of F:

$$F_c = F_{0.95,3,16} = 3.24.$$

Step 3. Computing the r-values:

$$r_i = \left[(k-1)F_c MS_{\text{error}} \sum_{j=1}^{k} \frac{c_{ij}^2}{n_j}\right]^{1/2} \qquad i = 1, 2$$

$$r_1 = [3(3.24)(12.7)(1/5)\{2^2 + (-1)^2 + (-1)^2\}]^{1/2} = 12.17$$

$$r_2 = [3(3.24)(12.7)(1/5)\{2^2 + (-3)^2 + (-1)^2 + 2^2\}]^{1/2} = 21.08.$$

Step 4. Performing hypothesis testing:
(a) reject H_1 since $|\hat{C}_1| = 31.20 > r_1 = 12.17$;
(b) fail to reject H_2 since $|\hat{C}_2| = 14.40 < r_2 = 21.08$.

4.3 STUDENTIZED RANGE TEST

The studentized range test [4] was proposed by Newman in 1939. The basic idea was either due to student (W. S. Gosset) or was an outgrowth of some of his ideas. Later, in 1952, Keuls [5] proposed the same test. Duncan [6] in 1942 advocated the use of the same studentized range test, but with a smaller conservative level of significance than the Newman–Keuls test.

The studentized range test is an approach for making pairwise comparisons of the k sample means $\bar{Y}_1, \bar{Y}_2, \ldots, \bar{Y}_k$ to determine if there are significant differences between the population means $\mu_1, \mu_2, \ldots, \mu_k$. The test has limited applicability since it requires the use of equal sample sizes in addition to the typical assumptions of the analysis of variance (section 3.1). However, when the assumptions of normality, equal sample sizes and equal variance are violated, a more attractive procedure would be Scheffe's method [1].

There are three different studentized range tests: Newman–Keuls' test [4, 5], Duncan's test [6] and Tukey's test [7]. These three tests are used to verify the null hypothesis $H_0: \mu_i = \mu_j$, based on the studentized range distribution [8]. The main difference among them is how the critical value for conducting the test is computed.

In this section the basic concepts underlying the development and application of the studentized range test and Newman–Keuls' procedure will be discussed. In section 4.4 other procedures will be succinctly reviewed and compared to one another.

4.3.1 Studentized range distribution

Let $z_1, z_2, \ldots, z_k$ be a random sample of size k from a normal density function with mean equal to zero and variance equal to one, and let y_i be the order statistic of z_i. Let v^2 be a chi-square variable with f degrees of freedom, and let $z_1, z_2, \ldots, z_k$ and v^2 be statistically independent. The random variable

$$Q_{kf} = \frac{y_k - y_1}{vf^{-1/2}} \tag{4.33}$$

is called the studentized range.

In order to find the distribution of the range $y_k - y_1$, consider first the general expression for the joint probability density function for the two order statistics y_1 and y_k [8]:

$$p(y_1, y_k) = k(k-1)[\Phi(y_k) - \Phi(y_1)]^{k-2}\phi(y_k)\phi(y_1) \tag{4.34}$$

where $\phi(z_i)$ is the probability density function of z_i and $\Phi(y_i)$ the cumulative density function of z_i. Thus the probability density function of the range $R = y_k - y_1$ can be obtained in terms of the following integral:

$$g(R) = k(k-1)\int_{-\infty}^{\infty} [\Phi(R+T) - \Phi(T)]^{k-2}\phi(R+T)\phi(T)\mathrm{d}T \tag{4.35}$$

where $T = y_1$.

Since the z_i terms and v^2 are independent, so are R and v^2. Consequently, the joint probability density function of R and v^2 is the product of their marginal probability density functions, g and w:

$$h(R, v^2) = g(R)w(v^2) \quad \text{for } 0 < R < \infty \quad 0 < v^2 < \infty. \tag{4.36}$$

The following transformations are relevant for the development of the range test:

$$Q = \frac{R}{vf^{-1/2}} \qquad u = \frac{v^2}{f}.$$

In order to find the probability density function $h(Q)$ we must know the Jacobian of the transformations given above and $m(Q, u)$. The Jacobian of the transformations is equal to $fu^{1/2}$; also,

$$m(Q, u) = g(Qu^{1/2})w(uf)fu^{1/2} \quad \text{for } 0 < Q < \infty \quad 0 < u < \infty. \tag{4.37}$$

Finally, the probability density function of the studentized range, $h(Q)$, is obtained by integrating equation (4.37) with respect to u:

$$h(Q) = \int_0^\infty g(Qu^{1/2})w(uf)fu^{1/2}\,\mathrm{d}u \tag{4.38}$$

where g is defined in equation (4.35) and w is a chi-square probability density function with f degrees of freedom. Then performing the corresponding substitution, equation (4.38) can be written as follows:

$$h(Q) = \int_0^\infty k(k-1)\int_{-\infty}^\infty [\Phi(Qu^{1/2} + T) - \Phi(T)]^{k-2}\phi(Qu^{1/2} + T)\phi(T)\mathrm{d}T$$
$$\times\left[\left(\Gamma\frac{f}{2}\right)^{-1}\left(\frac{1}{2}\right)^{f/2}(uf)^{f/2-1}\exp\left(-\frac{uf}{2}\right)fu^{1/2}\right]\mathrm{d}u. \tag{4.39}$$

No simplification of equation (4.39) is possible to facilitate the evaluation of the function $h(Q)$. This evaluation can be performed using numerical methods, but the computational work is rather extensive. However, the following integral

$$\int_{Q_{1-\alpha}}^\infty h(Q)\mathrm{d}Q = \alpha \tag{4.40}$$

has been tabulated for $\alpha = 0.01$, $\alpha = 0.05$ and $\alpha = 0.10$. Tables B.6 and B.7 give values of $Q_{1-\alpha}$ for $\alpha = 0.05$ and $\alpha = 0.01$, respectively.

A similar analysis can be conducted for the difference $y_j - y_i$ where $y_j > y_i$. In this case, it is possible to show that

$$Q_{rf} = \frac{y_j - y_i}{vf^{-1/2}} \tag{4.41}$$

follows a studentized range distribution with $r = j - i + 1$.

4.3.2 Basic testing procedure

Let $\bar{Y}_{(1)}, \bar{Y}_{(2)}, \ldots, \bar{Y}_{(k)}$ be the order statistics of the random variables $\bar{Y}_{.1}, \bar{Y}_{.2}, \ldots, \bar{Y}_{.k}$. Recall that

$$\frac{fS^2}{\sigma^2} \sim \chi_f^2 \tag{4.42}$$

where $S^2 = MS_{\text{error}}$ and $f = N - k$. The studentized range statistic for the range

$$R = \bar{Y}_{(j)} - \bar{Y}_{(i)}$$

is given by

$$Q_{rf} = (\bar{Y}_{(j)} - \bar{Y}_{(i)})\left(\frac{MS_{\text{error}}}{n}\right)^{-1/2} \tag{4.43}$$

Therefore, according to equation (4.43), if Q_c represents the value of Q_{rf} for which equation (4.40) is satisfied at the significance level α, the critical value of $R = \bar{Y}_{(j)} - \bar{Y}_{(i)}$ is obtained as

$$R_c = \left(\frac{MS_{\text{error}}}{n}\right)^{1/2} Q_c. \tag{4.44}$$

In equation (4.43), r is the number of steps between $\bar{Y}_{(j)}$ and $\bar{Y}_{(i)}$, that is, $r = j - i + 1$.

As a result of the previous discussion, the decision rule for the studentized range test would be to reject the hypothesis that $\mu_{(j)}$ is equal to $\mu_{(i)}$ if the observed range $\bar{Y}_{(j)} - \bar{Y}_{(i)}$ exceeds R_c. Otherwise, the hypothesis is accepted. Since the observed totals $T_{(1)}, T_{(2)}, \ldots, T_{(k)}$ are linked to the observed averages $\bar{Y}_{(1)}, \bar{Y}_{(2)}, \ldots, \bar{Y}_{(k)}$ by the relationship

$$T_{(i)} = n\bar{Y}_{(i)}$$

then it is also possible to express Q_{rf} as

$$Q_{rf} = \frac{T_{(j)} - T_{(i)}}{(nMS_{\text{error}})^{1/2}} \tag{4.45}$$

where $r = j - i + 1$ and $f = N - k$.

From equation (4.45), a critical value for the difference between the totals $T_{(j)}$ and $T_{(i)}$ is equal to $(nMS_{\text{error}})^{1/2}Q_c$. Therefore, the hypothesis that $\mu_{(i)}$ is equal to $\mu_{(j)}$ is rejected if the observed difference between the totals exceeds the value $(nMS_{\text{error}})^{1/2}Q_c$.

The observed totals can be recorded in a table which has k rows and k columns. The first row and the first column correspond to the smallest total, the second row and the second column correspond to the second total, and so on.

Once the values Q_c are obtained, the critical ranges can be computed and then the observed ranges can be checked against these critical values. The following procedure is suggested to perform the tests:

(a) The first test is on the difference in the upper right-hand corner. If the comparison is not significant, no additional tests are made.

(b) If the comparison is significant, tests are made successively on the entries of the first row, proceeding from right to left until the first nonsignificant difference is found. If the last significant entry occurs in column g, no additional tests will go beyond column g.
(c) The next test is made on the extreme right-hand entry in the second row. If the comparison is not significant, no additional tests are made.
(d) Tests are continued from right to left in the second row until either a nonsignificant difference is found or column g is reached, whichever happens first. If a nonsignificant difference is found first in column $h > g$, no additional tests are made beyond column h.
(e) For any row, proceeding from right to left, tests are made until a nonsignificant difference is found, or the point where the previous row was stopped is reached, whichever occurs first.

For the Newman–Keuls range test, it must be noted that the type I error rate is smaller and, therefore, provides a more conservative test than Duncan's. In other words, given the same number of treatment mean comparisons, it is more difficult to find a treatment mean pair significantly different using the Newman–Keuls method than with Duncan's procedure.

EXAMPLE 4.3

Five different types of fuses made by different manufacturers are going to be compared on the basis of service life. The durations in months, for random samples of size equal to seven, are given in Table 4.3. It is desired to test the hypothesis that the mean lives of these five types of fuses are equal.

In case that this hypothesis is rejected, the Newman–Keuls range test will be used with a level of significance equal to 0.05.

Table 4.3

A	*B*	*C*	*D*	*E*
1.25	2.25	6.75	3.25	5.50
1.50	4.50	4.25	3.50	3.50
2.50	4.75	6.25	4.75	7.25
1.30	3.25	7.50	3.50	4.75
2.70	1.50	4.50	5.50	4.00
1.20	3.75	4.75	3.75	6.25
1.80	4.50	8.00	7.25	3.75

ANOVA test

The totals of the five treatments being considered are shown below. In each case the order of the treatment is given in parentheses after arranging the totals according to their magnitudes.

	A ($j=1$)	B ($j=2$)	D ($j=3$)	E ($j=4$)	C ($j=5$)	Total
$T_{(j)}$	12.25	24.50	31.50	35.00	42.00	$T_{..}=145.25$
n_j	7	7	7	7	7	$N=35$

The hypothesis-testing procedure is summarized in Table 4.4. It can be verified that the corresponding critical value of the F-statistic is equal to 1.42. As a result of the ANOVA test indicated in Table 4.4, the hypothesis that there is no treatment effect is rejected. In order to see which significant differences caused this rejection, the Newman–Keuls range test will be used.

Newman–Keuls' range test

$T_{(1)}=12.25$, $T_{(2)}=24.50$, $T_{(3)}=31.50$, $T_{(4)}=35.00$, $T_{(5)}=42.00$. The observed ranges between these totals are shown in Table 4.5. Referring to Table 4.4,

Table 4.4 ANOVA results

Source	*df*	*SS*	*MS*	F_0
Treatments	4	73.15	18.29	11.02
Error	30	49.79	1.66	
Total	34	122.94		

Table 4.5 Ranges between totals

	$T_{(1)}$	$T_{(2)}$	$T_{(3)}$	$T_{(4)}$	$T_{(5)}$
$T_{(1)}$	–	12.25	19.25	22.75	29.75
$T_{(2)}$	–	–	7.00	10.50	17.50
$T_{(3)}$	–	–	–	3.50	10.50
$T_{(4)}$	–	–	–	–	7.00
$T_{(5)}$	–	–	–	–	–

Table 4.6 Results from the Newman–Keuls test for Example 4.3

	$\bar{Y}_{(1)}$	$\bar{Y}_{(2)}$	$\bar{Y}_{(3)}$	$\bar{Y}_{(4)}$	$\bar{Y}_{(5)}$
$\bar{Y}_{(1)}$	–	1.75*	2.75*	3.25*	4.25*
$\bar{Y}_{(2)}$	–	–	1.00	1.50	2.50*
$\bar{Y}_{(3)}$	–	–	–	0.50	1.50
$\bar{Y}_{(4)}$	–	–	–	–	1.00
$\bar{Y}_{(5)}$	–	–	–	–	–

* Indicates a significant difference between means.

note that $MS_{\text{error}} = 1.66$ and $n = 7$. For $r = 2, 3, 4, 5$ the values of $Q_{(r,\alpha,f)}$ can be found in Table B.6. The corresponding critical values for comparing treatment averages are computed using the relationship

$$R_c = \left(\frac{MS_{\text{error}}}{n}\right)^{1/2} Q_{(r,0.05,30)}.$$

The following results are obtained:

$$\begin{aligned} r = 2&: Q_{(2,0.05,30)} = 2.89 \qquad R_c = 1.41 \\ r = 3&: Q_{(3,0.05,30)} = 3.48 \qquad R_c = 1.70 \\ r = 4&: Q_{(4,0.05,30)} = 3.84 \qquad R_c = 1.87 \\ r = 5&: Q_{(5,0.05,30)} = 4.11 \qquad R_c = 2.00. \end{aligned}$$

As an illustration, the difference between the largest (C) and smallest (A) average will be considered. Note that in this case $R_c = 2.00$ and

$$\bar{Y}_{(5)} - \bar{Y}_{(1)} = \tfrac{1}{7}(T_{(5)} - T_{(1)}) = \tfrac{1}{7}(29.75) = 4.25.$$

Since this value exceeds the critical value $R_c = 2.00$, the difference between the effects of treatments A and C is declared significant. Proceeding in a similar fashion, and according to the systematic procedure previously outlined, it is concluded that the range between the following pairs of treatments (A, E), (A, D), (A, B) and (B, C) is significant, but those between (B, D), (B, E), (D, E), (C, D) and (C, E) are not. The test results are summarized as in Table 4.6.

4.4 MEAN SEPARATION PROCEDURES

The general methodology for the range tests was developed in section 4.3. As mentioned in that section, there are different kinds of range tests generally

Table 4.7 Mean separation procedures

Category	*Method*	*Critical values*	*Mean pairs*
I	(1) Multiple comparisons Least significant difference Tukey's test	1	$\frac{1}{2}k(k-1)$
	(2) Multiple range tests Newman–Keuls' test Duncan's test	2 or more	$\frac{1}{2}k(k-1)$
II	Dunnett's test	1	$k-1$

known as **mean separation procedures**. This section will summarize the most important aspects of these procedures.

Mean separation procedures [9] are designed to make a statistical inference concerning a given set of treatment means. Table 4.7 illustrates the two different categories under which mean separation procedures may be grouped. The first category consists of a group of studentized range-based procedures to test all possible pairs of mean differences. These procedures may be further subdivided into two types of methods, namely **multiple comparisons** and **multiple range tests**. As indicated in Table 4.7, in an experiment with k treatment means, $\frac{1}{2}k(k-1)$ comparisons can be conducted. The multiple comparison method uses a single critical value while the multiple range test procedure uses two or more critical values.

The purpose of the second category of mean separation procedures is to test mean differences between any treatment and a specified treatment, generally known as a control or standard treatment. As indicated in Table 4.7, in this case there are only $k-1$ paired comparisons.

4.4.1 Least significant difference method

The least significant difference (LSD) procedure, also known as Fisher's LSD, consists of making all possible $\frac{1}{2}k(k-1)$ standard t-tests for k treatment means. The LSD procedure is only used when the basic ANOVA test shows that at least one treatment mean is significantly different from the others, i.e. when the treatment source of variation is found to be significant by the F-statistic. The purpose of each of the pairwise treatment mean t-tests in the LSD procedure is to test the null hypothesis that the corresponding two treatment means are equal.

The critical value for conducting each test is the same for each individual comparison of mean pairs, and it depends on the number of degrees of freedom for the error, as well as the significance level of the test, α. It should be noted that α reflects the probability of a type I error for each individual comparison, and therefore cannot be applied to the entire group of individual comparisons since these are not mutually independent. Thus, the t-tests of all possible hypotheses are not independent.

The *LSD* method differs from the others in that it has a comparison-wise type I error, α, over all repetitions of the experiments. This means that the α risk will be inflated when using the *LSD* procedure. As the number of treatments increases, the type I error for the experiment becomes large. Due to this fact, a situation may arise whereby the F-statistic in the ANOVA is significant, yet the *LSD* procedure fails to find any pairs of treatment means which differ significantly from one another. This occurs since the F-statistic is considering all possible comparisons between treatment means simultaneously, not in a pairwise manner as the *LSD* procedure does.

The *LSD* looks at pairs of treatment means μ_i and μ_j such that if

$$|\bar{Y}_{i.} - \bar{Y}_{j.}| \geqslant t_{1-\alpha/2,N-k}\left[MS_{\text{error}}\left(\frac{1}{n_i}+\frac{1}{n_j}\right)\right]^{1/2}$$

then those pairs of treatment means are significantly different from one another. The *LSD* critical value is defined as

$$LSD = t_{1-\alpha/2,N-k}\left[MS_{\text{error}}\left(\frac{1}{n_i}+\frac{1}{n_j}\right)\right]^{1/2}$$

or in the case where you have a balanced design with an equal number of observations for each treatment, i.e. $n_1 = n_2 = \cdots = n$, the critical value of *LSD* raduces to

$$LSD = t_{1-\alpha/2,N-k}\left(\frac{2MS_{\text{error}}}{n}\right)^{1/2}.$$

Comparing the above result to that shown in equation (4.44), and setting $LSD = R_c$, it is concluded that $t_{1-\alpha/2,N-k} = Q_c(2)^{-1/2}$. In other words, the test can be conducted as a t-test with critical value $t_c = t_{1-\alpha/2,N-k}$, or as a range test with $r = 2$ and critical value $Q_c = 2^{1/2}t_c$.

It is easy to apply the *LSD* procedure, since one simply compares the observed absolute value of the difference between each pair of means. If the observed value exceeds *LSD*, then it can be determined that the population means μ_i and μ_j are different.

EXAMPLE 4.4

Consider the data in Example 4.3, concerning the service life response of five different types of fuses from separate manufacturers. The ANOVA methodology resulted in rejecting the null hypothesis that there is no treatment effect due to manufacturers. An analysis of the mean differences will be performed here in terms of the *LSD*.

Note that $MS_{\text{error}} = 1.66$ and $\bar{Y}_{(i)} = T_{(i)}n^{-1}$. In this example, the critical

Table 4.8 Results from *LSD* method for Example 4.4

	$\bar{Y}_{(1)}$	$\bar{Y}_{(2)}$	$\bar{Y}_{(3)}$	$\bar{Y}_{(4)}$	$\bar{Y}_{(5)}$
$\bar{Y}_{(1)}$	–	1.75*	2.75*	3.25*	4.25*
$\bar{Y}_{(2)}$	–	–	1.00	1.50	2.50*
$\bar{Y}_{(3)}$	–	–	–	0.50	1.50
$\bar{Y}_{(4)}$	–	–	–	–	1.00
$\bar{Y}_{(5)}$	–	–	–	–	–

* Indicates a significant difference between means.

value for any individual comparison of mean differences is given by equation (4.44) after setting $r = 2$, with $Q_{(r,\alpha,f)} = Q_{(2,0.05,30)} = 2.89$ obtained from Table B.6.

$$R_c = \left(\frac{MS_{\text{errpr}}}{n}\right)^{1/2} Q_{(r,\alpha,f)} = \left(\frac{1.66}{7}\right)^{1/2} Q_{(2,0.05,30)} = 1.41.$$

Therefore, all pairs of treatment means whose difference exceeds the critical value are considered as significant. The test results are summarized as shown in Table 4.8.

4.4.2 Tukey's range procedure

Tukey's procedure [7] is a special version of the range test that assumes that the parameter r (number of steps between averages) is equal to k for each pair of means. This assumption implies that all mean pairs are positioned at the end points of a numerical scale extending from the smallest to the largest average (or total), whereas the LSD procedure assumes they are positioned adjacent to each other. This extreme ordering requires a larger critical difference, and for this reason Tukey's test is said to be more conservative than the LSD procedure. Tukey's test has an experiment-wise type I error, which means that an experiment with only one incorrect decision is as undesirable as an experiment with many incorrect decisions.

The test procedure in Tukey's method is the same as that in the LSD method, except that the value of r used for calculating the critical value is k instead of 2. Notice in Example 4.5 below that Tukey's method gives a smaller number of significant mean pairs than the LSD procedure when both methods are applied to the same data.

EXAMPLE 4.5

Consider again Example 4.3. The critical value for Tukey's range procedure with $k = 5$ is equal to

Table 4.9 Results from the Tukey's range procdure for Example 4.5

	$\bar{Y}_{(1)}$	$\bar{Y}_{(2)}$	$\bar{Y}_{(3)}$	$\bar{Y}_{(4)}$	$\bar{Y}_{(5)}$
$\bar{Y}_{(1)}$	–	1.75	2.75*	3.25*	4.25*
$\bar{Y}_{(2)}$	–	–	1.00	1.50	2.50*
$\bar{Y}_{(3)}$	–	–	–	0.50	1.50
$\bar{Y}_{(4)}$	–	–	–	–	1.00
$\bar{Y}_{(5)}$	–	–	–	–	–

* Indicates a significant difference between means.

$$\left(\frac{MS_{\text{error}}}{n}\right)^{1/2} Q_{(5,0.05,30)} = \left(\frac{1.66}{7}\right)^{1/2} (4.11) = 2.00.$$

The results of the test are given in Table 4.9. As is expected, only four mean pairs are considered significant while the LSD method concluded that seven mean pairs were significantly different.

4.4.3 Duncan's multiple range test

Duncan's method [6] is similar to Newman–Keuls' procedure in the sense that it uses several critical values for conducting the test, i.e. one critical value for each subset of mean pairs associated with the same number of steps $r = 2, 3, \ldots, k$. However, Duncan's procedure applies a different 'error rate' $\alpha_r = 1 - (1 - \alpha)^r$ for each subset associated with r steps $(r = 2, 3, \ldots, k)$ in calculating the critical value, while Newman–Keuls' procedure uses the same α for all subsets. Tables B.8 and B.9 show critical values for Duncan's multiple range test, considering $\alpha = 0.01$ and 0.05, respectively. These critical values have already incorporated the error rate α_r.

Comparing Tables B.8 and B.9 to Tables B.6 and B.7, it is easy to appreciate that the incorporation of the α_r values results in smaller critical values for Duncan's test for $r > 2$. Duncan refers to the α_r as special protection levels against finding false significant differences. Example 4.6 illustrates the procedure for Duncan's multiple range test.

EXAMPLE 4.6

Consider Example 4.3 again. The critical values for R_c in this case are determined using

$$R_c = \left(\frac{MS_{\text{error}}}{n}\right)^{1/2} Q'_{(r,0.05,30)}$$

Table 4.10 Results from Duncan's test for Example 4.6

	$\bar{Y}_{(1)}$	$\bar{Y}_{(2)}$	$\bar{Y}_{(3)}$	$\bar{Y}_{(4)}$	$\bar{Y}_{(5)}$
$\bar{Y}_{(1)}$	–	1.75*	2.75*	3.25*	4.25*
$\bar{Y}_{(2)}$	–	–	1.00	1.50*	2.50*
$\bar{Y}_{(3)}$	–	–	–	0.50	1.50*
$\bar{Y}_{(4)}$	–	–	–	–	1.00
$\bar{Y}_{(5)}$	–	–	–	–	–

* Indicates a significant difference between means.

where the required critical values of the Q-statistic come from Table B.9. The following critical values are obtained:

$$r = 2: Q'_{(2,0.05,30)} = 2.89 \qquad R_c = 1.41$$
$$r = 3: Q'_{(3,0.05,30)} = 3.04 \qquad R_c = 1.48$$
$$r = 4: Q'_{(4,0.50,30)} = 3.13 \qquad R_c = 1.52$$
$$r = 5: Q'_{(5,0.05,30)} = 3.20 \qquad R_c = 1.56.$$

Comparing the critical values $Q_{r\alpha f}$ used in Example 4.3 to the values $Q'_{r\alpha f}$ used in the current example, it can be seen that $Q'_{(r,0.05,30)} < Q_{(r,0.05,30)}$ for $r = 3, 4, 5$.

Following the same testing procedure presented in Example 4.3, it is concluded that significant mean differences exist between the treatments in each of the following pairs: (A, C), (A, E), (A, D), (A, B), (B, D), (B, E) and (C, E), as shown in Table 4.10.

4.4.4 Dunnett's procedure

This procedure is appropriate for determining significant differences between any treatment mean and the specified control treatment. Dunnett's procedure [10] makes use of an experiment-wise type I error. The basic testing procedure is shown below, where μ_s represents the specified treatment mean:

(a) Null hypothesis:

$$H_0: \mu_s = \mu_j \quad j \neq s.$$

(b) Alternative hypotheses:

$$H_1: \mu_s > \mu_j$$
$$H'_1: \mu_s < \mu_j$$
$$H''_1: \mu_s \neq \mu_j.$$

(c) Test statistic:

$$D_j = Y_j - Y_s.$$

(d) Critical value:

$$D = \left(\frac{2MS_{\text{error}}}{n}\right)^{1/2} d_{(k-1,\alpha,f)}$$

$$D' = -\left(\frac{2MS_{\text{error}}}{n}\right)^{1/2} d_{(k-1,\alpha,f)}$$

$$D'' = \left(\frac{2MS_{\text{error}}}{n}\right)^{1/2} d_{(k-1,\alpha/2,f)}.$$

(e) Conclusion: reject H_0 if $D_j \geqslant D, D_j \leqslant D'$ or $|D_j| \geqslant D''$.

In the above relationships, $d_{(k-1,\alpha/2,f)}$ is the value from Tables B.10 and $d_{(k-1,\alpha,f)}$ the value from Table B.11 for making $k-1$ one-sided or two-sided comparisons, excluding the control treatment, with f degrees of freedom for the MS_{error}.

EXAMPLE 4.7

Consider Example 4.3 again. Assume $\bar{Y}_{(3)}$ is the control treatment. Apply a two-sided Dunnett's procedure with $\alpha = 0.05$ level of significance. From Table B.10, the value of $d_{(5-1,0.025,30)} = 2.58$.

(a) Null hypothesis:

$$H_0: \mu_3 = \mu_j, j = 1, 2, 4, 5.$$

(b) Alternative hypothesis:

$$H_1: \mu_3 \neq \mu_j.$$

(c) Test statistic:

$$\bar{Y}_j - \bar{Y}_3.$$

(d) Critical value:

$$D'' = \left(\frac{2MS_{\text{error}}}{n}\right)^{1/2} d_{(5-1,0.025,30)}$$

$$= \left(\frac{2(1.66)}{7}\right)^{1/2} (2.58)$$

$$= 1.78$$

(e) Comparisons:

$$|\bar{Y}_{(3)} - \bar{Y}_{(1)}| = 2.75 > D''$$
$$|\bar{Y}_{(3)} - \bar{Y}_{(2)}| = 1.00 < D''$$
$$|\bar{Y}_{(3)} - \bar{Y}_{(4)}| = 0.50 < D''$$
$$|\bar{Y}_{(3)} - \bar{Y}_{(5)}| = 1.50 < D''.$$

(f) Conclusion: reject $H_0: \mu_3 = \mu_1$.

EXERCISES

1. Give examples of (a) orthogonal contrasts, (b) nonorthogonal contrasts for an experiment with three treatments and sample sizes equal to 4, 10 and 7, respectively.
2. (a) Prove the orthogonality conditions for contrasts C_1 and C_2.
   (b) Develop the formulas for contrasts defined on totals. Give examples.
3. The Kenton Food Company wishes to tests four different package designs for a new breakfast cereal. Ten stores, with approximately equal volumes, were selected as the experimental units. These 10 stores were randomly assigned to the package designs, and the number of cases sold was recorded for a specified study period. Other relevant conditions beside package design, such as price, amount and location of shelf space, and special promotional efforts were kept the same for all stores. The results obtained are given in Table 4.11.
   (a) Perform an analysis of variance.
   (b) Interpret the results.
4. Suppose that the characteristics of each package design in Exercise 3 can be described as indicated in Table 4.12.
   (a) Define a set of orthogonal contrasts to compare, using averages:
      (i) the two 3-color designs;
      (ii) the two 5-color designs;
      (iii) the 3 color designs with the 5-color designs.
   (b) Perform the corresponding analysis of variance using the results given in Exercise 3.

Table 4.11

Package design			
1	*2*	*3*	*4*
12	14	19	24
18	12	17	30
	13	21	

Table 4.12

Design	*Characteristics*
1	3-color, with cartoons
2	3-color, without cartoons
3	5-color, with cartoons
4	5-color, without cartoons

5. In the case discussed in Exercise 3, we wish to estimate the following contrasts and perform an analysis of variance using Scheffe's method:
   (a) the two 3-color designs;
   (b) the two 5-color designs;
   (c) the 3-color designs with the 5-color designs.
6. Consider the following hypothetical results for an experiment with five treatments: assume that $MS_{\text{error}} = 2$.

Treatments	A	B	C	D	E
$T_{.j}$	7	20	18	24	14
n_j	4	4	4	4	4

   Perform a range test using a 1% level of significance.
7. Use Scheffe's method to test if the mean life of a fuse type A is equal to the average of the expected lives of types B and C in the experiment mentioned in Exercise 5, Chapter 3.
8. Find 95% confidence limits for the mean of the second treatment in Exercise 5, Chapter 3.
9. Perform the studentized range test for Exercise 5, Chapter 3, using a 1% level of significance and considering ranges between averages instead of totals.
10. The owner of a lawn and garden supply store is currently considering four brands of lawn fertilizer based on growth rate. He conducted an experiment by applying each brand to six randomly selected yards. Initial tests indicate that there are significant differences between brands ($\alpha = 0.05$). Given the information below, use the LSD procedure to determine which brands are significantly different. Assume that $MS_{\text{error}} = 4.76$.

$$T_1 = 181.8 \qquad T_3 = 183.0$$
$$T_2 = 198.0 \qquad T_4 = 214.2.$$

11. Apply Tukey's procedure to the experimental data in Exercise.10. Briefly discuss why some differences in the results may occur.
12. Given the experimental data below, use Duncan's procedure to determine any significant differences between pairs of means ($\alpha = 0.05$).

$$\bar{Y}_1 = 1.94 \qquad \bar{Y}_4 = 3.14$$
$$\bar{Y}_2 = 2.63 \qquad \bar{Y}_5 = 2.71.$$
$$\bar{Y}_3 = 3.52$$

13. In the experimental situation given in Exercise 22, Chapter 3, it is desired to test if there is a significant difference between the largest and smallest

averages, using the Newman–Keuls range test and considering a level of significance equal to 0.05.

(a) Find the maximum range allowed between the two averages before the null hypothesis is rejected, taking into consideration the number of steps.

(b) Find the maximum range allowed between the two averages before the null hypothesis is rejected, ignoring the number of steps.

REFERENCES

1. Scheffe, H. (1953) A method for judging all contrasts in the analysis of variance. *Biometrika*, **40**, 87–104.
2. Seber, G. A. F. (1977) *Linear Regression Analysis*, John Wiley & Sons, New York.
3. Miller, R. G. Jr (1986) *Simultaneous Statistical Inference*, McGraw-Hill, New York.
4. Newman, D. (1939) The distribution of the range in samples from a normal population, expressed in terms of an independent estimate of standard deviation. *Biometrika*, **31**.
5. Keuls, M. (1952) The use of the studentized range in connection with an analysis of variance. *Euphytica*, **1**, 112–22.
6. Duncan, D. B. (1955) Multiple range and multiple *F* tests. *Biometrics*, **11**, 1–42.
7. Turkey, J. W. (1949) Comparing individual means in the analysis of variance. *Biometrics*, **5**, 99–114.
8. Mood, A. M. and Graybill, F. A. (1963) *Introduction to the Theory of Statistics*, 2nd edn, McGraw-Hill, New York.
9. Lentner, M. and Bishop, T. (1986) *Experimental Design and Analysis*, Valley Book Company, Blacksburg.
10. Dunnett, C. W. (1955) A multiple comparison procedure for comparing several treatments with a control. *Jour. Amer. Stat. Assoc.*, **50**, 1096–121.

FURTHER READING

Pearson, E. S. and Hartley, H. O. (1956) *Biometrica Tables for Statisticians Volume 1*, 3rd edn, Cambridge University Press, Cambridge.

5 Analysis of single-factor experiments using randomized block designs

5.1 INTRODUCTION

Perhaps the most common local control strategy in industrial experimentation is the use of blocks. A proper subset of the experimental conditions of a randomized experiment having an error variation which is expected to be less than the variation in the entire group of treatments is referred to as a block. As a precaution against systematic variation from one trial to another, it is desirable to arrange the treatments within each block in a random order. When this is done, the associated experimental design is called a **randomized block design**. As mentioned in Chapter 1, the purpose of local control is to increase the homogeneity of the experimental material to reduce the random error.

An illustration of the use of blocking is given as follows. The manufacturing manager of a plant wants to buy a machine for performing a particular operation of a production process. In this process the machines under test need to be tended by any of the five available operators. There are four companies that manufacture the type of machine needed and each would allow the manager to use its machine for a few days. The quantity produced per day is chosen as the measure of effectiveness in the selection of the company from which the machine will be bought. Following the concepts developed in Chapter 3, a totally randomized experiment could be designed considering the four manufacturers as the levels of a single factor (machine). In this experiment one operator could be assigned to each machine and be allowed to work on that machine for five days. However, this design is not really satisfactory, since the skills of the operators chosen may affect the results of the experiment. A more appropriate experiment would involve the use of all of the five operators working on each machine during a day. If the results associated with each operator are obtained at random and put together within a block, this second experimental design will allow us to study the effects of the operators in addition to the effect of the machines.

In the experiment to be studied in this chapter there is still a single factor under consideration. The blocks simply represent one restriction on complete randomization due to the environment in which the experment is conducted. The data for the analysis can be arranged in a two-way classification table, where rows represent blocks and columns represent treatments. Depending on the size of the block used, there are two basic types of randomized block designs:

(a) *Complete block design.* Each block contains all the treatments.
(b) *Incomplete block design.* The size of at least one block is less than the number of treatments in the experiment. There are two types of randomized block designs with incomplete blocks:
   (i) *Balanced.* All blocks have the same size and the number of times any pair of treatments appears together, in a block, in the design is constant. Furthermore, a balanced incomplete block design is said to be **symmetric** if the number of treatments is equal to the number of blocks.
   (ii) *Unbalanced.* The number of blocks containing any pair of treatments is not constant. It may differ from one pair to another.

Some examples of treatment layouts are given in Tables 5.1–5.3 to clarify the previous definitions. Table 5.1 shows a complete block design, Table 5.2 gives a balanced incomplete block design and Table 5.3 shows an unbalanced incomplete design. In Table 5.1 there are four treatments and three blocks. As can be seen, treatments A, B, C and D are contained in each block. In the incomplete block designs shown in Tables 5.2 and 5.3 there are five treatments A, B, C, D, E and the size of the block is equal to four. In the balanced

Table 5.1 Complete block design

	Positions			
	A	B	C	D
Blocks	B	A	C	D
	D	C	B	A

Table 5.2 Balanced incomplete design

	Positions			
	A	B	C	D
	A	B	C	E
Blocks	A	B	D	E
	A	C	D	E
	B	C	D	E

Table 5.3 Unbalanced incomplete design

	Positions			
	B	A	E	D
	A	B	C	D
Blocks	B	A	C	D
	E	D	C	A
	D	C	A	B

Table 5.4 Analysis techniques

Design	*Analysis*
Complete	Two-way ANOVA: section 5.2
Balanced incomplete	Special ANOVA (adjusted): section 5.3
Unbalanced incomplete	General least-square method: section 5.6

incomplete design shown in Table 5.2 the number of times any pair of treatments appears together (in a block) is constant and equal to three. For a general unbalanced design, this number is not constant, as illustrated in the design shown in Table 5.3.

Table 5.4 indicates the type of analysis technique that can be used in each of the three possible cases illustrated in Tables 5.1–5.3.

5.2 ANALYSIS OF VARIANCE FOR COMPLETE BLOCKS

Consider the fixed-effect model discussed in Chapter 3, with the additional restriction that the experimental material is grouped into homogeneous blocks. If β_i represents the effect due to the ith block, the corresponding model can be formulated as

$$Y_{ij} = \mu + \beta_i + \tau_j + \varepsilon_{ij} \tag{5.1}$$

where $i = 1, 2, \ldots, n$ and $j = 1, 2, \ldots, k$. In the above model

$$\beta_i = \mu_{i.} - \mu \tag{5.2}$$

where $\mu_{i.}$ is the expected value of the population from which observations are taken for the ith block, and

$$\tau_j = \mu_{.j} - \mu \tag{5.3}$$

where $\mu_{.j}$ is the expected value of the observation in the jth treatment.

A new assumption, in addition to those considered in Chapter 3, is that the effect of each treatment is the same in all blocks. In order to explain this assumption, the estimated responses associated with the ith observation in each of two treatments j and j', $j \neq j'$, will be considered. These responses can be written as

$$Y_{ij} = \mu + (\mu_{i.} - \mu) + (\mu_{.j} - \mu) \tag{5.4}$$

and

$$Y_{ij'} = \mu + (\mu_{i.} - \mu) + (\mu_{.j'} - \mu). \tag{5.5}$$

Similarly, for the i'th observation, $i \neq i'$, in each of the treatments j and j', is given by

$$Y_{i'j} = \mu + (\mu_{i'.} - \mu) + (\mu_{.j} - \mu) \tag{5.6}$$

and

$$Y_{i'j'} = \mu + (\mu_{i'.} - \mu) + (\mu_{.j} - \mu). \tag{5.7}$$

Let $A_{ii'j}$ and $A_{ii'j'}$ be defined as

$$A_{ii'j} = E[Y_{ij}] - E[Y_{i'j}] \tag{5.8}$$

$$A_{ii'j'} = E[Y_{ij'}] - E[Y_{i'j'}]. \tag{5.9}$$

As can be seen from equations (5.2)–(5.7), $A_{ii'j} = A_{ii'j'} = \beta_{i.} - \beta_{i'.}$ for all i, i', j and j'. Thus, in this case the interaction effect between blocks and treatments does not exist. When $A_{ii'j} \neq A_{ii'j'}$, there exists an interaction effect between blocks and treatments.

The fundamental equation of ANOVA for the randomized block design is now developed. From equation (5.1), it can be seen that the error term is equal to

$$\varepsilon_{ij} = Y_{ij} - \mu_{i.} - \mu_{.j} + \mu. \tag{5.10}$$

Therefore, the total deviation of Y_{ij} from the grand mean μ can be broken down into the following three components:

$$Y_{ij} - \mu = (\mu_{i.} - \mu) + (\mu_{.j} - \mu) + (Y_{ij} - \mu_{i.} - \mu_{.j} + \mu). \tag{5.11}$$

Based on equation (5.11), the following relationship can be considered for the estimates of the corresponding unknown population parameters:

$$Y_{ij} - \bar{Y}_{..} = (\bar{Y}_{i.} - \bar{Y}_{..}) + (\bar{Y}_{.j} - \bar{Y}_{..}) + (Y_{ij} - \bar{Y}_{i.} - \bar{Y}_{.j} + \bar{Y}_{..}). \tag{5.12}$$

In addition to the sums of squares defined in Chapter 3, it is possible to consider the sum of squares for blocks and to reformulate the sum of squares for the experimental error. The corresponding results are shown as follows:

$$SS_{\text{total}} = \sum_i \sum_j (Y_{ij} - \bar{Y}_{..})^2 \tag{5.13}$$

$$SS_{\text{block}} = \sum_i \sum_j (\bar{Y}_{i.} - \bar{Y}_{..})^2 \tag{5.14}$$

$$SS_{\text{treatment}} = \sum_i \sum_j (\bar{Y}_{.j} - \bar{Y}_{..})^2 \tag{5.15}$$

$$SS_{\text{error}} = \sum_i \sum_j (Y_{ij} - \bar{Y}_{i.} - \bar{Y}_{.j} + \bar{Y}_{..})^2. \tag{5.16}$$

Following a procedure similar to that explained in Chapter 3 for the one-way model, it is possible to derive the fundamental equation of the ANOVA for randomized block designs:

$$SS_{\text{total}} = SS_{\text{block}} + SS_{\text{treatment}} + SS_{\text{error}}. \tag{5.17}$$

The number of degrees of freedom of the total sum of squares is equal to $(nk - 1)$; this number can be partitioned into three components, that is $(n - 1)$ for blocks, $(k - 1)$ for treatments and $(n - 1)(k - 1)$ for the error term. It can be verified that this number of degrees of freedom for the error is indeed equal to

$$(nk - 1) - (n - 1) - (k - 1).$$

As indicated previously, blocking is used to reduce the experimental error by making the experimental material more homogeneous. This can be appreciated by considering the one-way and two-way ANOVA models for the ith observation of the jth treatment:

$$Y_{ij} + \mu + \tau_j + \quad \varepsilon_{ij} \qquad \text{(one way)}$$

$$Y_{ij} = \mu + \tau_j + \beta_i + \varepsilon_{ij} \quad \text{(two-way)}.$$

Note that

$$\varepsilon_{ij}(\text{one-way}) = \beta_i + \varepsilon_{ij} \quad \text{(two-way)} \tag{5.18}$$

It can be verified that this equality implies

$$SS_{\text{error}}(\text{one-way}) = SS_{\text{block}} + SS_{\text{error}} \quad \text{(two-way)}. \tag{5.19}$$

The mean squares associated with the three independent sources of variation included in equation (5.17) are defined as

$$MS_{\text{block}} = \frac{SS_{\text{block}}}{(n-1)} \tag{5.20}$$

$$MS_{\text{treatment}} = \frac{SS_{\text{treatment}}}{k-1} \tag{5.21}$$

$$MS_{\text{error}} = \frac{SS_{\text{error}}}{(n-1)(k-1)} \tag{5.22}$$

The expected value of the block mean square can be found as follows. It will be assumed that the variance of the random error is σ_e^2, and that the treatment effects are fixed. From equation (5.14) it is first concluded that

$$SS_{\text{block}} = k \sum_i (\bar{Y}_{i.} - \bar{Y}_{..})^2. \tag{5.23}$$

The expected value of the above sum of squares is given by

$$\begin{aligned} E[SS_{\text{block}}] &= kE\left[\sum_i (\bar{Y}_{i.} - \bar{Y}_{..})^2\right] \\ &= k\sum_i E[(\bar{Y}_{i.} - \bar{Y}_{..})^2] \\ &= k\sum_i (E[\bar{Y}_{i.} - \bar{Y}_{..}])^2 + k\sum_i V[\bar{Y}_{i.} - \bar{Y}_{..}] \\ &= k\sum_i \beta_i^2 + nkV[\bar{Y}_{i.} - \bar{Y}_{..}]. \end{aligned} \tag{5.24}$$

The variance term $V\{\bar{Y}_{i.} - \bar{Y}_{..}\}$ shown on the right-hand side of equation (5.24) is obtained as follows. First it is noted that

$$\begin{aligned} \bar{Y}_{i.} - \bar{Y}_{..} &= \bar{Y}_{i.} - \left(\frac{1}{n}\right)(\bar{Y}_{1.} + \bar{Y}_{2.} + \cdots + \bar{Y}_{n.}) \\ &= -\left(\frac{\bar{Y}_{1.}}{n}\right) - \left(\frac{\bar{Y}_{2.}}{n}\right) - \cdots + \frac{(n-1)\bar{Y}_{i.}}{n} - \cdots - \left(\frac{\bar{Y}_{n.}}{n}\right). \end{aligned} \tag{5.25}$$

Because of the assumption of independence, the variance of the sum is equal to the sum of variances. Also, $V[\bar{Y}_{i.}] = \sigma_e^2/k^{-1}$, for all i. Therefore, from equation (5.25), the variance term under consideration becomes equal to

$$\begin{aligned} V[\bar{Y}_{i.} - \bar{Y}_{..}] &= \left(\frac{\sigma_e^2}{k}\right)\left(\frac{(n-1)}{n^2} + \frac{(n-1)^2}{n^2}\right) \\ &= \left(\frac{(n-1)}{nk}\right)\sigma_e^2. \end{aligned} \tag{5.26}$$

Finally, equation (5.24) yields the result shown as follows:

$$E[SS_{\text{block}}] = \text{k}\sum_i \beta_i^2 + (n-1)\sigma_e^2. \tag{5.27}$$

An exactly similar argument leads to the expected value of the sum of squares for treatments:

$$E[SS_{\text{treatments}}] = n\sum_j \tau_j^2 + (k-1)\sigma_e^2. \tag{5.28}$$

Since $E[MS]$ is equal to $E[SS]$ divided by the corresponding number of degrees of freedom, it can be concluded that

$$E[MS_{\text{block}}] = \frac{k}{(n-1)}\sum_i \beta_i^2 + \sigma_e^2 \tag{5.29}$$

and

$$E[MS_{\text{treatment}}] = \frac{n}{(k-1)}\sum_j \tau_j^2 + \sigma_e^2. \tag{5.30}$$

For the total sum of squares the argument is again similar:

$$\begin{aligned} E[SS_{\text{total}}] &= E\left[\sum_i \sum_j (Y_{ij} - \bar{Y}_{..})^2\right] \\ &= \sum_i \sum_j E[Y_{ij} - \bar{Y}_{..}]^2 \\ &= \sum_i \sum_j E[Y_{ij} - \bar{Y}_{..}]^2 + \sum_i \sum_j V[Y_{ij} - \bar{Y}_{..}] \\ &= \sum_i \sum_j (\beta_i + \tau_j)^2 + (nk - 1)\sigma_e^2. \end{aligned} \tag{5.31}$$

Now it is noted that

$$\sum_i \sum_j (\beta_i + \tau_j)^2 = k\sum_i \beta_i^2 + n\sum_j \tau_j^2 + \sum_i \sum_j \beta_i \tau_j$$

$$\sum_{i \neq j} \sum_j \beta_i \tau_j = 0. \tag{5.32}$$

Since $\sum_i \sum_j \beta_i \tau_j = (\sum_i \beta_i)(\sum_j \tau_j)$, $\sum_i \beta_i = 0$ and $\sum_j \tau_j = 0$, equation (5.31) can be rewritten as

$$E[SS_{\text{total}}] = k\sum_i \beta_i^2 + n\sum_j \tau_j^2 + (nk - 1)\sigma_e^2.$$

From the above results, it is concluded that

$$\begin{aligned} E[SS_{\text{error}}] &= E[SS_{\text{total}}] - E[SS_{\text{block}}] - E[SS_{\text{treatment}}] \\ &= (nk - 1 - n + 1 - k + 1)\sigma_e^2 \\ &= (n - 1)(k - 1)\sigma_e^2 \end{aligned} \tag{5.33}$$

and therefore

$$E[MS_{\text{error}}] = \sigma_e^2. \tag{5.34}$$

The procedure to test the hypothesis H_0: $\beta_i = 0$ for all i is as follows. Under the assumption that the null hypothesis is true, the expected value of the mean square for blocks is given by

$$E[MS_{\text{block}}] = \sigma_e^2. \tag{5.35}$$

The result implies that

$$\frac{SS_{\text{block}}}{\sigma_e^2} \sim \chi^2_{(n-1)}. \tag{5.36}$$

Similarly, the result $E[MS_{\text{error}}] = \sigma_e^2$ (which is true regardless of the null hypothesis) implies that

$$\frac{SS_{\text{error}}}{\sigma_e^2} \sim \chi^2_{(n-1)(k-1)}. \tag{5.37}$$

Therefore, if the null hypothesis is assumed to be true,

$$F_{(n-1,(n-1)(n-1))} = \frac{MS_{\text{block}}}{MS_{\text{error}}}. \tag{5.38}$$

After computing the value of the F-statistic based on a random sample that includes one observation for each block–treatment combination, the F-value is checked against the critical value for the specified degrees of freedom and significance level. If the observed F-value is larger than the critical F-value the null hypothesis is rejected. Otherwise, the null hypothesis cannot be rejected.

Even though the variation between blocks may be of no direct interest, a large MS value associated with blocks would mean that the use of blocks has been effective in reducing the error mean square. This reduction makes the experiment more sensitive.

If the blocks are large enough, such that each treatment may be tested more than once per block, the hypothesis that there is no interaction between blocks and treatments could also be tested. If this interaction is statistically significant, the relative performance of a treatment would not be the same in all blocks.

Following a similar procedure the hypothesis H_0: $\tau_j = 0$ for all j may be tested. In this case the F-statistic would be defined as

$$F_{((k-1),(n-1)(k-1))} = \frac{MS_{\text{treatment}}}{MS_{\text{error}}}. \tag{5.39}$$

As in Chapter 3, the computations needed to perform an ANOVA can be simplified by rewriting the sums of squares in terms of the sample totals. The corresponding expressions are given as follows:

(a) For blocks:

$$SS_{\text{block}} = \sum_i \frac{T_{i\cdot}^2}{k} - \frac{T_{\cdot\cdot}^2}{N}. \tag{5.40}$$

(b) For treatments:

$$SS_{\text{treatment}} = \sum_j \frac{T_{\cdot j}^2}{n} - \frac{T_{\cdot\cdot}^2}{N}. \tag{5.41}$$

(c) For the total:

$$SS_{\text{total}} = \sum_i \sum_j Y_{ij}^2 - \frac{T_{\cdot\cdot}^2}{N}. \tag{5.42}$$

(d) For the error:

$$SS_{\text{error}} = SS_{\text{total}} - SS_{\text{block}} - SS_{\text{treatment}}. \tag{5.43}$$

Table 5.5 Data format

	$j=1$	$j=2$	$\cdots$	$j=k$	
$i=1$	Y_{11}	Y_{12}	$\cdots$	Y_{1k}	$T_{1\cdot}$
$i=2$	Y_{21}	Y_{22}	$\cdots$	Y_{2k}	$T_{2\cdot}$
$\vdots$	$\vdots$	$\vdots$	$\vdots$	$\vdots$	$\vdots$
$i=n$	Y_{n1}	Y_{n2}	$\cdots$	Y_{nk}	$T_{n\cdot}$
	$T_{.1}$	$T_{.2}$	$\cdots$	$T_{.k}$	$T_{..}$

Table 5.6 Two-way ANOVA table

Source	*df*	*SS*	*MS*
Treatments	$(k-1)$	$\sum_j \frac{T_{.j}^2}{n} - \frac{T_{..}^2}{N}$	$\frac{SS_{\text{treatment}}}{(k-1)}$
Blocks	$(n-1)$	$\sum_i \frac{T_{i.}^2}{k} - \frac{T_{..}^2}{N}$	$\frac{SS_{\text{block}}}{(n-1)}$
Error	$(k-1)(n-1)$	$SS_{\text{total}} - SS_{\text{treatment}} - SS_{\text{block}}$	$\frac{SS_{\text{error}}}{(n-1)(k-1)}$
Total	$N-1$	$\sum_i \sum_j Y_{ij}^2 - \frac{T_{..}^2}{N}$	

In the above equations $T_{i.}$ is the total of the observations in the ith block, $T_{.j}$ is the total for the jth treatment and $T_{..}$ is the grand total. Normally the data are arranged according to the format shown in Table 5.5, which allows for an easy calculation of the block and treatment totals. The ANOVA results developed in this section can be summarized as shown in Table 5.6.

EXAMPLE 5.1

In order to illustrate the design of a complete block experiment, consider a manufacturing process in which the yield of a chemical reaction is affected by a particular catalyst [1]. This catalyst may be prepared in any of four distinct methods, A, B, C and D. The process is pictorially described in Figure 5.1. It is desired to design a set of experiments in order to test if there are significant differences among the four possible methods of preparing the catalyst.

Each catalyst may be tested over an entire month. This would, however, allow possible variations due to plant efficiency and other causes to affect

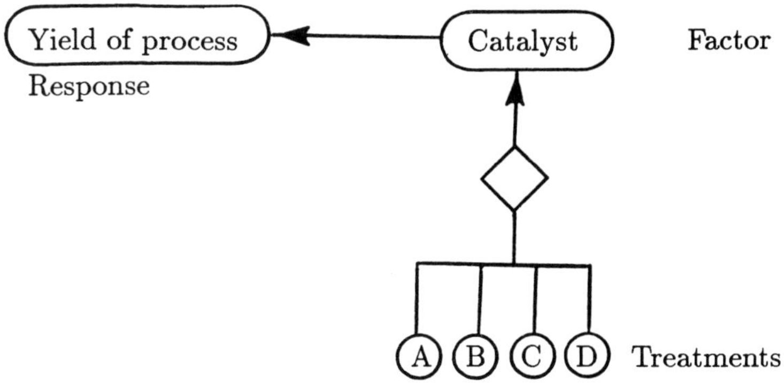

Figure 5.1 Pictorial representation of process.

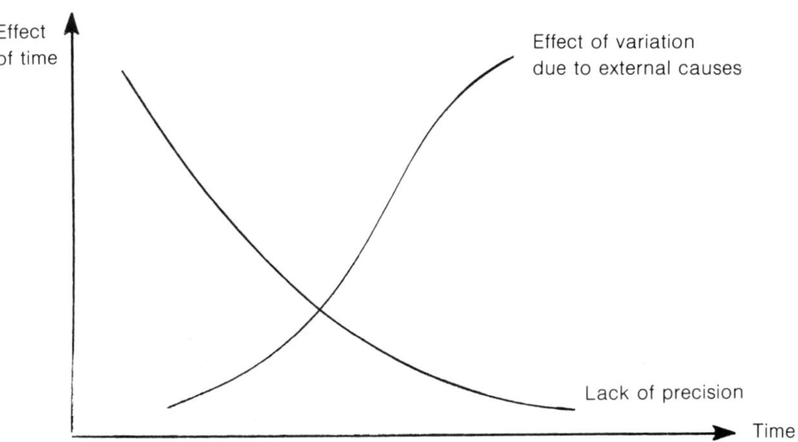

Figure 5.2 Effects associated with the length of testing period.

the results of the experiments. By reducing the test time of each catalyst, the variation from month to month will be reduced. Conversely, as illustrated in Figure 5.2, the precision of the test will be reduced, due to the reduction in the test time of each catalyst. However, the precision of the testing procedure can be enhanced by blocking the test time into one-month intervals, with each catalyst tested during a different week of the month. The randomized block design shown in Table 5.7 may be used. Each block has four positions, with each position considered to be one week; during the first week of month 1 a catalyst prepared by method A will be tested, during the second week catalyst B will be tested, and so on.

Table 5.7 Complete block design

		Treatments			
		Week 1	*Week 2*	*Week 3*	*Week 4*
Blocks	Month 1	A	B	C	D
	Month 2	C	A	B	D
	Month 3	A	D	C	B
	Month 4	C	D	B	A
	Month 5	B	A	C	D
	Month 6	A	D	B	C

EXAMPLE 5.2

An organic chemical is produced using any of five different blends [1]. The actual yield is below the theoretical yield due to a loss of product during a filtration process, as depicted in Figure 5.3. It is desired to design an experiment to investigate any differences in the yield due to the blend used.

The experimenter has decided to obtain three observations for each blend. The filtration process can produce five batches in one day. So the 15 batches of products were divided into three blocks of five. Within a block, the order in which the five blends are tested is completely randomized. The chosen design is shown in Table 5.8.

The experimental response measured is the amount lost expressed as a percentage; the corresponding data are shown in Table 5.9, and the ANOVA results is Table 5.10. From the results shown in Table 5.10, it is concluded that the variation among treatments, or blends, is significant, while that among blocks is not, at a level of significance $\alpha = 0.10$.

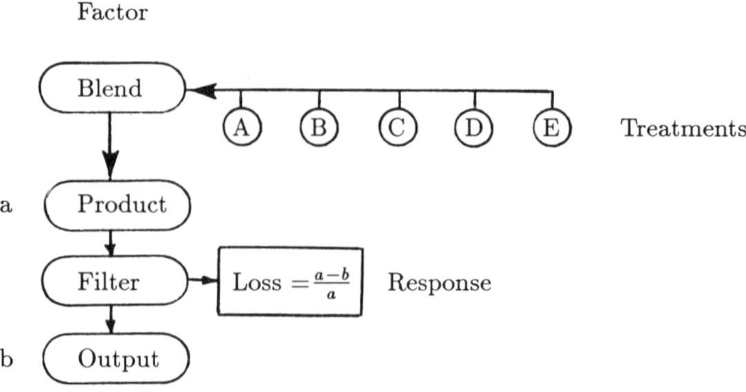

Figure 5.3 Filtration process for Example 5.2.

Table 5.8 Complete block design

	Blend				
	1	*2*	*3*	*4*	*5*
Block I	B	A	C	E	D
Block II	A	E	B	C	D
Block III	B	D	C	E	A

Table 5.9 Responses for Example 5.2

	Treatment (blend)					
	A	*B*	*C*	*D*	*E*	$T_{i.}$
Block I	16.9	18.2	17.0	15.1	18.3	85.5
Block II	16.5	19.2	18.1	16.0	18.3	88.1
Block III	17.5	17.1	17.3	17.8	19.8	89.5
$T_{.j}$	50.9	54.5	52.4	48.9	56.4	$T_{..} = 263.1$

Table 5.10 Two-way ANOVA

Source	*df*	*SS*	*MS*	*F*	*Critical values*
Treatments	4	11.56	2.89	3.32	$F_c = F_{0.90}(4, 8) = 2.81$
Block	2	1.65	0.82	0.94	$F_c = F_{0.90}(2, 8) = 3.11$
Error	8	6.99	0.87		
Total	14	20.20			

5.3 BALANCED INCOMPLETE BLOCK DESIGNS

In some experiments using randomized block designs, it may not be possible to run all the treatment combinations in each block. Situations like this generally occur because of shortages of funds, time, experimental facilities, or the physical size of the block. For example, in the experiment described in Example 5.2, for testing blocks it may happen that because of time restrictions the filtration process can produce only three batches per day. Therefore each blend cannot be tested on each day. For this type of problem it is still possible to use randomized block designs in which every treatment is not present in every block. These designs are known as randomized incomplete block designs.

When all treatment comparisons are equally important, the treatments used in each block should be selected in a balanced manner, that is, any pair of treatments occurring together the same number of times as any other pair. In this case the design is called a balanced incomplete block design. The following notation will be used in this section:

t = total number of treatments
b = number of blocks
k = block size ($k < t$)
r = number of blocks containing any treatment ($r <$ b)
λ = number of times each pair of treatments occurs together ($\lambda < r$)

The above notation is illustrated in Table 5.11 for a balanced incomplete block design with $k = 3$, $b = 10$, $\lambda = 3$, $r = 6$, $t = 5$.

In order to express λ in terms of r, k and t the following considerations are useful. With a block size equal to k, treatment j may be paired with $k - 1$ other treatments within a single block. If there are r replications, treatment j will appear in r different blocks. Thus, the total number of pairs in which treatment j appears within some common block will be $r(k - 1)$. Also, in a balanced design each treatment is paired within some common block with each of $(t - 1)$ treatments λ times.

Therefore

$$\lambda(t - 1) = r(k - 1)$$

or, equivalently,

$$\lambda = \frac{r(k - 1)}{(t - 1)}. \tag{5.44}$$

Table 5.11 Balanced incomplete bloak design

		Treatments		
Blocks	1	A	B	C
	2	A	B	E
	3	A	D	E
	4	B	C	D
	5	C	D	E
	6	A	B	D
	7	A	C	D
	8	A	C	E
	9	B	C	E
	10	B	D	E

Table 5.12 Minimal number of blocks in a balanced design

t	k	r	b
4	2	3	6
5	2	4	10
5	3	6	10
6	2	5	15
6	3	5	10
6	4	10	15
7	2	6	21
7	3	3	7
7	4	4	7

Assuming that the integers t and k are known, the minimal number of blocks required for a balanced design is obtained as the solution to the following model:

$$\text{Minimize } b = \frac{rt}{k} \tag{5.45}$$

subject to

$$\lambda = \frac{r(k-1)}{(t-1)}$$

where r, b and λ are restricted to be positive integers.

Table 5.12 gives the minimum number of blocks needed in a design when t and k are known. The results shown in this table were obtained by solving the above optimization problem for each case.

EXAMPLE 5.3

It is desired to obtain a balanced design for six treatments and blocks of size equal to three. The minimal number of blocks that can be found from the solution of the simple model is shown below:

$$\text{Minimize } b = 2r$$

subject to

$$\lambda = \tfrac{2}{5}r$$

where r, b and λ are constrained to be positive integers. By inspection, it is possible to see that the optimal solution is $r = 5$, $\lambda = 5$ and $b = 10$. Note that these are the same results given in Table 5.12 for $t = 6$ and $k = 3$.

EXAMPLE 5.4

The standard procedure for measuring the life of automobile tires is to run the tire on a test car under normal road conditions [1]. If there are several types of tires to be tested, the tires must be used on the same car in order to reduce the large variation between different automobiles. They must also be rotated frequently to eliminate differences due to the position on the car. Since only four tires may be tested per car, the design is somewhat limited. Usually the experimenter can arrange the work and only consider two to four treatments at a time. If, however, there are five tire brands A, B, C, D and E to be considered, and all comparisons are equally important, then a balanced design, with five blocks, will be needed. Each block has four positions. The design is shown in Table 5.13, where the blocks correspond to cars.

Large experimental error is usually found in road tests. This may be reduced by manufacturing tires having more than one compound. Comparisons between different parts of the same tire would seem to be more reliable than comparisons between different tires. For example, suppose that there are four compounds, or treatments, A, B, C and D, and a three-part tread is to be employed. Table 5.14 shows a suitable balanced incomplete block design for this case with the four tires considered as blocks having three positions. It

Table 5.13 Balanced incomplete block design: five cars used

		Position			
		1	*2*	*3*	*4*
	1	A	B	C	D
	2	A	B	C	E
Block	3	A	B	D	E
	4	A	C	D	E
	5	B	C	D	E

Table 5.14 Balanced incomplete block design: one car used

		Position		
		1	*2*	*3*
	1	A	B	C
	2	A	B	D
Block	3	A	C	D
	4	B	C	D

is finally noticed that in the design of Table 5.13 five cars are used, whereas in the design on Table 5.14 only one car is used.

5.4 SPECIAL ANOVA FOR BALANCED INCOMPLETE BLOCK DESIGNS

The statistical model for a randomized complete block design is

$$Y_{ij} = \mu + \beta_i + \tau_j + \varepsilon_{ij}$$

and the estimates t_j and b_i are such that for each block $\sum_j t_j = 0$, and for each treatment $\sum_i b_i = 0$. This is, however, not true for the case of an incomplete design, since the design is not orthogonal. As will be shown in this section, in balanced incomplete designs the treatment totals must be adjusted because blocks and treatments are confounded.

To develop the adjusted ANOVA methodology it is possible to proceed as follows. The linear model under consideration is formulated as follows:

$$Y_{ijm} = \mu + \beta_i + \tau_j + \varepsilon_{ijm} \tag{5.46}$$

where $i = 1, 2, \ldots, b$, $j = 1, 2, \ldots, t$ and $m = 1$, if the observation corresponds to block i and treatment j; $m = 0$, otherwise.

First, it is noted that a treatment total such as $T_{.j}$ estimates the sum of three terms, as shown as follows:

$$T_{.j} \doteq r\mu + r\tau_j + \sum_{i \in (j)_{\mathrm{B}}} \beta_i \tag{5.47}$$

In equation (5.47) $\doteq$ means 'estimates'. The notation $(j)_{\mathrm{B}}$ in the last term on the right side is the sum of the block effects in which treatment j appears. For any two treatments A and B, equation (5.47) can be used to obtain the following results:

$$T_{.\mathrm{A}} \doteq r\mu + r\tau_a + \sum_{i \in (a)_{\mathrm{B}}} \beta_i$$

$$T_{.\mathrm{B}} \doteq r\mu + r\tau_b + \sum_{i \in (b)_{\mathrm{B}}} \beta_i.$$

Therefore

$$T_{.\mathrm{A}} - T_{.\mathrm{B}} \doteq r(\tau_a - \tau_b) + \left(\sum_{i \in (a)_{\mathrm{B}}} \beta_i - \sum_{i \in (b)_{\mathrm{B}}} \beta_i \right). \tag{5.48}$$

From the above result it can be seen that the difference in treatment effects depends on the effects of those blocks containing treatments A and B and hence it can be concluded that the balanced incomplete block design (BIBD) is not orthogonal.

The normal equations for a BIBD are formulated as shown in equations

(5.49)–(5.51):

$$bkm + r\sum_{j=1}^{t}\hat{t}_j + k\sum_{i=1}^{b} b_i = T_{..} \tag{5.49}$$

$$km + \sum_{j=1}^{t} n_{sj}\hat{t}_j + kb_s = T_{s.} \quad s = 1, 2, \ldots, b \tag{5.50}$$

$$rm + r\hat{t}_p + \sum_{i=1}^{b} n_{ip}b_i = T_{.p} \quad p = 1, 2, \ldots, t \tag{5.51}$$

where $n_{ij} = 1$ if the jth treatment occurs in the ith block and $n_{ij} = 0$ otherwise. As an illustration of the meaning of n_{ij}, the BIBD shown in Table 5.2 will be considered. For this design $s = 1, 2, 3, 4, 5$ and $p = 1, 2, 3, 4, 5$ for A, B, C, D, E, respectively:

$$\begin{array}{ll} n_{11} = n_{21} = n_{31} = n_{41} = 1 & n_{51} = 0 \\ n_{12} = n_{22} = n_{32} = n_{52} = 1 & n_{42} = 0 \\ n_{13} = n_{23} = n_{43} = n_{53} = 1 & n_{33} = 0 \\ n_{14} = n_{34} = n_{44} = n_{54} = 1 & n_{24} = 0 \\ n_{25} = n_{35} = n_{45} = n_{55} = 1 & n_{15} = 0 \end{array}$$

Note that

$$\sum_i n_{ij} = 4 = r$$

for $j = 1, 2, 3, 4, 5$ where r is the number of treatment replications and it is equal to $\sum_i n_{ip}$. From equation (5.50), the quantity $(m + b_s)$ can be computed as

$$m + b_s = \frac{1}{k}\left(T_{s.} - \sum_j n_{sj}\hat{t}_j\right). \tag{5.52}$$

Therefore, from equation (5.51) it is concluded that

$$r\hat{t}_p + \sum_i n_{ip}(m + b_i) = T_{.p}. \tag{5.53}$$

Substituting equation (5.52) into equation (5.53) the following result is obtained:

$$r\hat{t}_p + \sum_{i=1}^{b} n_{ip}\frac{1}{k}\left(T_{i.} - \sum_{j=1}^{t} n_{ij}\hat{t}_j\right) = T_{.p}$$

or, equivalently,

$$r\hat{t}_p - \frac{1}{k}\sum_{j=1}^{t}\sum_{i=1}^{b} n_{ip}n_{ij}\hat{t}_j = T_{.p} - \frac{1}{k}\sum_{i=1}^{b} n_{ip}T_{i.}$$

Considering $\hat{t}_p$ as a common factor of the second term on the left-hand side, the above equation can be reformulated as

$$r\hat{\tau}_p - \frac{1}{k}\sum_{j=1}^{b} n_{ip}n_{ip}\hat{\tau}_p - \frac{1}{k}\sum_{j=1,j\neq p}^{t}\sum_{i=1}^{b} n_{ip}n_{ij}\hat{\tau}_j = T_{.p} - \frac{1}{k}\sum_{i=1}^{b} n_{ip}T_{i.}$$

From the definitions of λ and n_{ij} it is possible to show that

$$\lambda = \sum_{i=1}^{b} n_{ip}n_{ij} \tag{5.54}$$

if $p \neq j$. Since $n_{ij} = 0$ or 1, then $n_{ip}^2 = n_{ip}$. This result can be used to obtain the following relationship:

$$\left(r - \frac{r}{k}\right)\hat{\tau}_p - \frac{\lambda}{k}\sum_{j=1,j\neq p}^{t} \hat{\tau}_j = T_{.p} - \frac{1}{k}\sum_{i=1}^{b} n_{ip}T_{i.} \tag{5.55}$$

Since $\sum_{j=1}^{t}\hat{\tau}_j = 0$ for each block, then

$$\sum_{j=1,j\neq p}^{t} \hat{\tau}_j = -\hat{\tau}_p.$$

If the above result is now substituted into equation (5.55), this equation would become

$$\left(r - \frac{r}{k} + \frac{\lambda}{k}\right)\hat{\tau}_p = T_{.p} - \frac{1}{k}\sum_{i=1}^{b} n_{ip}T_{i.} \quad p = 1, 2, \ldots, t.$$

Or, equivalently, after considering equation (5.44) along with the previous result:

$$\frac{\lambda t}{k}\hat{\tau}_p = T_{.p} - \frac{1}{k}\sum_{i=1}^{b} n_{ip}T_{i.} \tag{5.56}$$

For any treatment p, let a term Q_p be defined as

$$Q_p = kT_{.p} - \sum_{(j)} n_{ip}T_{i.}$$

Equivalently,

$$Q_p = kT_{.p} - \sum_{i\in(p)_{\mathrm{B}}} T_{i.}$$

The Q_p terms are usually referred to as the adjusted treatment totals. It can be verified that they always add up to zero.

Using the above relationship it can be seen that equation (5.56) yields the result $Q_p = \lambda\hat{\tau}_p$. In conclusion,

$$\hat{\tau}_p = \frac{Q_p}{\lambda t}.$$

The expected value and the variance of Q_j can be obtained as follows.

Substituting equations (5.50) and (5.51) in equation (5.56), and using equation (5.46), Q_p can be written as

$$Q_p = k\left(r\mu + r\tau_p + \sum_{i=1}^{b} n_{ip}\beta_i + \sum_{im}\varepsilon_{ipm}\right) - \sum_{i=1}^{b} n_{ip}\left(k\mu + \sum_{j=1}^{t} n_{ij}\tau_j + k\beta_i + \sum_{jm}\varepsilon_{ijm}\right).$$

Using the result $\sum_i n_{ip} = r$, and the above equation, the adjusted totals can be written as

$$Q_p = kr\mu + kr\tau_p + k\sum_{i=1}^{b} n_{ip}\beta_i + k\sum_{im}\varepsilon_{ipm} - kr\mu - \sum_{i=1}^{b}\sum_{j=1}^{t} n_{ip}n_{ij}\tau_j$$

$$- k\sum_{i=1}^{b} n_{ip}\beta_i - \sum_{ijm} n_{ip}\varepsilon_{ijm}$$

$$= kr\tau_p - \sum_{i=1}^{b}\sum_{j=1}^{t} n_{ip}n_{ij}\tau_j + k\sum_{im}\varepsilon_{ip} - \sum_{ijm} n_{ip}\varepsilon_{ijm}.$$

Now the identity

$$\sum_{j=1}^{t} n_{ij}\tau_j = n_{ip}\tau_p + \sum_{j=1, j\neq p}^{t} n_{ij}\tau_j$$

and equation (5.54) can be used to conclude that

$$Q_p = kr\tau_p - \sum_{i=1}^{b} n_{ip}n_{ip}\tau_p - \sum_{j=1, j\neq p}^{t} \lambda\tau_j + k\sum_{im}\varepsilon_{ipm} - \sum_{ijm} n_{ip}\varepsilon_{ijm}.$$

Since $\sum_i n_{ip} = r$, the coefficient of τ_p becomes $r(k-1)$. But, from equation (5.45), $r(k-1) = \lambda(t-1)$. Therefore

$$Q_p = \lambda(t-1)\tau_p - \sum_{(j=1, j\neq p)}^{t} \lambda\tau_j + k\sum_{i} n_{ip}\varepsilon_{ip} - \sum_{ijm} n_{ip}\varepsilon_{ijm}.$$

That is,

$$Q_p = \lambda t(\tau_p - \bar{\tau}) + \left(k\sum_{im}\varepsilon_{ipm} - \sum_{ijm} n_{ip}\varepsilon_{ijm}\right). \tag{5.57}$$

Since the errors are distributed with mean 0, the expected value of Q_p is found to be

$$E[Q_p] = \lambda t(\tau_p - \bar{\tau}). \tag{5.58}$$

Additionally, since the errors are distributed with variance σ^2,

$$V[Q_p] = E\left[\left(k\sum_{i} n_{ip}\varepsilon_{ip} - \sum_{ij} n_{ij}n_{ip}\varepsilon_{ij}\right)^2\right] = kr(k-1)\sigma^2 \tag{5.59}$$

The covariance of any two treatment effect estimates can be found from the above equations. It can be verified that [2] the covariance of Q_p and

$Q_{p'}$, for $p \neq p'$, is given by

$$\operatorname{cov}(Q_p Q_{p'}) = -k\lambda\sigma^2.$$

As a result of these developments, we can conclude the following results [2]. The linear minimum-variance unbiased estimate of any contrast $C = \sum_j c_j \tau_j$ is given by $(\lambda t)^{-1} \sum c_j Q_j$. The variance of the estimate is

$$k\sigma^2(\lambda t)^{-1} \sum c_j^{\,2}.$$

If the errors are normally distributed, the estimate is maximum likelihood and is the best unbiased estimate. For deriving the confidence interval of a contrast of the τ'_j, it can be verified that the following quantity is distributed as $N(0, 1)$:

$$z = \sum c_j \hat{\tau}_j - \sum c_j \tau_j \left[\sigma^2 \sum_j c_j^2 \left(\frac{k}{\lambda t} \right) \right]^{-1/2}$$

If the variance σ^2 is not known, it can be estimated by the MS_{error}. In such a case, the estimate

$$t = \left(\sum_j c_j \hat{\tau}_j - \sum_j c_j \tau_j \right) \left[MS_{\text{error}} \sum_j c_j^{\,2} \left(\frac{k}{\lambda t} \right) \right]^{-1/2}$$

is distributed as a t-distribution with $(bk - t - b + 1)$ degrees of freedom. A $100(1 - \alpha)\%$ confidence interval is

$$\sum_j c_j \hat{\tau}_j - \theta \leqslant \sum_j c_j \hat{\tau}_j \leqslant \sum_j c_j \hat{\tau}_j + \theta \tag{5.60}$$

where

$$\theta = t_{1-\alpha/2} \left(\frac{MS_{\text{error}} \sum c_j^{\,2} k}{\lambda t} \right)^{1/2}. \tag{5.61}$$

It can be shown that the sum of squares of the adjusted treatments is equal to

$$SS_{\text{treatment(adjusted)}} = \sum_{j=1}^{t} \frac{Q_j^2}{(k\lambda t)}. \tag{5.62}$$

The ANOVA table to test the hypothesis that all treatments effects are negligible is given in Table 5.15. The mean square for block effects is discarded in Table 5.15 because the block totals are influenced by differences between treatments, and only the SS for treatments was adjusted. The value of the statistic SS_{error} is obtained from the fundamental equation of ANOVA:

$$SS_{\text{error}} = SS_{\text{total}} - SS_{\text{block}} - SS_{\text{treatment(adjusted)}}. \tag{5.63}$$

It is also possible to study block effects when the design is symmetric, that is, if $b = t$. In that case, the adjusted totals for blocks are found to be equal to

$$Q'_i = rT_{i.} - \sum_{j \in (i)_T} T_{.j} \tag{5.64}$$

and

$$SS_{\text{block(adjusted)}} = \sum_{i=1}^{b} \frac{(Q'_i)^2}{(r\lambda b)}. \tag{5.65}$$

The special ANOVA table with adjusted block totals used for testing the hypothesis that all block effects are negligible is shown in Table 5.16. In Tables 5.15 and 5.16, G represents the correction term $G = T^2/N$. In Table 5.16 the mean square for treatments is discarded. The following example shows an application of the procedures to test treatment and block effects.

Table 5.15 Special ANOVA table (adjusted for treatments)

Source	*df*	*SS*	*MS*
Treatments (adjusted)	$t-1$	$\sum_j \frac{Q_j^2}{(k\lambda t)}$	$\frac{SS_{\text{treatment}}}{(t-1)}$
Blocks	$b-1$	$\sum_i \frac{T_{i.}^2}{k} - G$	Discard
Error	$N-t-b+1$	SS_{error}	$\frac{SS_{\text{error}}}{(N-t-b+1)}$
Total	$N-1$	$\sum_i \sum_j Y_{ij}^2 - G$	

Table 5.16 Special ANOVA table (adjusted for blocks)

Source	*df*	*SS*	*MS*
Treatments	$t-1$	$\sum_j \frac{T_{.j}{}^2}{n_j} - G$	Discard
Blocks (adjusted)	$b-1$	$\sum_i \frac{Q_i^2}{(r\lambda b)}$	$\frac{SS_{\text{block}}}{(b-1)}$
Error	$N-t-b+1$	SS_{error}	$\frac{SS_{\text{error}}}{(N-t-b+1)}$
Total	$N-1$	$\sum_i \sum_j Y_{ij}^2 - G$	

EXAMPLE 5.5

A plant manager wants to buy machines for one of his manufacturing processes. There are four companies A, B, C, and D making such machines, and each has allowed the manager to use a machine for a limited time. There are four operators assigned to the process, but, due to time limits, each can work on only three machines. Tables 5.17 and 5.18 show the experimental design and the corresponding data, respectively.

In the design shown in Table 5.17 the treatments are the machines, and the blocks are operators. Positions refer to the order in which machines A through D were used by operators 1 through 4. Each operator used a machine for a constant period of time, and the machines were assigned at random. With the above design two analyses will be performed to test differences between machines and between operators.

Machines

For this part of the analysis it is necessary to adjust for the treatment effects. First, compute the sum of squares for blocks from equation (5.40):

$$SS_{\text{block}} = \frac{(2400)^2 + (2810)^2 + (2580)^2 + (2440)^2}{3} - \frac{(10\,230)^2}{12} = 34\,291.7.$$

Table 5.17 Balanced incomplete block design

		Position		
		1	*2*	*3*
Block	1	A	B	C
	2	A	C	D
	3	B	C	D
	4	A	B	D

Table 5.18 Responses for Example 5.5

Operator	*Machines* *A*	*B*	*C*	*D*	$T_{i.}$
1	780	820	800	–	2400
2	950	–	920	940	2810
3	–	880	880	820	2580
4	840	780	–	820	2440
$T_{.j}$	2570	2480	2600	2580	$T_{..} = 10\,230$

Using equation (5.42), the sum of squares for the total can be computed as follows:

$$SS_{\text{total}} = \sum_i \sum_j Y_{ij}^2 - G$$

$$= 8\,760\,900 - 8\,721\,075 = 39\,825.$$

To proceed with the analysis, compute the adjusted treatment totals ($j = 1$ for A, $j = 2$ for B, $j = 3$ for C and $j = 4$ for D):

$$Q_1 = 3(2570) - (2400 + 2810 + 2440) = 60$$
$$Q_2 = 3(2480) - (2400 + 2580 + 2440) = 20$$
$$Q_3 = 3(2600) - (2400 + 2810 + 2580) = 10$$
$$Q_4 = 3(2580) - (2810 + 2580 + 2440) = -90.$$

From equation (5.62)

$$SS_{\text{treatment(adjusted)}} = \sum_j \frac{(Q_j)^2}{(k\lambda t)}$$
$$= \frac{(60)^2 + (20)^2 + (10)^2 + (-90)^2}{(3 \times 2 \times 4)}$$
$$= 508.33.$$

Therefore

$$SS_{\text{error}} = SS_{\text{total}} - SS_{\text{block}} - SS_{\text{treatment(adjusted)}}$$
$$= 5025.$$

The ANOVA results are given in Table 5.19. Based on these results it is concluded that there is no significant effect due to machines, using a level of significance $\alpha = 0.05$.

Operators

For this part of the analysis it is necessary to adjust for the block effects. The sum of squares for treatments is obtained from equation (5.41):

$$SS_{\text{treatment}} = \frac{(2570)^2 + (2480)^2 + (2600)^2 + (2580)^2}{3} - 8\,721\,075$$
$$= 2825.$$

The adjusted block totals are given by

$$Q'_1 = 3(2400) - (2570 + 2480 + 2600) = -450$$
$$Q'_2 = 3(2810) - (2570 + 2600 + 2580) = 680$$
$$Q'_3 = 3(2580) - (2480 + 2600 + 2580) = 80$$
$$Q'_4 = 3(2440) - (2570 + 2480 + 2580) = -310.$$

Table 5.19 Special ANOVA (adjusted for treatments)

Source	*df*	*SS*	*MS*	*F*	*Significant?*
Machines(adj.)	3	508.3	169.43	0.169	No
Operators	3	34 291.7	–		
Error	5	5 025.0	1005.00		
Total	11	39 825.0			

Table 5.20 Special ANOVA (adjusted blocks)

Source	*df*	*SS*	*MS*	*F*	*Significant?*
Machines	3	2 825	–		
Operators(adj.)	3	31 975	10 658.33	10.6	Yes
Error	5	5 025	1 005.00		
Total	11	39 825			

Using equation (5.65):

$$SS_{\text{block(adj)}} = \sum_i \frac{(Q_i')^2}{(r\lambda b)}$$

$$= \frac{(-450)^2 + (680)^2 + (80)^2 + (-310)^2}{(3 \times 2 \times 4)}$$

$$= 31\,975.$$

Therefore

$$SS_{\text{error}} = 39\,825 - 31\,975 - 2825$$

$$= 5025$$

as already known from the study of the machine effect. The ANOVA results are summarized in Table 5.20, where the operator effect is shown to be significant, using $\alpha = 0.05$.

5.5 AFTER-ANOVA TESTS

Sometimes it may be desired to study contrasts for a balanced incomplete design that has $b = t$. Using the following equations, contrasts between treatments and between blocks can be examined:

(a) For treatments:

$$SS_C = \frac{C^2}{(\sum_j c_j^2)(k\lambda t)} \tag{5.66}$$

$$C = \sum_j c_j Q_j \tag{5.67}$$

$$\sum_j c_j = 0. \tag{5.68}$$

(b) For blocks:

$$SS_C = \frac{C^2}{(\sum_i c_i^2)(r\lambda b)} \tag{5.69}$$

$$C = \sum_i c_i Q'_i \tag{5.70}$$

$$\sum_i c_i = 0. \tag{5.71}$$

Example 5.6 illustrates the use of the above results.

EXAMPLE 5.6

Using Example 5.5 it is possible to examine contrasts between operators 1 and 4, 2 and 3, and between a combination of 1 and 4 and a combination of 2 and 3. The following contrasts will be examined:

$$\begin{aligned} C_1 &= Q'_1 \qquad\qquad\qquad - Q'_4 \\ C_2 &= \qquad\quad Q'_2 - Q'_3 \\ C_3 &= Q'_1 - Q'_2 - Q'_3 + Q'_4. \end{aligned}$$

This is equivalent to testing the following hypotheses:

$$\begin{aligned} H_0^1&: \beta_1 = \beta_4 \\ H_0^2&: \beta_2 = \beta_3 \\ H_0^3&: \beta_1 + \beta_4 = \beta_2 + \beta_3. \end{aligned}$$

The contrasts and sums of squares under consideration can be estimated as follows, using equations (5.69) and (5.70):

$$C_1 = -450 + 310 = -140$$

$$SS_{c1} = \frac{(-140)^2}{(1^2 + 1^2)(2 \times 3 \times 4)} = 408.33$$

$$C_2 = 680 - 80 = 600$$

$$SS_{c2} = \frac{(600)^2}{(2)(24)} = 7500$$

Table 5.21 Special ANOVA for contrasts of blocks

Source	*df*	*SS*	*MS*	*F*	*Significant?*
C_1	1	408.33	408.33	0.41	No
C_2	1	7 500.00	7 500.00	7.46	Yes
C_3	1	24 066.60	24 066.60	23.95	Yes
Machines	3	2 825.00	–	–	
Error	5	5 025.00	1 005.00		
Total	11	39 825.00			

$$C_3 = -450 - 680 - 80 - 310 = -1520$$

$$SS_{c3} = \frac{(-1520)^2}{(4)(24)} = 24\,066.6.$$

The ANOVA results are summarized in Table 5.21. From Table 5.21 it is concluded that operators 2 and 3 perform differently, and that their combined performance is different from that of operators 1 and 4 together.

5.6 LEAST-SQUARES SIGNIFICANCE TEST

In this section, the least-squares methodology will be applied to block designs. First the no-interaction model will be treated in section 5.6.1. Section 5.6.2 will consider a BIBD example. Finally, the model for two-way ANOVA with interactions will be covered in section 5.6.3. For a detailed study of the topic, references [2] and [3] are recommended.

5.6.1 Two-way ANOVA with no interactions

Under the assumption that there is no interaction between blocks and treatments, the ANOVA model can be written as

$$Y_{ij} = \mu + \beta_i + \tau_j + \varepsilon_{ij}$$

where $i = 1, \ldots, n$ and $j = 1, \ldots, k$. If m, b_i, and t_j are the least-squares estimates for μ, β_i and τ_j, respectively, the regression model can be written as

$$Y_{ij} = m + b_i + t_j + e_{ij}. \tag{5.72}$$

The procedure is to minimize the sum of squares for the error, just as in Chapter 3. That is,

$$SS_{\text{error}} = \sum_i \sum_j (Y_{ij} - m - b_i - t_j)^2. \tag{5.73}$$

The normal equtions are obtained by setting the partial derivatives of

SS_{error} with respect to m, b and t equal to zero. This results in

$$T_{..} = Nm + k\sum_i b_i + n\sum_j t_j \tag{5.74}$$

$$T_{i.} = km + kb_i + \sum_j t_j \quad \text{for } i = 1, 2, \ldots, n \tag{5.75}$$

$$T_{.j} = nm + \sum_i b_i + nt_j \quad \text{for } j = 1, 2, \ldots, k. \tag{5.76}$$

The normal equations given in equations (5.74)–(5.76) are not linearly independent. However, after including the additional conditions $\sum_i b_i = 0$ and $\sum_j t_j = 0$, a unique solution can be obtained.

If it is desired to test the hypothesis that all treatments are not significant, that is H_0: $\tau_j = 0$ for all j, the reduced model can be written as

$$Y_{ij} = m' + b_i' + e_{ij}'. \tag{5.77}$$

On the other hand, if it is desired to test whether the block effects are not significant, the reduced model would be

$$Y_{ij} = m'' + t_j'' + e_{ij}''. \tag{5.78}$$

After formulating and solving the normal equations associated with the reduced models given in equations (5.77) and (5.78), the procedure from Chapter 3 can be used to obtain the following results:

$$SS_{\text{reg}}(m, b_i, t_j) = mT_{..} + \sum_i b_i T_i + \sum_j t_j T_{.j} \tag{5.79}$$

$$SS_{\text{reg}}(m', b_i') = m'T_{..} + \sum_i b_i' T_{i.} \tag{5.80}$$

$$SS_{\text{treatment}} = SS_{\text{reg}}(m, b_i, t_j) - SS_{\text{reg}}(m', b_i') \tag{5.81}$$

$$SS_{\text{reg}}(m'', t_j'') = m''T_{..} + \sum_j t_j'' T_{.j} \tag{5.82}$$

$$SS_{\text{block}} = SS_{\text{reg}}(m, b_i, t_j) - SS_{\text{reg}}(m'', t_j''). \tag{5.83}$$

EXAMPLE 5.7

Consider an experiment designed to test the quality of four different brands of automobile tires [4]. The tires will be tested using a real car, under normal driving conditions. Four cars will be used, each having one tire of each of the four brands. In order to account for differences due to the position of the tires on the car, the positions of the tires will be randomized. The results in Table 5.22, which have been coded, are available.

The normal equations are given as follows:

$$-15 = 16m + 4b_1 + 4b_2 + 4b_3 + 4b_4 + 4t_1 + 4t_2 + 4t_3 + 4t_4$$
$$4 = 4m + 4b_1 + t_1 + t_2 + t_3 + t_4$$

$$\begin{aligned}
-1 &= 4m \quad +4b_2 \quad +t_1 + t_2 + t_3 + t_4 \\
-5 &= 4m \quad +4b_3 \quad +t_1 + t_2 + t_3 + t_4 \\
-13 &= 4m \quad +4b_4 + t_1 + t_2 + t_3 + t_4 \\
5 &= 4m + b_1 + b_2 + b_3 + b_4 + 4t_1 \\
-3 &= 4m + b_1 + b_2 + b_3 + b_4 \quad + 4t_2 \\
-9 &= 4m + b_1 + b_2 + b_3 + b_4 \quad + 4t_3 \\
-8 &= 4m + b_1 + b_2 + b_3 + b_4 \quad + 4t_4.
\end{aligned}$$

Solving the above system of equations after including the additional conditions $\sum_i b_i = 0$ and $\sum_j t_j = 0$, the following results are obtained:

$$\begin{array}{llll}
m = -0.94 & & & \\
b_1 = 1.94 & b_2 = 0.69 & b_3 = -0.31 & b_4 = -2.31 \\
t_1 = 2.19 & t_2 = 0.19 & t_3 = -1.31 & t_4 = -1.06.
\end{array}$$

The sum of squares for the regression on m, b and t is given by equation (5.79):

$$\begin{aligned}
SS_{\text{reg}}(m, b_i, t_j) &= mT_{..} + \sum_i b_i T_{i.} + \sum_j t_j T_{.j} \\
&= (-15)(-0.94) + 4(1.94) - 1(.069) - 5(-0.31) - 13(-2.31) \\
&\quad + 5(2.19) - 3(0.19) - 9(-1.31) - 8(-1.06) \\
&= 83.40.
\end{aligned}$$

Thus

$$\begin{aligned}
SS_{\text{error}} &= \sum_i \sum_j Y_{ij}^2 - SS_{\text{reg}}(m, b_i, t_j) \\
&= 95 - 83.40 \\
&= 11.60.
\end{aligned}$$

Table 5.22 Coded responses for Example 5.6

Car	*Brand* A	B	C	D	$T_{i.}$
I	4	1	−1	0	4
II	1	1	−1	−2	−1
III	0	0	−3	−2	−5
IV	0	−5	−4	−4	−13
$T_{.j} =$	5	−3	−9	−8	$-15 = T_{..}$
$\sum Y_{ij}^2 =$	17	27	27	24	$95 = \sum_i \sum_j Y_{ij}^2$

The reduced model to test the significance of the treatments is given by

$$Y_{ij} = m' + b'_i + e'_{ij}.$$

The least-squares estimates can be found to be

$$m'_1 = -0.94$$
$$b'_1 = 1.94 \qquad b'_2 = 0.69 \qquad b'_3 = -0.31 \qquad b'_4 = -2.31.$$

The regression sum of squares due to the reduced model is obtained from equation (5.80):

$$\begin{aligned} SS_{\text{reg}}(m', b'_i) &= m'T_{..} + \sum b'_i T_{i.} \\ &= (-15)(-0.94) + 4(1.94) - 1(0.69) \\ &\quad - 5(-0.31) - 13(-2.31) \\ &= 52.75. \end{aligned}$$

Therefore, using equation (5.66),

$$\begin{aligned} SS_{\text{treatment}} &= SS_{\text{reg}}(m, b_i, t_j) - SS_{\text{reg}}(m', b'_i) \\ &= 83.40 - 52.75 \\ &= 30.65. \end{aligned}$$

Similarly, the reduced model to examine the block effects is given by

$$Y_{ij} = m'' + t''_j + e''_{ij}.$$

The least-squares estimates can be found to be $m'' = -0.94$, $t''_1 = 2.19$, $t''_2 = 0.19$, $t''_3 = 1.31$, $t''_4 = -1.06$. Using these results, it can be verified that the regression sum of squares for the reduced model under consideration is equal to

$$\begin{aligned} SS_{\text{reg}}(m'', t''_j) &= m''T_{..} + \sum_j t''_j T_{.j} \\ &= 44.75. \end{aligned}$$

Finally, the sum of squares for the block effects can be computed from equation (5.83):

$$\begin{aligned} SS_{\text{block}} &= SS_{\text{reg}}(m, b_i, t_j) - SS_{\text{reg}}(m'', t''_j) \\ &= 38.65. \end{aligned} \tag{5.84}$$

The ANOVA results for this example are as follows:

(a) For treatments, $F_0 = MS_{\text{treatment}}/MS_{\text{error}} = 7.93$ and the critical value of the F-statistic is equal to 3.86 using $\alpha = 0.05$. Therefore, the hypothesis that the treatment effects are negligible is rejected at the 5% level of significance.

(b) For blocks, $F_0 = MS_{\text{block}}(MS_{\text{error}})^{-1} = 10.00$ and the critical value of the F-statistic is equal to 3.86. Therefore, the hypothesis that the block effects are negligible is rejected at the 5% level of significance.

5.6.2 Balanced incomplete block design

The normal equations for a BIBD were given in equations (5.49)–(5.51). These equations can be solved using the additional conditions $\sum_i b_i = 0$ and $\sum_j \hat{t}_j = 0$. In Example 5.8 given below, we will use the symbol t_j instead of $\hat{t}_j$ to simplify the formulation of the normal equations.

EXAMPLE 5.8

The approach can be applied to a balanced incomplete design. This example will demonstrated it on the data from Example 5.5. The normal equations follow:

$$
\begin{array}{lllllllllll}
T_{..}: & 12m + 3b_1 & +3b_2 & +3b_3 & +3b_4 & +3t_1 & +3t_2 & +3t_3 & +3t_4 & = 10\,230 \\
T_{1.}: & 3m + 3b_1 & & & & +t_1 & +t_2 & +t_3 & & = 2400 \\
T_{2.}: & 3m & +3b_2 & & & +t_1 & & +t_3 & +t_4 & = 2810 \\
T_{3.}: & 3m & & +3b_3 & & & +t_2 & +t_2 & +t_4 & = 2580 \\
T_{4.}: & 3m & & & +3b_4 & +t_1 & +t_2 & & +t_4 & = 2440 \\
T_{.1}: & 3m + b_1 & +b_2 & & +b_4 & +3t_1 & & & & = 2570 \\
T_{.2}: & 3m + b_1 & & +b_3 & +b_4 & & +3t_2 & & & = 2480 \\
T_{.3}: & 3m + b_1 & +b_2 & +b_3 & & & & +3t_3 & & = 2600 \\
T_{.4}: & 3m & +b_2 & +b_3 & +b_4 & & & & +3t_4 & = 2580 \\
 & \quad\; b_1 & +b_2 & +b_3 & +b_4 & & & & & = 0 \\
 & & & & & t_1 & +t_2 & +t_3 & +t_4 & = 0
\end{array}
$$

After solving the above equations, the same procedures developed in section 3.3.2 can be used. It is left as an exercise for the student to verify the following results:

(a) Solution of original normal equations:

$m = 852.5 \quad b_1 = -56.25 \quad b_2 = 85 \quad b_3 = 10 \quad b_4 = -38.75$

$t_1 = 7.5 \quad t_2 = 2.5 \quad t_3 = 1.25 \quad t_4 = -11.25$

$$SS_{\text{ref}}(m, b_i, t_j) = 8\,755\,875.$$

(b) Solution of reduced normal equations for testing treatment effects:

$m' = 852.5 \quad b'_1 = -52.5 \quad b'_2 = 84.17 \quad b'_3 = 7.5 \quad b'_4 = -39.17$

$$SS_{\text{reg}}(m', b'_i) = \frac{26\,266\,100}{3} = 8755\,366.67.$$

(c) Solution of reduced normal equations for testing block effects:

$m'' = 852.5 \quad t''_1 = 4.16 \quad t''_2 = -25.83 \quad t''_3 = 14.17 \quad t''_4 = 7.5$

$$SS_{\text{reg}}(m'', t''_j) = 8723\,900.$$

Using these results, we find that $SS_{\text{block}} = 31\,975$ and $SS_{\text{treatment}} = 508.33$.

5.6.3 Two-way ANOVA with interactions

There are many experiments in which it is possible to analyze the expected change in the response variable due to blocks and treatments. The linear model that we have used for the analysis of a randomized block design is given by (equation (5.1)):

$$Y_{ij} = \mu + \beta_i + \tau_j + \varepsilon_{ij}.$$

This model is said to be additive since it does not have an interaction term. The interaction term is not needed because it is assumed that

$$E[Y_{ij}] - E[Y_{i'j}] = E[Y_{ij'}] - E[Y_{i'j'}] = 0 \quad \text{for all } i, i', j, j'.$$

While the above simple additive model is often useful, there are situations where it is inadequate. Let γ_{ij} denote the interaction effect due to treatment j and block i. In this case, a more appropriate nonadditive model is

$$Y_{ij} = \mu + \beta_i + \tau_j + \gamma_{ij} + \varepsilon_{ij}. \tag{5.85}$$

In order to analyze the interaction effects, it will be necessary to collect several observations for each possible combination of treatments and blocks. Let n_{ij} be the number of replicates in the cell corresponding to treatment j and block i. If $n_{ij} = 1$ for the entire experiment, then the best strategy is to consider the additive model (no-interaction model).

Consider the special case with k levels of the factor (treatments), n blocks and r replications in each cell. It is possible to analyze if a significant difference between replications exists. This will be explained in the next example.

If m, b_i, t_j and g_{ij} are the least-squares estimators for $\mu, \beta_i, \tau_j, \gamma_{ij}$, the model becomes

$$Y_{ij} = m + b_i + t_j + g_{ij} + \varepsilon_{ij}.$$

Taking the partial derivatives of

$$SS_{\text{error}} = \sum_i \sum_j \sum_k (Y_{ijk} - m - b_i - t_j - g_{ij})^2 \tag{5.86}$$

with respect to m, b_i, t_j and g_{ij} and equating them to zero, the following normal equations are obtained:

$$T_{...} = r\left(nkm + k\sum_i b_i + n\sum_j t_j + \sum_i \sum_j g_{ij} \right) \tag{5.87}$$

$$T_{i..} = r\left(km + kb_i + \sum_j t_j + \sum_i \sum_j g_{ij} \right) \quad \text{for } i = 1, 2, \ldots, n \tag{5.88}$$

$$T_{.j.} = r\left(nm + \sum_i b_i + nt_j + \sum_i g_{ij} \right) \quad \text{for } j = 1, 2, \ldots, k \tag{5.89}$$

$$T_{ij.} = r(m + b_i + t_j + g_{ij}) \quad \text{for } i = 1, 2, \ldots, n \text{ and } j = 1, 2, \ldots, k. \tag{5.90}$$

It can be easily verified that the above equations are linearly dependent. It can also be shown that a solution to these equations is $m=0$, $b_i=0$ for all i, $t_j=0$ for all j and $g_{ij}=T_{ij.}r^{-1}$. Therefore

$$SS_{\text{reg}}(m,b_i,t_j,g_{ij})=\sum_i\sum_j g_{ij}T_{ij.}$$

$$=\sum_i\sum_j\frac{T_{ij.}^2}{r}. \qquad (5.91)$$

In order to test the hypothesis $H_0:\gamma_{ij}=0$ for all i and j, the reduced model is

$$Y_{ijk}=\mu+\beta_i+\tau_j+\varepsilon_{ijk}.$$

Note that this is the same model as equation (5.54) but with replications. The normal equations are

$$T_{...}=r\left(nkm'+k\sum_i b_i'+n\sum_j t_j'\right) \qquad (5.62)$$

$$T_{i..}=r\left(km'+kb_i'+\sum_j t_j'\right) \quad \text{for } i=1,2,\ldots,n \qquad (5.93)$$

$$T_{.j.}=r\left(nm'+\sum_i b_i'+nt_j'\right) \quad \text{for } j=1,2,\ldots,k. \qquad (5.94)$$

The additional conditions

$$\sum_i b_i'=0 \qquad \sum_j t_j'=0$$

can be considered to find the following solution:

$$m'=\bar{Y}_{...}$$
$$b_i'=\bar{Y}_{i..}-\bar{Y}_{...}$$
$$t_j'=\bar{Y}_{.j.}-\bar{Y}_{...}$$

Therefore

$$SS_{\text{reg}}(m',b',t')=mT_{...}+\sum_i b_i'T_{i..}+\sum_j t_j'T_{.j.}$$

$$=\sum_i\frac{T_{i..}^2}{kr}+\sum_j\frac{T_{.j.}^2}{nr}-\frac{T_{...}^2}{N}$$

where $N=nkr$.

In order to test the hypothesis $H_0:\tau_j=0$ for all j, the reduced model to be considered is

$$Y_{ijk}=\mu+\beta_i+\varepsilon_{ijk}.$$

The normal equations associated with this model are

$$T_{...} = r\left(nkm'' + k\sum_i b_i''\right)$$

$$T_{i..} = r(km'' + kb_i'') \quad i = 1, 2, \ldots, n.$$

After substituting $\sum_i b_i'' = 0$ the solution becomes $m'' = \bar{Y}_{...}, b_i = \bar{Y}_{i..} - \bar{Y}_{...}$, for all i. Therefore

$$SS_{\text{reg}}(m'', b_i'') = m''T_{...} + \sum_i T_{i..}$$

$$= \sum_i \frac{T_{i..}^2}{kr}. \tag{5.95}$$

To test the hypothesis $H_0: \beta_i = 0$ for all i, the reduced model to be considered is

$$Y_{ijk} = \mu + \tau_j + \varepsilon_{ijk}.$$

The normal equations associated with the above model are

$$T_{...} = r\left(nkm''' + n\sum_j t_j'''\right) \tag{5.96}$$

$$T_{.j.} = r(nm''' + nt_j''') \quad j = 1, 2, \ldots, k. \tag{5.97}$$

The solution of these normal equations, after including the condition $\sum_j t_j''' = 0$, is found to be

$$m''' = \frac{T_{...}}{rnk}$$

$$t_j''' = \frac{T_{.j.}}{nr} - m.$$

Therefore

$$SS_{\text{reg}}(m''', t_j''') = mT_{...} + \sum_j t_j''' T_{.j.}$$

$$= \sum_j \frac{T_{.j.}^2}{nr}. \tag{5.98}$$

To construct the ANOVA table we first calculate the following sum of squares:

$$SS_{\text{treatment}} = SS_{\text{reg}}(m, b', t') - SS_{\text{reg}}(m, b'')$$
$$SS_{\text{blocks}} = SS_{\text{reg}}(m, b', t') - SS_{\text{reg}}(m, t''')$$
$$SS_{\text{interaction}} = SS_{\text{reg}}(m, b, t, g) - SS_{\text{reg}}(m, b', t')$$
$$SS_{\text{total}} = \sum_i \sum_j \sum_k Y_{ijk}^2 - \frac{T_{...}^2}{N}.$$

The above sums of squares have degrees of freedom equal to $k-1$, $n-1$, $(k-1)(n-1)$ and $N-1$, respectively. The value of SS_{error} and its degrees of freedom are calculated by subtraction. Example 5.9 illustrates the methodology.

EXAMPLE 5.9

It is desired to analyze the effects of temperature in an oven on the time required to make bread. Three identical ovens are available at the bakery store in order to speed the results. The experimental conditions and the results are given in Table 5.23.

Several replications were run; however, oven C broke down after carrying out the first treatment. Also other conditions forced the experimenter to consider different numbers of replications in each experimental condition. In spite of this inconvenience it was possible to analyze the data as a two-way classification model and test the results. The model under consideration is given as (equation (5.85)):

$$Y_{ij} = \mu + \beta_i + \tau_j + \gamma_{ij} + \varepsilon_{ij}.$$

In a set of balanced data each experimental condition would have $n = 3$ observations; however, with unbalanced data each cell may have a different sample size. Some cells may even have no observations. Table 5.24 shows

Table 5.23 Responses for Example 5.9

		Replications		
Oven	*Temperature (°F)*	*1*	*2*	*3*
A	90	2.1	1.9	2
A	110	1.1	1.5	–
B	90	2.5	–	–
B	110	1.6	1.1	–
C	90	2.0	2.4	–
C	110	–	–	–

Table 5.24 Sample sizes for example 5.9

i	$j=1$	$j=2$	$n_{i.}$
1	3	2	5
2	1	2	3
3	2	0	2
$n_{.j}$	6	4	$N = 10$

Table 5.25 Total time per experimental condition

Oven	*90 °F*	*110 °F*	$T_{.j.}$
A	6.0	2.6	8.6
B	2.5	2.7	5.2
C	4.4	–	4.4
$T_{i..}$	12.9	5.3	$T_{...} = 18.2$

the number of trials in each experimental condition. The total time is recorded in Table 5.25.

The normal equations are

$$\begin{aligned}
10m + 5b_1 + 3b_2 + 2b_3 + 6t_1 + 4t_2 + 3g_{11} + 2g_{12} + g_{21} + 2g_{22} + 2g_{31} &= 18.2\\
5m + 5b_1 + 3t_1 + 2t_2 + 3g_{11} + 2g_{12} &= 8.6\\
3m + 3b_2 + t_1 + 2t_2 + g_{21} + 2g_{22} &= 5.2\\
2m + 2b_3 + 2t_1 + 2g_{31} &= 4.4\\
6m + 3b_1 + b_2 + 2b_3 + 6t_1 + 3g_{11} + g_{21} + 2g_{31} &= 12.9\\
4m + 2b_1 + 2b_2 + 4t_2 + 2g_{12} + 2g_{22} &= 5.3\\
3m + 3b_1 + 3t_1 + 3g_{11} &= 6.0\\
2m + 2b_1 + 2t_2 + 2g_{12} &= 2.6\\
m + b_2 + t_1 + g_{21} &= 2.5\\
2m + 2b_2 + 2t_2 + 2g_{22} &= 2.7\\
2m + 2b_3 + 2t_1 + 2g_{31} &= 4.4.
\end{aligned}$$

A solution of this system of equations, as indicated in section 5.6.3, is

$$m = 0 \qquad b_1 = b_2 = b_3 = 0 \qquad t_1 = t_2 = 0$$
$$g_{11} = 2 \qquad g_{21} = 2.5 \qquad g_{31} = 2.2 \qquad g_{12} = 1.3 \qquad g_{22} = 1.35.$$

Therefore

$$SS_{\text{reg}}(m, b_i, t_j, g_{ij}) = \sum_i \sum_j g_{ij} T_{ij.} = 35.$$

The reduced model for testing $H_0: \gamma_{ij} = 0$ for $i = 1, 2, 3$ and $j = 1, 2$ is formulated as follows:

$$Y_{ijk} = \mu + \beta_i + \tau_j + \varepsilon_{ijk}.$$

The normal equations associated with this model are

$$\begin{aligned}
10m + 5b'_1 + 3b'_2 + 2b'_3 + 6t'_1 + 4t'_2 &= 18.2\\
5m + 5b'_1 + 3t'_1 + 2t'_2 &= 8.6
\end{aligned}$$

$$\begin{aligned}
3m \quad\quad + 3b'_2 \quad\quad + t'_1 + 2t'_2 &= 5.2\\
2m \quad\quad\quad + 2b'_3 + 2t'_1 \quad &= 4.4\\
6m + 3b'_1 + b'_2 + 2b'_3 + 6t'_1 \quad &= 12.9\\
4m + 2b'_1 + 2b'_2 \quad\quad\quad + 4t'_2 &= 5.3.
\end{aligned}$$

The additional conditions in this case are

$$\begin{aligned}
5b'_1 + 3b'_2 + 2b'_3 \quad\quad &= 0\\
6t'_1 + 4t'_2 &= 0.
\end{aligned}$$

After including these conditions the solution is

$$m = 1.82 \qquad b'_1 = -0.1, \qquad b'_2 = 0.1429 \qquad b'_3 = 0.0357$$
$$t'_1 = 0.3443 \qquad t'_2 = -0.5764.$$

Therefore

$$SS_{\text{reg}}(m, b'_i, t'_j) = mT_{...} + \sum_i b'_i T_{i..} + \sum_j t'_j T_{.j.} = 34.87.$$

The reduced model for testing $H_0\colon \beta_i = 0$ for $i = 1, 2, 3$ is formulated as follows:

$$Y_{ijk} = \mu + \tau_j + \varepsilon_{ijk}.$$

The normal equations associated with this model are

$$\begin{aligned}
10m + 6t'''_1 + 4t'''_2 &= 18.2\\
6m + 6t'''_1 \quad\quad &= 12.9\\
4m \quad\quad + 4t'''_2 &= 5.3.
\end{aligned}$$

The only additional condition needed is $6t'''_1 + 4t'''_2 = 0$. The solution of these equations after considering the additional conditions is

$$m = 1.82 \qquad t'''_1 = 0.33 \quad t'''_2 = -0.495.$$

Therefore

$$SS_{\text{reg}}(m, t'''_j) = mT_{...} + \sum_j t'''_j T_{.j.} = 34.7575.$$

The reduced model for testing $H_0\colon \tau_j = 0$, for $j = 1, 2$ is formulated as follows:

$$Y_{ijk} = \mu + \beta_i + \varepsilon_{ijk}.$$

The reduced equations associated with this model are

$$\begin{aligned}
10m + 5b''_1 + 3b''_2 + 2b''_3 &= 18.2\\
5m + 5b''_1 \quad\quad\quad &= 8.6\\
3m \quad\quad + 3b''_2 \quad &= 5.2\\
2m \quad\quad\quad + 2b''_3 &= 4.4.
\end{aligned}$$

Again, the additional condition needed is $5b''_1 + 3b''_2 + 2b''_3 = 0$. After

Table 5.26 ANOVA results

Source	*df*	*SS*	*MS*	*F*
Treatments	1	1.3846	1.3846	10.885
Blocks	2	0.1125	0.05625	0.4422
Interaction	2	0.13	0.065	0.511
Error	4	0.5088	0.1272	
Total	9	2.136		

considering this condition, the solution of these equations is

$$m = 1.82 \qquad b''_1 = -0.1 \qquad b''_2 = -0.0867 \qquad b''_3 = 0.38.$$

Therefore

$$SS_{\text{reg}}(m, b''_i) = mT_{..} + \sum_i b''_i T_{i..} = 33{:}485.$$

With the above results, we can now get the ANOVA results shown in Table 5.26, after computing the following sums of squares:

$$\begin{aligned}
SS_{\text{treat}} &= SS_{\text{reg}}(m, b'_i, t'_j) - SS_{\text{reg}}(m, b''_i) = 34.87 - 33.485 = 1.3846\\
SS_{\text{block}} &= SS_{\text{reg}}(m, b'_i, t'_j) - SS_{\text{reg}}(m, b'''_j) = 34.87 - 34.757 = 0.1125\\
SS_{\text{interaction}} &= SS_{\text{reg}}(m, b_i, t_j, g_{ij}) - SS_{\text{reg}}(m, b'_i, t'_j) = 35 - 34.87 = 0.13\\
SS_{\text{total}} &= \sum_i \sum_j \sum_k Y^2_{ijk} - \frac{T^2_{...}}{N} = 35.26 - 33.124 = 2.136\\
SS_{\text{error}} &= 0.5088.
\end{aligned}$$

From Table 5.26 it is concluded that the hypothesis that the treatment effect is negligible should be rejected at 5% level of significance.

5.7 MISSING VALUES

For a single-factor completely randomized design, the ANOVA can be run with unequal n_j. In this case, there are no problems with missing data; however, in the case of a block design missing observations can cause the design to become unbalanced and, therefore, not orthogonal.

For this reason, the missing observations, if they are not too many, can be estimated before conducting an ANOVA study. Orthogonality is a property which ensures that the effects of all sources of variations included in the model can be estimated independently of one another, as a result of having additive sums of squares. A randomized lock design is orthogonal when $\sum_j t_j = 0$ for all blocks, and $\sum_i b_i = 0$ for all treatments. Thus, in the case of missing values, the design is no longer orthogonal, unless the missing data points are estimated. In all cases, however, the number of degrees of freedom is equal to the number of real observations minus one. Consider a randomized block design with k treatments and n blocks. Suppose that the observation in cell (m, p) is missing. The sum of squares of the error is given by

$$SS_{\text{error}} = SS_{\text{total}} - SS_{\text{block}} - SS_{\text{treatment}}. \tag{5.99}$$

Using equations (5.41)–(5.43), the error sum of squares in equation (5.99) can be written as a function of the missing observation Y_{mp} as follows, where $i \neq m$ and $j \neq p$:

$$SS_{\text{error}}(Y_{mp}) = \sum_i \sum_j Y_{ij}^2 + Y_{mp}^2 + \frac{(\Sigma_i \Sigma_j Y_{ij} + Y_{mp})^2}{nk} - \frac{(\Sigma_i T_{i.}^2 + (Y_{mp} + T_{m.})^2)}{k} - \frac{(\Sigma_j T_{.j}^2 + (Y_{mp} + T_{.p})^2)}{n}. \tag{5.100}$$

Usually a good estimate Y_{mp} is the value that minimizes the error sum of squares. From elementary calculus,

$$\frac{d(SS_{\text{error}})}{d(Y_{mp})} = 0 \tag{5.101}$$

is a necessary condition to minimize the error sum of squares. Solving equation (5.101) for Y_{mp} the estimate is

$$Y_{mp} = \frac{(nT_{m.} + kT_{.p} - \Sigma_i \Sigma_j y_{ij})}{(k-1)(n-1)}. \tag{5.102}$$

The value of Y_{mp} given by equation (5.102) can now be used to recompute the sums of squares for treatments, blocks and random error. The corresponding degrees of freedom are shown in Table 5.27.

An exact solution to this problem can be found using the least-squares significance test as illustrated in Example 5.9. For the case of several missing values, the present discussion remains valid, and is easily extended by

Table 5.27 Degrees of freedom

SS	df
Total	$nk-2$
Blocks	$n-1$
Treatments	$k-1$
Error	$(n-1)(k-1)-1$

considering partial derivatives with respect to each estimate of a missing observation. This procedure is also illustrated in Example 5.10 for a case with two missing observations.

EXAMPLE 5.10

Consider again the problem given in Example 5.2, involving the manufacturing of an organic chemical using one of five different blends. Assume that batches 1 and 15 were not available. Table 5.28 summarizes the results, with missing values represented by x and y.

ANOVA estimating missing observations

The estimation of the missing values follows:

$$\begin{aligned} T_{1.} &= y + 67.3 & T_{.2} &= y + 36.3 \\ T_{2.} &= 88.1 & T_{.3} &= 52.4 \\ T_{3.} &= x + 72 & T_{.4} &= 48.9 \\ T_{.1} &= x + 33.4 & T_{.5} &= 56.4. \end{aligned}$$

It can be verified that

$$\begin{aligned} SS_{\text{error}} &= SS_{\text{total}} - SS_{\text{block}} - SS_{\text{treatment}} \\ &= \frac{(8x^2 + 8y^2 - 311.2x - 312y + 2xy)}{15} + R \end{aligned}$$

where R is the sum of the terms not containing x or y. To estimate the missing observations the partial derivatives of SS_{error} must be taken with respect to x and y, set equal to zero, and solved for x and y. The corresponding system of equations is given by $16x - 311.2 + 2y = 0$ and $16y - 312 + 2x = 0$. Solving these equations, the result is $x = 17.28$ and $y = 17.34$. Therefore the missing values are estimated as

$$\begin{aligned} &\text{Batch } 1: 17.34 \approx 17.3 \\ &\text{Batch } 15: 17.28 \approx 17.3. \end{aligned}$$

Table 5.28 Yield of process for manufacturing an organic chemical

		Blends				
		A	*B*	*C*	*D*	*E*
	1	16.9	y	17.0	15.1	18.3
Blocks	2	16.5	19.2	18.1	16.0	18.3
	3	x	17.1	17.3	17.8	19.8

Table 5.29 Data on yield of manufacturing process after estimating missing observation

		A	*B*	*C*	*D*	*E*	$T_{i.}$
	1	16.9	17.3	17.0	15.1	18.3	84.6
Blocks	2	16.5	19.2	18.1	16.0	18.3	88.1
	3	17.3	17.1	17.3	17.8	19.8	89.3
	$T_{.j}=$	50.7	53.6	52.4	48.9	56.4	$262.0=T_{..}$

Table 5.30 ANOVA table estimating missing values

Source	*df*	*SS*	*MS*	*F*	*Significant?*
Blocks	2	2.39	1.20	1.10	No
Blends	4	10.86	2.72	2.50	No
Error	6	6.54	1.09		
Total	12	19.79			

Note that the original values given in Example 5.2 were 18.2 and 17.5 respectively. The analysis can be completed using the estimated values, as shown in Tables 5.29 and 5.30. From the results given in Table 5.29, $\sum_i\sum_j Y_{ij}^2=4596.06$ and $N=15$. The ANOVA results for this example using $\alpha=0.10$ are summarized in Table 5.30.

As can be seen in Table 5.30, the effect of the blend is not significant. It should be noted that in Example 5.2 this effect was not found negligible at the same level of significance.

ANOVA using least-squares procedure

This procedure gives an exact solution, whereas the solution found via estimation was only approximate. The normal equations for this analysis are given as follows:

$$
\begin{aligned}
T_{..}&: 13m+4b_1+5b_2+4b_3+2t_1+2t_2+3t_3+3t_4+3t_5=227.4\\
T_{1.}&: 4m+4b_1+t_1+t_3+t_4+t_5=67.3\\
T_{2.}&: 5m+5b_2+t_1+t_2+t_3+t_4+t_5=88.1\\
T_{3.}&: 4m+4b_3+t_2+t_3+t_4+t_5=72.0\\
T_{.1}&: 2m+b_1+b_2+2t_1=33.4\\
T_{.2}&: 2m+b_2+b_3+3t_2=36.3\\
T_{.3}&: 3m+b_1+b_2+b_3+3t_3=52.4\\
T_{.4}&: 3m+b_1+b_2+b_3+3t_4=48.9
\end{aligned}
$$

$$
\begin{aligned}
T_{.5}:\ & 3m + b_1 + b_2 + b_3 + 3t_5 = 56.4 \\
& b_1 + b_2 + b_3 = 0 \\
& t_1 + t_2 + t_3 + t_4 + t_5 = 0
\end{aligned}
$$

The solution of the normal equations yields the following results:

$$m = 17.4681$$
$$b_1 = -0.54021 \qquad b_2 = 0.15185 \qquad b_3 = 0.38836.$$

Reduced normal equations for blocks:

$$
\begin{aligned}
& 13m' + 4b'_1 + 5b'_2 + 4b'_3 = 227.4 \\
\mathrm{i} = 1:\ & 4m' + 4b'_1 = 67.3 \\
i = 2:\ & 5m' + 5b'_2 = 88.1 \\
i = 3:\ & 4m' + 4b'_3 = 72.0 \\
& b'_1 + b'_2 + b'_3 = 0
\end{aligned}
$$

The solution of the above equations yields the following results:

$$m' = 17.482$$
$$b'_1 = -0.657 \qquad b'_2 = 0.138 \qquad b'_3 = 0.518.$$

Reduced normal equations for treatments:

$$
\begin{aligned}
& 13m'' + 2t''_1 + 2t''_2 + 3t''_3 + 3t''_4 + 3t''_5 = 227.4 \\
j = 1:\ & 2m'' + 2t''_1 = 33.4 \\
j = 2:\ & 2m'' + 2t''_2 = 36.3 \\
j = 3:\ & 3m'' + 3t''_3 = 52.4 \\
j = 4:\ & 3m'' + 3t''_4 = 48.9 \\
j = 5:\ & 3m'' + 3t''_5 = 56.4 \\
& t''_1 + t''_2 + t''_3 + t''_4 + t''_5 = 0.
\end{aligned}
$$

The solution of the above equations yields the following results:

$$m'' = 17.483$$
$$t''_1 = -0.783 \qquad t''_2 = 0.667 \qquad t''_3 = -0.017$$
$$t''_4 = -1.183 \qquad t''_5 = 1.317.$$

Table 5.31 ANOVA table without estimating missing values

Source	*df*	*SS*	*MS*	*F*	*Significant?*
Blocks	2	1.638	0.819	0.637	No
Blends	4	10.378	2.595	2.018	No
Error	6	7.713	1.286		
Total	12	19.729			

The ANOVA results are summarized in Table 5.31, where a level of significance $\alpha = 0.10$ has been chosen. As can be seen in this table, both hypotheses are rejected, as they were in Table 5.30. Again, it is noted that if the two observations were not missing, treatment effects would have been found significant (as shown in Table 5.10).

EXERCISES

1. The desulfurization of coal from five sources was tested using a randomized block design. Fifteen batches, three from each source, were divided into three blocks in random order, as shown in Table 5.32. Conduct a statistical analysis for the variation between sources and blocks.
2. In Exercise 1 suppose that only enough coal for two batches was available from source A. Delete the value in block 1 for source A, keeping all others the same. Repeat the analysis, estimating the missing observation.
3. In Exercise 1, assume that coal data are missing for source A in block 1 and source C in block 3. Do an ANOVA using the least-squares significance test.
4. Show how and why a balanced incomplete design must be adjusted if an ANOVA test is used.
5. What is the minimal number of blocks required to have a balanced incomplete design with block size four and six treatments in total?
6. Explain how to perform a special ANOVA method for symmetrical incomplete block designs.

Table 5.32

Block	*Batch*	*Source*	*Weight loss(%)*
	1	A	16.5
	2	E	18.3
1	3	B	19.2
	4	C	18.1
	5	D	16.0
	6	B	18.2
	7	A	16.9
2	8	C	17.0
	9	E	18.3
	10	D	15.1
	11	B	17.1
	12	D	17.8
3	13	C	17.3
	14	E	19.8
	15	A	17.5

Table 5.33

Pen	*Feed*	*Initial weight*	*Growth rate*
	A	48	9.94
1	B	48	10.00
	C	48	9.75
	B	32	9.24
2	C	28	8.66
	A	32	9.48
	C	33	7.63
3	A	35	9.32
	B	41	9.34
	C	50	10.37
	A	48	10.56
4	B	46	9.68
	B	37	9.67
5	A	32	8.82
	C	30	8.57

7. How can you tell if a block design is or is not orthogonal, by inspecting the normal equations?
8. Table 5.33 contains data on the initial weights, in pounds, and the growth rates, in pounds per week, of 15 pigs, classified according to pen and type of feed given. Examine the difference between the three types of feed A, B and C in their effect on the growth of pigs, correcting for differences in initial weight (source: reference [5]).
9. Use the least-squares significance test on the data given in Exercise 1.
10. Explain what condition is required on the number of observations per cell if it is necessary to investigate the significance of the interaction effect between treatments and blocks in a completely randomized block design.
11. Consider the results in Table 5.34 for a design with treatments A, B, C, D and E and blocks 1, 2, 3, 4 and 5. Do the corresponding ANOVA.

Table 5.34

	A	*B*	*C*	*D*	*E*
1	10	9	14	20	–
2	12	11	15	–	17
3	10	12	–	18	15
4	13	–	12	19	18
5	–	10	15	17	17

Table 5.35

Material composition	*Machine* 1	2	3	4	5
A	24, 28, 30	–	30, 27	23, 21	31
B	21	23, 25	23	29, 22	28, 27
C	26, 24	–	22, 26	28, 26	23
D	24	21, 24	–	22, 24	26, 21

12. Test the significance of the block effect using the least-squares significance test, using the data given in Exercise 11.
13. Solve exercise 3, estimating the missing observations, rather than using the least-squares significance test.
14. In a manufacturing plant, it is desired to test if the maintenance tool wear is influenced by the material composition of the tools. Since the tools are used randomly in the plant according to machine breakdowns, the quality control engineer decides to conduct a test to verify the hypothesis whether the tool materials have an effect on the tool life or not. The results of tool wear measured in microns are given in Table 5.35 (replicates are separated by commas).

    Define the treatments and blocks. Perform an ANOVA test to find whether treatment effects, block effects or interactions are significant.

REFERENCES

1. Davies, O. L. *et al.* (1956) *The Design and Analysis of Industrial Experiments*, 2nd edn, Imperial Chemical Industrial Ltd, London.
2. Graybill, F. A. (1961) *An Introduction of Linear Statistical Models*, Vol. 1. McGraw-Hill, New York.
3. Searle, S. R. (1971) *Linear Models*, John Wiley & Sons, Inc., New York.
4. Hicks, C. R. (1973) *Fundamental Concepts in the Design of Experiments*, 2nd edn, Holt, Rinehart and Winston, New York.
5. Chadravarti, I. M., Laha, R. G. and Roy, J. (1967) *Handbook of Applied Statistics*, Vol. II, John Wiley & Sons, Inc., New York.

FURTHER READING

Montgomery, D. C. (1991) *Design and Analysis of Experiments*, 3rd edn, John Wiley & Sons, New York.

6 Latin, Graeco-Latin and Youden squares

In a randomized block experiment the homogeneity of the experimental material is enhanced by applying the treatments over compact blocks of relatively similar experimental conditions, thus reducing the residual variation (which is assigned to random error) and making the comparisons more sensitive. For example, an experiment to compare fuel economy of different brands of automobiles may require more than one driver. As explained in Chapter 5, a balanced experiment with each brand as a treatment and each driver as a block will allow the separation of the effect of different driving habits from the effect due to automobile brands.

When additional restrictions on randomization exist, the experiment must be further subdivided to increase the homogeneity of the experimental conditions. In a fabric wear-testing machine with four positions and four treatments, for example, the results obtained after a run of standard length may vary due to experimental errors, differences among fabric (treatments) and differences among positions. Therefore, the comparisons between different materials will be more precise if all fabrics are tested in the same position. In this case, however, for all fabrics to be tested the experiment must consist of four runs of the machine. As there may exist variation from run to run, comparisons between different materials may be affected by run differences. It is physically impossible to compare every pair of treatments in both the same run and the same position, but it is possible to ensure that each treatment is tested the same number of times in every position and also in every run. The experiment in this case is subdivided not only by fabrics (treatments) and positions but also by machine runs. In fact, if different machine operators could affect results, additional groupings for homogeneity could be possible, each possibly increasing the sensitivity of the experiment. As more restrictions on randomizations are considered, it is necessary, however, to ensure that the random error retains a sufficient number of degrees of freedom.

6.1 LATIN SQUARES

When an experimental design uses two subdivisions or 'blocking variables' for enhancing the homogeneity of the experimental material, an arrangement of trials called a **Latin square** may be used. In the fabric wear example previously discussed, such a design ensures that each fabric is tested once in every run and once in every position. The design is orthogonal and each source of variation can be independently separated and investigated.

A Latin square is a square containing r rows and r columns, and consequently r^2 cells. Each cell contains one of r letters, corresponding to r treatments, and each letter occurs once and only once in each row and each column. The total variation among the results is made up of four separate sources:

(a) between rows (runs or blocks)
(b) between columns (positions on machine)
(c) between letters (treatments)
(d) error (residual).

As an illustration, a Latin square design for the sample problem under consideration is given in Table 6.1. It is assumed that runs, positions and treatments affect the results independently of one another. In practice, this means that either the interactions are not important or do not exist. In the design shown in Table 6.1 there are four treatments, four blocks and four positions. Each letter A, B, C, D represents a single treatment.

6.1.1 Analysis of variance

Let β_i, τ_j and γ_k be the effect of the ith block, the jth treatment and the kth position, respectively. The main null hypothesis to be tested is that no effect exists due to treatments, that is:

$$H_0: \tau_j = 0 \quad j = 1, 2, 3, \ldots, r$$

where it is assumed that $j = 1$ implies treatment A, $j = 2$ implies treatment B, and so on. Additionally, two other null hypotheses can be tested:

$$H'_0: \beta_i = 0 \quad i = 1, 2, 3, \ldots, r$$
$$H''_0: \gamma_k = 0 \quad k = 1, 2, 3, \ldots, r.$$

Table 6.1

		Positions		
Blocks	*1*	*2*	*3*	*4*
1	C	D	A	B
2	D	A	B	C
3	B	C	D	A
4	A	B	C	D

H_0' states that no block has a significant effect, and H_0'' that no position has a significant effect. These three hypotheses can be tested with only r^2 observations, instead of r^3 as would be needed in a full design including all possible level combinations.

Consider the Latin square previously given. Let T_A and T_B be the observed totals for treatments A and B, respectively. Therefore

$$T_A \doteq (\tau_A + \mu + \beta_1 + \gamma_3) + (\tau_A + \mu + \beta_2 + \gamma_2) + (\tau_A + \mu + \beta_3 + \gamma_4) + (\tau_A + \mu + \beta_4 + \gamma_1)$$

and

$$T_B \doteq (\tau_B + \mu + \beta_1 + \gamma_4) + (\tau_B + \mu + \beta_2 + \gamma_3) + (\tau_B + \mu + \beta_3 + \gamma_1) + (\tau_B + \mu + \beta_4 + \gamma_2)$$

Therefore

$$T_A - T_B \doteq 4\tau_A - 4\tau_B$$

or in general,

$$T_A - T_B \doteq r\tau_A - r\tau_B. \tag{6.1}$$

Since $\bar{Y}_A = T_A/r$ and $\bar{Y}_B = T_B/r$ equation (6.1) can be rewritten as

$$\bar{Y}_A - \bar{Y}_B \doteq \tau_A - \tau_B. \tag{6.2}$$

According to equation (6.2), differences between any two treatment effects can be estimated separately and independently. This is also true for block and position effects. For any r, there are $(r-1)!r!$ possible Latin square arrangements. For the case $r = 4$, for example, there are 144 possible squares. The particular square used in a specific application can be thought of as having been chosen at random from a collection of 144 squares. Tables with Latin square designs for different values of r can be obtained from sources such as the books by Fisher and Yates [1] and by Federer [2].

In Chapter 3, the mathematical model for a single-factor experiment with no restrictions on randomization was formulated as

$$Y_{ij} = \mu + \tau_j + \varepsilon_{ij} \tag{6.3}$$

where μ is the common effect in all cells, Y_{ij} the result for the ith observation in the jth treatment, τ_j the jth treatment effect and ε_{ij} the corresponding random error.

In Chapter 5, subdividing the experiment into complete randomized blocks, with block effects independent of treatment effects, resulted in

$$Y_{ij} = \mu + \beta_i + \tau_j + \varepsilon_{ij} \tag{6.4}$$

where β_i is the ith block effect. Equation (6.4) can be rearranged as

$$Y_{ij} = \mu + \tau_j + (\beta_i + \varepsilon_{ij}). \tag{6.5}$$

The error term of equation (6.3), which contained all variation not accounted for by treatment differences, has been subdivided into $\beta_i + \varepsilon_{ij}$ in equation (6.5) to include block effects.

A Latin square design further subdivides the error term in equation (6.4) into a position effect and a reduced random error. The following mathematical model describes the experiment for a design with r treatments, r rows and r columns:

$$Y_{ijk} = \mu + \beta_i + \tau_j + \gamma_k + \varepsilon_{ijk} \tag{6.6}$$

where the index i refers to rows (blocks), j to treatments (Latin letters) and k to columns (positions).

Due to the orthogonality of the design, the sum of squares of the total can be proved to be given by

$$SS_{\text{total}} = SS_{\text{block}} + SS_{\text{treatment}} + SS_{\text{position}} + SS_{\text{error}}. \tag{6.7}$$

The sum of the squares SS_{position} of equation (6.7) can be obtained as

$$SS_{\text{position}} = \sum_i \sum_j \sum_k (\bar{Y}_{..k} - \bar{Y}_{...})^2 \tag{6.8}$$

or

$$SS_{\text{position}} = \sum_k \frac{T_{..k}^2}{r} - \frac{T_{...}^2}{N} \tag{6.9}$$

where $T_{..k}$ is the observed total for the kth position.

Table 6.2 ANOVA table for a Latin square design

Source	*df*	*SS*	*MS*
Treatments	$(r-1)$	$\sum_j \frac{T_{.j.}^2}{r} - \frac{T_{..}^2}{N}$	$\frac{SS_{\text{treatment}}}{(r-1)}$
Blocks	$(r-1)$	$\sum_i \frac{T_{i..}^2}{r} - \frac{T_{..}^2}{N}$	$\frac{SS_{\text{block}}}{(r-1)}$
Positions	$(r-1)$	$\sum_k \frac{T_{k..}^2}{r} - \frac{T_{..}^2}{N}$	$\frac{SS_{\text{position}}}{(r-1)}$
Error	$(r-1)(r-2)$	$SS_{\text{total}} - SS_{\text{treatment}} - SS_{\text{block}} - SS_{\text{position}}$	$\frac{SS_{\text{error}}}{(r-1)(r-2)}$
Total	$r^2 - 1$	$\sum_i \sum_k Y_{ijk}^2 - \frac{T_{..}^2}{N}$	

From equation (6.7), the sum of the squares of the error is given by

$$SS_{\text{error}} = SS_{\text{total}} - SS_{\text{block}} - SS_{\text{position}} - SS_{\text{treatment}}. \tag{6.10}$$

Note that

$$df_{\text{error}} = (r^2 - 1) - (r - 1) - (r - 1) - (r - 1).$$

After simplifying the above equation, the following result is obtained:

$$df_{\text{error}} = (r - 1)(r - 2). \tag{6.11}$$

As can be seen in equation (6.10), the additional restriction on the randomization of the experiment further reduces the experimental error. On the other hand, as shown in equation (6.11), the number of degrees of freedom of error is also reduced, and hence there is less precision in estimating the error variance (mean square).

The ANOVA for a Latin square design can be summarized as in Table 6.2. In this table, $T_{i..}$, $T_{.j.}$ and $T_{..k}$ are the observed totals for the ith row, the jth treatment and the kth column. Also, $N = r^2$.

EXAMPLE 6.1

A job-lot machine shop manager desires to select the best turning method, from a total of five, which will produce the maximum number of acceptable aluminium alloy parts; these combinations of tool and speed are labeled A, B, C, D and E. It is suspected that lathe and operator may affect quality; thus, the manager assigns the methods for each lathe and operator shown in Table 6.3.

The design is a Latin square with five rows, five columns and five treatments (letters). The number in parentheses is the daily production of parts with acceptable machining tolerances and surface finish. The corresponding

Table 6.3

		Operator					
	No.	*1*	*2*	*3*	*4*	*5*	*Total*
Lathe	I	A(15)	D(12)	C(14)	B(11)	E(10)	62
	II	E(11)	C(15)	B(13)	A(14)	D(12)	65
	III	D(11)	B(12)	A(16)	E(10)	C(13)	62
	IV	C(13)	A(15)	E(10)	D(13)	B(12)	63
	V	B(11)	E(10)	D(12)	C(12)	A(16)	61
Total		61	64	65	60	63	313

Table 6.4 Analysis of variance of Example 6.1, $\alpha = 0.05$

Source	*df*	*SS*	*MS*	*F*	F_c	*Significant?*
Treatments	4	70.64	17.660	28.48	3.26	Yes
Lathes	4	1.84	0.460	0.66	3.26	No
Operators	4	3.44	0.860	1.24	3.26	No
Error	12	8.32	0.693			
Total	24	84.24				

totals are given as follows:

$$T_{i..} = 62, 65, 62, 63, 61 \quad \text{(lathes)}$$
$$T_{.j.} = 76, 59, 67, 60, 51 \quad \text{(treatments)}$$
$$T_{..k} = 61, 64, 65, 60, 63 \quad \text{(operators).}$$

Also

$$T_{...} = 313 \qquad \sum_i \sum_k Y_{ijk}^2 = 4003 \qquad G = \frac{(313)^2}{25} = 3918.76.$$

Therefore

$$SS_{\text{total}} = 4003 - 3918.76 = 84.24$$
$$SS_{\text{treatment}} = \tfrac{1}{5}(76^2 + 59^2 + 67^2 + 60^2 + 51^2) - 3918.76 = 70.64$$
$$SS_{\text{lathe}} = \tfrac{1}{5}(62^2 + 65^2 + 62^2 + 63^2 + 61^2) - 3918.76 = 1.84$$
$$SS_{\text{operator}} = \tfrac{1}{5}(61^2 + 64^2 + 65^2 + 60^2 + 63^2) - 3918.76 = 3.44$$

The ANOVA can be summarized as in Table 6.4.

Thus, the machining method has a significant effect on daily production, while lathe and operator effects are not significant. In this case, the manager should use method A on all machines with all operators in order to produce the greatest number of acceptable parts per day.

EXAMPLE 6.2

A railroad maintenance-of-way engineer wishes to investigate the daily increase in number of crosstie replacements to be achieved by giving a six-man gang a backhoe and a hi-rail gang truck which can run on both roads and rails [3]. The engineer devises an experiment with four treatments:

A: normal equipment
B: normal equipment plus backhoe with operator
C: normal equipment plus hi-rail truck
D: normal equipment plus backhoe (with operator) and hi-rail truck.

Table 6.5

Weather	*Working conditions*			
	Poor	*Fair*	*Good*	*Very good*
Rain	A (5.17)	B (9.33)	C (8.50)	D(12.00)
Drizzle	D(10.33)	A (7.00)	B(10.00)	C (7.83)
Moderate	C (8.00)	D(12.67)	A (6.83)	B(11.17)
Hot sun	B(11.50)	C (8.50)	D(11.67)	A (7.02)

Table 6.6 Analysis of variance for Example 6.2, $\alpha = 0.05$

Source	*df*	*SS*	*MS*	*F*	F_c	*Significant?*
Treatments	3	64.10	21.37	30.97	4.76	Yes
Conditions	3	1.31	0.44	0.64	4.76	No
Weather	3	3.25	1.08	1.57	4.76	No
Error	6	4.17	0.69			
Total	15	72.83				

Besides equipment, working conditions (location and traffic delays) as well as weather will affect gang production, which has been averaging seven ties replaced per man per day without the extra equipment. Consequently, a 4×4 Latin square design is used (values in parentheses are ties per man per day). Using the data in Table 6.5, the following results are obtained:

$$T_{i..} = 35.00, 35.16, 38.67, 38.69 \quad \text{(working conditions)}$$
$$T_{.j.} = 26.02, 42.00, 32.83, 46.67 \quad \text{(treatments)}$$
$$T_{..k} = 35.00, 37.50, 37.00, 38.02 \quad \text{(weather)}$$

Also

$$T_{...} = 147.52 \qquad \sum_i \sum_k Y_{ijk}^2 = 1432.96 \qquad G = \frac{(147.52)^2}{16} = 1360.13.$$

The ANOVA is summarized in Table 6.6.

As an exercise, the reader should verify that the backhoe with operator causes the greatest incremental increase in productivity (almost as great as the two combined). If productivity increases justify the costs, however, both machines should be added.

6.1.2 Replication of Latin square designs

When small Latin squares are used, it is often desirable to replicate them; there are three different cases of replications of Latin squares. In the following discussion rows and columns are generally referred to as 'blocking variables', although up to this point rows have represented actual blocks and columns have been used to indicate block positions.

(a) Case I

The Latin square design is replicated using the same blocking variables. The ANOVA for the replication of the Latin squares is based on the following model:

$$Y_{ijkm} = \mu + \beta_i + \tau_j + \gamma_k + \omega_m + \varepsilon_{ijkm} \tag{6.12}$$

where $i = 1, 2, \ldots, r$, $j = 1, 2, \ldots, r$, $k = 1, 2, \ldots, r$ and $m = 1, 2, \ldots, n$. In equation (6.12), ω_m is the effect of the replication. The fundamental equation of the ANOVA for this experiment can be written as

$$SS_{\text{total}} = SS_{\text{block}} + SS_{\text{treatment}} + SS_{\text{position}} + SS_{\text{replication}} + SS_{\text{error}} \tag{6.13}$$

where

$$SS_{\text{position}} = \sum_k \frac{T^2_{..k.}}{rn} - \frac{T^2_{....}}{N} \tag{6.14}$$

$$SS_{\text{block}} = \sum_i \frac{T^2_{i...}}{rn} - \frac{T^2_{....}}{N} \tag{6.15}$$

$$SS_{\text{treatment}} = \sum_j \frac{T^2_{.j..}}{rn} - \frac{T^2_{....}}{N} \tag{6.16}$$

$$SS_{\text{replication}} = \sum_m \frac{T^2_{...m}}{r^2} - \frac{T^2_{....}}{N} \tag{6.17}$$

$$SS_{\text{total}} = \sum_i \sum_j \sum_k \sum_m Y^2_{ijkm} - \frac{T^2_{....}}{N}. \tag{6.18}$$

The number of degrees of freedom of the total sum of squares is equal to $nr^2 - 1$; this number can be partitioned into five components, namely: $(r-1)$ for blocks, treatments and positions and $(n-1)$ for replications. The number of degrees of freedom for the error is $(r-1)[n(r+1)-3]$ which is obtained from the following equation:

$$df_{\text{error}} = (nr^2 - 1) - (r-1) - (r-1) - (r-1) - (n-1) = (r-1)[n(r+1)-3].$$

The ANOVA for case I is summarized in Table 6.7.

(b) Case II

The Latin square design is replicated by holding one of the blocking variables unchanged, that is, the replication is performed either by having the same

Table 6.7 ANOVA table for case I

Source	*df*	*SS*	*MS*
Treatments	$(r-1)$	$\sum_j \frac{T^2_{.j..}}{rn} - \frac{T^2_{....}}{N}$	$\frac{SS_{\text{treatment}}}{(r-1)}$
Blocks	$(r-1)$	$\sum_i \frac{T^2_{i...}}{rn} - \frac{T^2_{....}}{N}$	$\frac{SS_{\text{block}}}{(r-1)}$
Positions	$(r-1)$	$\sum_k \frac{T^2_{..k.}}{rn} - \frac{T^2_{....}}{N}$	$\frac{SS_{\text{position}}}{(r-1)}$
Replications	$(n-1)$	$\sum_m \frac{T^2_{..m}}{r^2} - \frac{T^2_{....}}{N}$	$\frac{SS_{\text{replication}}}{(n-1)}$
Error	$(r-1)$ $[n(r+1)-3]$	$SS_{\text{total}} - SS_{\text{treatment}} - SS_{\text{block}} - SS_{\text{position}} - SS_{\text{replication}}$	$\frac{SS_{\text{error}}}{(r-1)[n(r+1)-3]}$
Total	nr^2-1	$\sum_i \sum_j \sum_k \sum_m Y^2_{ijkm} - \frac{T^2_{....}}{N}$	

Table 6.8 ANOVA table for case II

Source	*df*	*SS*	*MS*
Treatments	$(r-1)$	$\sum_j \frac{T^2_{.j..}}{rn} - \frac{T^2_{....}}{N}$	$\frac{SS_{\text{treatment}}}{(r-1)}$
Blocks	$n(r-1)$	$\sum_i \frac{T^2_{i...}}{r} - \frac{T^2_{....}}{N} - SS_{\text{replication}}$	$\frac{SS_{\text{block}}}{n(r-1)}$
Positions	$(r-1)$	$\sum_k \frac{T^2_{..k.}}{rn} - \frac{T^2_{....}}{N}$	$\frac{SS_{\text{position}}}{(r-1)}$
Replications	$(n-1)$	$\sum_m \frac{T^2_{...m}}{r^2} - \frac{T^2_{....}}{N}$	$\frac{SS_{\text{replication}}}{(n-1)}$
Error	$(r-1)(nr-2)$	$SS_{\text{total}} - SS_{\text{treatment}} - SS_{\text{block}} - SS_{\text{position}} - SS_{\text{replication}}$	$\frac{SS_{\text{error}}}{(r-1)(nr-2)}$
Total	nr^2-1	$\sum_i \sum_j \sum_k \sum_m Y^2_{ijkm} - \frac{T^2_{....}}{N}$	

rows (blocks) and varying the columns (positions) or vice versa. Since in this case only one of the two blocking variables remains the same for all replicates, there can be two kinds of replications.

Table 6.8 shows the ANOVA formulas for the case where the rows (blocks) are replicated. In this case the ANOVA table is essentially that shown for case I, although the sum of squares of blocks and degrees of freedom for error need to be recalculated. When the columns (positions) are replicated while the blocks remain unchanged, the ANOVA table is again essentially the same as the table for case I, except that the sum of squares for the positions and the degrees of freedom for the error will be changed. In this case, the sum of squares for the positions and the degrees of freedom for the error are given by

$$SS_{\text{position}} = \sum_{k=1}^{nr} \frac{T^2_{..k.}}{r} - \frac{T^2_{....}}{N} - SS_{\text{replication}} \tag{6.19}$$

with $n(r-1)$ degrees of freedom. Additionally, the number of degrees of freedom for the error is given by

$$df_{\text{error}} = (r-1)(nr-2).$$

(c) Case III

The Latin square design is replicated with different rows (blocks) and different columns (positions). The ANOVA table for the Latin square case III is shown in Table 6.9.

Table 6.9 ANOVA table for case III

Source	*df*	*SS*	*MS*
Treatments	$(r-1)$	$\sum_j \frac{T^2_{.j..}}{rn} - \frac{T^2_{....}}{N}$	$\frac{SS_{\text{treatment}}}{(r-1)}$
Blocks	$n(r-1)$	$\sum_i \frac{T^2_{i...}}{r} - \frac{T^2_{....}}{N} - SS_{\text{replication}}$	$\frac{SS_{\text{block}}}{n(r-1)}$
Positions	$n(r-1)$	$\sum_k \frac{T^2_{..k.}}{r} - \frac{T^2_{....}}{N} - SS_{\text{replication}}$	$\frac{SS_{\text{position}}}{n(r-1)}$
Replications	$(n-1)$	$\sum_m \frac{T^2_{...m}}{r^2} - \frac{T^2_{....}}{N}$	$\frac{SS_{\text{replication}}}{(n-1)}$
Error	$(r-1)$ $[n(r-1)-1]$	$SS_{\text{total}} - SS_{\text{treatment}}$ $- SS_{\text{block}} - SS_{\text{position}}$ $- SS_{\text{replication}}$	$\frac{SS_{\text{error}}}{(r-1)[n(r-1)-1]}$
Total	$nr^2 - 1$	$\sum_i \sum_j \sum_k \sum_m Y^2_{ijkm} - \frac{T^2_{....}}{N}$	

EXAMPLE 6.3

Table 6.10 gives results of an experiment for three replications. The experiments were carried out in a 4×4 Latin square relating to the testing of rubber-covered fabric in the Martindale wear tester ([4], p. 64). This machine consists of four rectangular brass plates on each of which is fastened an abrading surface of special quality emery paper. Four weighted bushes into which the test samples of fabric are fixed rest on the emery surface. The mechanical device moves the bushes over the surface of the emery, thus abrading the test specimens. The loss in weight after a given number of cycles is used as criterion of resistance to abrasion. In the actual experiment, four materials A, B, C, D were tested together in each of the four runs of the machine. The entries in Table 6.10 show the loss of weight in 0.1 mg in a run of standard length.

For case I to be possible, it will be assumed that each position in the machine has three heads to hold the test specimen. This is to permit all the three replications to be carried out simultaneously. Furthermore, for case II it will be assumed that only the runs are replicated. For case III it needs to be assumed that each replication is carried out on a different machine.

For the data in Table 6.10 the following results are easily obtained:

$$G = \frac{(11\,301)^2}{(16)(3)} = 2660\,679.188$$

$$T_{i...} = 2761, 3004, 2817, 2719 \quad \text{(row totals)}$$

$$T_{..k.} = 2729, 3010, 2784, 2778 \quad \text{(position totals)}$$

$$T_{.j..} = 3129, 2603, 2853, 2716 \quad \text{(treatment totals)}$$

$$T_{...m} = 3852, 3727, 3722 \quad \text{(replication totals)}$$

$$T_{....} = 11\,301.$$

Table 6.10 Results of wear testing experiment

Replication 1	A251	B241	D227	C229
	D234	C273	A294	B226
	C235	D236	B218	A268
	B195	A270	C230	D225
Replication 2	A232	B236	D211	C213
	D245	C265	A263	B220
	C239	D229	B211	A260
	B209	A258	C225	D211
Replication 3	A247	B235	D218	C221
	D229	C270	A260	B225
	C228	D231	B202	A260
	B185	A266	C225	D220

Table 6.11 ANOVA table for case I

Source	*df*	*SS*	*MS*	*F*
Treatments	3	12 863.73	4287.91	57.757
Blocks	3	3 953.063	1317.688	17.749
Positions	3	3 944.23	1314.74	17.709
Replications	2	678.125	339.063	4.567
Error	36	2 672.664	74.241	
Total	47	24 111.817		

Therefore using equations (6.14)–(6.18), the appropriate sum of squares can be calculated:

$$SS_{\text{total}} = 24\,111.817$$
$$SS_{\text{treatment}} = \tfrac{1}{12}(3129^2 + 2603^2 + 2853^2 + 2716^2) - G = 12\,863.730$$
$$SS_{\text{block}} = \tfrac{1}{12}(2761^2 + 3004^2 + 2817^2 + 2719^2) - G = 3953.063$$
$$SS_{\text{position}} = \tfrac{1}{12}(2729^2 + 3010^2 + 2784^2 + 2778^2) - G = 3944.230$$
$$SS_{\text{replication}} = \tfrac{1}{16}(3852^2 + 3727^2 + 3722^2) - G = 678.125$$
$$SS_{\text{error}} = 24\,111.817 - 12\,863.730 - 3953.063 - 3944.230 - 678.125$$
$$= 2672.664.$$

The ANOVA can be summarized as shown in Table 6.11.

In case II all the sums of squares are the same as in case I except for block and error. The sum of squares for the blocks is calculated using the expression shown in Table 6.8 (considering 12-row totals) and is given by

$$SS_{\text{block}} = 10\,662\,099/4 - G - SS_{\text{replication}} = 4162.437$$
$$SS_{\text{error}} = 24\,111.817 - 12\,863.729 - 4162.437 - 3944.229 - 678.125$$
$$= 2463.297.$$

The ANOVA table for case II is shown in Table 6.12.

Table 6.12 ANOVA table for case II

Source	*df*	*SS*	*MS*	*F*
Treatments	3	12 863.729	4287.91	52.22
Blocks	9	4 162.437	462.492	2.633
Positions	3	3 944.229	1314.74	16.012
Replications	2	678.125	339.063	4.129
Error	30	2 463.297	82.1	
Total	47	24 111.817		

Table 6.13 ANOVA table for case III

Source	*df*	*SS*	*MS*	*F*
Treatments	3	12 863.729	4287.91	52.22
Blocks	9	4 162.437	462.492	5.633
Positions	9	4 442.937	493.66	6.03
Replications	2	678.125	339.063	4.129
Error	24	1 964.588	81.858	
Total	47	24 111.817		

In case III the total sum of squares, the sum of squares for replications, the sum of squares for treatments and the sum of squares for blocks are the same as for case II. The new sums of squares are obtained from Table 6.9, as follows:

$$SS_{\text{position}} = \frac{10\,663\,201}{4} - G - SS_{\text{replication}} = 4442.937$$

$$SS_{\text{error}} = 24\,111.817 - 12\,863.729 - 4162.437 - 4442.937 - 678.125$$
$$= 1964.588.$$

The ANOVA results are shown in Table 6.13.

6.1.3 Missing values

Although the exact solution to the problem of missing observations can be obtained by the general regression significance test, an approximate solution is also possible.

When the missing observations are estimated, the resulting complete Latin square is orthogonal and can be solved in the normal manner explained in section 6.1.1. It must be realized, however, that the number of degrees of freedom of the total is reduced by the number of estimated observations, and that the results are only approximate.

Let Y_{mpq} be the estimate of the missing observations in row m, treatment p and column q, and let R, C and T be the totals of known values in the row, column and treatments containing Y_{mpq}. Additionally, let S be the sum of all known values, and r be the number of rows, columns or treatments.

The sum of squares of the error is given by equation (6.10):

$$SS_{\text{error}} = SS_{\text{total}} - SS_{\text{block}} - SS_{\text{position}} - SS_{\text{treatment}}.$$

The error sum of squares can be rewritten as a function of the missing observation Y_{mpq} as follows:

$$SS_{\text{error}}(Y_{mpq}) = \sum\sum\sum Y_{ijk}^2 + Y_{mpq}^2 + \frac{2(S+Y_{mpq})^2}{r^2} - \frac{[\sum T_{i..}^2 + (Y_{mpq}+R)^2]}{r}$$

$$- \frac{[\sum T_{.j.}^2 + (Y_{mpq}+C)^2]}{r} - \frac{[\sum T_{..k}^2 + (Y_{mpq}+T)^2]}{r}. \tag{6.20}$$

In equation (6.20) all summations are computed for observed values or totals. A necessary condition to minimize the error sum of squares is $\mathrm{d}SS_{\text{error}}/\mathrm{d}Y_{mpq} = 0$. After setting the first derivative equal to zero and solving for Y_{mpq}, the result given in equation (6.21) is obtained:

$$Y_{mpq} = \frac{[r(R+C+T) - 2S]}{[(r-1)(r-2)]}. \tag{6.21}$$

6.1.4 Orthogonal squares

Two squares are said to be orthogonal if after being superimposed the following conditions are achieved:

(a) Each letter of either square occurs exactly once in each row and each column.
(b) Each letter of the first square is paired exactly once in any single cell with each letter of the second square.

As an illustration, consider the two 4 × 4 Latin squares in Tables 6.14 and 6.15. The arrangement in Table 6.16 is obtained after superimposing the two Latin squares. Since each letter of the first square is combined with each letter of the second square exactly once, the given Latin squares are orthogonal. As indicated by Fisher and Yates [1], there are three orthogonal squares of dimension 4 × 4, four orthogonal squares of dimension 5 × 5, none of dimension 6 × 6, and six of dimension 7 × 7, and so on. In Fisher and Yates's publication we can find mutually orthogonal Latin squares of various sizes.

A total of r treatments can be tested using r^2 experimental units grouped according to $m < r$ restrictions on randomization (or ways of grouping) using

Table 6.14

	I	*II*	*III*	*IV*
1	A	B	C	D
2	B	A	D	C
3	C	D	A	B
4	D	C	B	A

Table 6.15

	I	*II*	*III*	*IV*
1	Z	Y	X	W
2	X	W	Z	Y
3	W	X	Y	Z
4	Y	Z	W	X

Table 6.16

	I	*II*	*III*	*IV*
1	AZ	BY	CX	DW
2	BX	AW	DZ	CY
3	CW	DX	AY	BZ
4	DY	CZ	BW	AX

$m-1$ mutually orthogonal Latin squares of size $r \times r$. For example, if $m=3$ and $r=4, 16$ pigs can be classified according to 4 types of feed, 4 litters, 4 weight groups and 4 feeding pens, superimposing two mutually orthogonal Latin squares: (a) one square for treatments A, B, C, D; (b) one square for litters W, X, Y, Z. After this, the columns of the resulting square can be associated with weight groups (I, II, III, IV) and the rows as litters $(1,2,3,4)$. In this case the previous arrangement resulting from superimposing two orthogonal Latin squares can be used.

The ANOVA for this kind of design is carried out by computing in the usual fashion one sum of squares with $r-1$ degrees of freedom for each way of grouping, thus accounting for $m(r-1)$ degrees of freedom in all. The treatment sum of squares has $r-1$ degrees of freedom. The error sum of squares is obtained by subtraction of the above components from the total sum of squares and has $(r-1)(r-m)$ degrees of freedom.

The design resulting from the superimposition of two Latin squares is well known as a Graeco-Latin square design since one square usually contains Latin letters and the other Greek letters. This design will be studied in section 6.2.

6.2 GRAECO-LATIN SQUARES

A Graeco-Latin square is an arrangement of r Greek letters and r Latin letters in a square of r rows and r columns, such that each Greek letter and each Latin letter appears exactly once in each row and each column, and each Greek–Latin combination appears exactly once. With this arrangement, it is possible to test the significance of treatments (Latin letters) under the presence of three restrictions on randomization (rows, columns and Greek letters).

Table 6.17 is an example of a Graeco–Latin square where α, λ, ϕ, δ are the levels of the third restriction. Notice that this is the same type of arrangement as two orthogonal Latin squares, which is necessary if we are to assume the orthogonality of effects.

The ANOVA for a Graeco-Latin square design is based on the following

Table 6.17

	I	*II*	*III*	*IV*
1	Aα	Bλ	Cϕ	Dδ
2	Bϕ	Aδ	Dα	Cλ
3	Cδ	Dϕ	Aλ	Bα
4	Dλ	Cα	Bδ	Aϕ

model:

$$Y_{ijkm} = \mu + \beta_i + \tau_j + \gamma_k + \omega_m + \varepsilon_{ijkm}. \tag{6.22}$$

In equation (6.22), ω_m is the effect of a third restriction on the randomization. The fundamental equation of the ANOVA for this experiment can be written as

$$SS_{\text{total}} = SS_{\text{block}} + SS_{\text{treatment}} + SS_{\text{position}} + SS_{\text{w}} + SS_{\text{error}}. \tag{6.23}$$

Note that equation (6.23) is similar to equation (6.7), with the exception that the SS_{w} term does not appear in equation (6.7). This term can be computed as

$$SS_{\text{w}} = \sum_k \frac{T^2_{\dots m}}{r} - \frac{T^2_{\dots\dots}}{r^2}. \tag{6.24}$$

In equation (6.24), $T_{\dots m}$ is the observed total for the mth Greek letter and $T_{\dots\dots}$ is the observed grand total.

By construction a Graeco-Latin square represents a balanced design. Assuming that the effects due to rows, columns, Latin letters and Greek letters are additive, it is possible to test separately and independently each of the four hypotheses with only r^2 observations instead of r^4 as would be needed in a full design including all level combinations. Let $\beta_i, \tau_j, \gamma_k$ and ω_m be the effect due to the ith row, the jth treatment, the kth column and the mth Greek letter. Using this notation, the hypotheses now to be tested can be formulated as follows:

$$\begin{aligned} H_0&: \tau_j = 0 \quad j = 1, 2, \dots, r \\ H'_0&: \beta_i = 0 \quad i = 1, 2, \dots, r \\ H''_0&: \gamma_k = 0 \quad k = 1, 2, \dots, r \\ H'''_0&: \omega_m = 0 \quad m = 1, 2, \dots, r. \end{aligned}$$

H_0 states that no treatment is significant, H'_0 that no row effect is significant, H''_0 that no column effect is significant and H'''_0 that no Greek letter effect (a third restriction on randomization) is significant.

The Graeco-Latin square mathematical model is derived from the Latin square model by splitting the Latin square's error term (ε_{ijk}) into the ω_m

Table 6.18 ANOVA for a Graeco-Latin square design

Source	*df*	*SS*	*MS*
Treatments	$(r-1)$	$\sum_j \frac{T^2_{.j..}}{r} - \frac{T^2_{..}}{N}$	$\frac{SS_{\text{treatment}}}{(r-1)}$
Blocks	$(r-1)$	$\sum_i \frac{T^2_{i...}}{r} - \frac{T^2_{..}}{N}$	$\frac{SS_{\text{block}}}{(r-1)}$
Positions	$(r-1)$	$\sum_k \frac{T^2_{..k.}}{r} - \frac{T^2_{..}}{N}$	$\frac{SS_{\text{position}}}{(r-1)}$
Greek letters	$(r-1)$	$\sum_m \frac{T^2_{...m}}{r} - \frac{T^2_{..}}{N}$	$\frac{SS_{\text{Greek}}}{(r-1)}$
Error	$(r-1)(r-3)$	$SS_{\text{total}} - SS_{\text{treatment}} - SS_{\text{block}} - SS_{\text{position}} - SS_{\text{Greek}}$	$\frac{SS_{\text{error}}}{(r-1)(r-3)}$
Total	r^2-1	$\sum_i \sum_k Y^2_{ijkm} - \frac{T^2_{..}}{N}$	

effect of the third restriction on randomization and a reduced random error (ε_{ijkm})

The ANOVA for a Graeco-Latin square design can be summarized as in Table 6.18. In this table, $T_{i...}$, $T_{.j..}$, $T_{..k.}$ and $T_{...m}$ are the observed totals for the ith row, the jth treatment (Latin letters), the kth column and the mth Greek letter.

EXAMPLE 6.4

A manufacturer produces control valves by turning brass castings supplied by four outside suppliers. He suspects that one or more of the four turret lathes used to machine the castings is out of tolerance, as the number of out-of-tolerance valves has been increasing; but he does not want to shut down any more lathes than necessary for readjustment and repairs. Thus, he devises the following single-day experiment to determine if there is, in fact, a difference between lathes. As he suspects that the lathe operator, and his state of fatigue, can affect results, he changes each operator to a new lathe every two hours, and also compares the number of defective parts for each two-hour time period. Finally, he wishes to check whether casting

Table 6.19

Lathe	$C\beta(16)$	$B\gamma(12)$	$D\delta(17)$	$A\alpha(11)$
	$B\alpha(15)$	$C\delta(14)$	$A\gamma(15)$	$D\beta(14)$
	$A\delta(12)$	$D\alpha(6)$	$B\beta(14)$	$C\gamma(13)$
	$D\gamma(9)$	$A\beta(9)$	$C\alpha(8)$	$B\delta(9)$

suppliers have any effect on quality. Thus, the four factors under consideration are:

(a) turret lathes (rows)
(b) operators (columns)
(c) two-hour time periods (Greek letters)
(d) suppliers (Latin letters).

The design to be used is the 4 × 4 Graeco-Latin square shown in Table 6.19.

Let Y_{ijkm} be the observation corresponding to the ith lathe, the jth supplier, the kth operator and the mth period. The responses shown in Table 6.19 in parentheses are number of rejects.

The following totals are obtained:

$$T_{i...} = 56, 58, 45, 35 \quad \text{(lathes)}$$
$$T_{.j..} = 47, 50, 51, 46 \quad \text{(suppliers)}$$
$$T_{..k.} = 52, 41, 54, 47 \quad \text{(operators)}$$
$$T_{...m} = 40, 53, 49, 52 \quad \text{(time periods)}$$
$$T_{....} = 194$$
$$\sum_i \sum_k Y_{ijkm}^2 = 2504$$
$$G = \frac{(194)^2}{16} = 2352.25.$$

The ANOVA results are shown in Table 6.20.

Table 6.20 Analysis of variance for Example 6.4, $\alpha = 0.05$

Source	*df*	*SS*	*MS*	*F*	F_c	*Significant?*
Lathes	3	85.25	28.42	7.939	9.28	No
Operators	3	25.25	8.42	2.352	9.28	No
Suppliers	3	4.25	1.42	0.397	9.28	No
Time periods	3	26.25	8.25	2.444	9.28	No
Error	3	10.75	3.58			
Total	15	151.75				

Note that the critical F-value which corresponds to $n_1 = 3, n_2 = 3, \alpha = 0.10$, from an F-table, is 5.39, which makes machine effects significant at the 10% confidence level. Further studies may be warranted if quality control continues to be a problem.

EXAMPLE 6.5

The object of the experiment is to ascertain whether a modified annealing procedure could be introduced into the production of light gauge domestic copper tube [5]. Essential conditions are that the tensile strength of the product should be uniform, with a minimum of 17 ton in^{-2} (262 MPa), which has been achieved hitherto by annealing the tubes fully and then applying a light final draft as shown in Figure 6.1. A possible alternative procedure is to anneal the material at a lower temperature than that required for full annealing, thereby eliminating the final drawing operation. The experiment is designed to investigate this alternative.

In the present method, the purpose of the annealing operation is to soften the material for the final drafting, while in the proposed method the annealing operation will eliminate the imperfections in the surface caused by drafting.

In deciding the form of tests, it was necessary to consider possible causes of variation in the results, including variation in the material itself and variations in temperature over the annealing furnace. Material variations were studied by taking samples of eight tubes at random on each of eight days spread over a period of three weeks, thus allowing normal process variations to be covered adequately.

In order to minimize temperature effects the furnace used was chosen to give as uniform a temperature as possible. The 64 tubes were held in the furnace in a jig having 8 horizontal rows and 8 vertical columns of holes, i.e. an 8 × 8 square (Figure 6.2). Tests were made at various nominal temperatures to determine that most likely to lead to favorable results. The construction of the furnace indicated that temperature variations, if present, would be mainly horizontal or vertical, and no appreciable interaction was

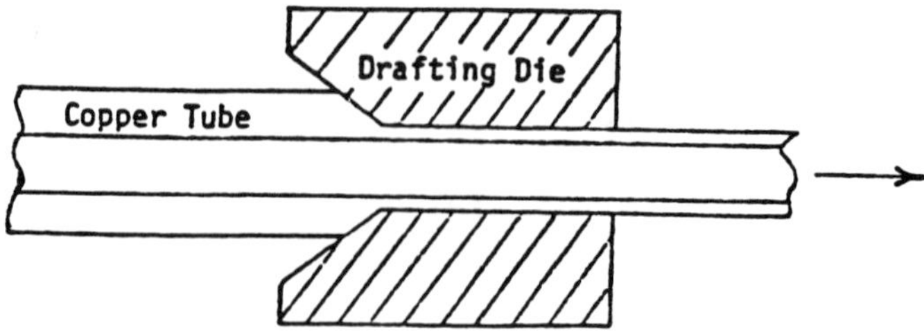

Figure 6.1 Drafting operation.

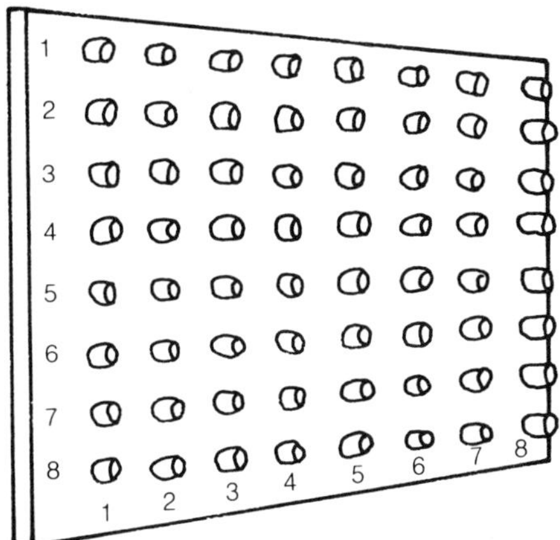

Figure 6.2 Tube-holding jig.

expected between rows and columns. This assumption cannot always be made, because the variation in temperature of some furnaces is radial.

A Graeco-Latin square can be employed in the experiment. In this design 'rows' correspond to the actual rows in the rectangular jig, 'columns' to the actual columns, 'Latin letters' to days to manufacturing (lots) and 'Greek letters' to the number assigned to an individual tube within the sample of eight. The numbers 1 to 8 could, if required, represent order of manufacturing during the day, different pretreatments, or different after-treatments to tubes.

6.3 YOUDEN SQUARES

In general, a Youden square is defined as a balanced incomplete block design in which each treatment appears exactly the same number of times in each position of the blocks. A particularly attractive Youden square design is a symmetrical incomplete block design in which each treatment occurs once and only once in each position of the block. W. J. Youden [6, 7] developed and made extensive use of this design in agricultural experimentation. Although a Youden square with $b = t$ is always a Latin square with one or more rows or columns or diagonals missing, the converse is not always true. A Latin square from which more than one row (column or diagonal) is missing is not necessarily a Youden square since removing rows from a Latin square may destroy the balance. It is, however, possible to construct Youden squares from all symmetrical balanced incomplete block designs. The construction of Youden squares is discussed in section 6.3.2.

Table 6.21 is an example of a Youden square. Note that it corresponds

Table 6.21

Blocks	*Positions* 1	2	3
1	A	B	C
2	D	A	B
3	C	D	A
4	B	C	D

to a Latin square with a missing column. In this case, the missing column has treatments D, C, B and A in blocks 1, 2, 3 and 4, respectively.

Using the notation of Chapter 5 for balanced incomplete block designs, $t = 4$ treatments, $b = 4$ blocks, $k =$ block size of 3, $r = 3$ replications of each treatment and $\lambda = 2$ (times each pair of treatments occurs together).

Situations calling for a Youden square design could arise in cases where one factor has less than r levels, such as a test of four materials with four operators but only three milling machines, or in cases of lost data.

6.3.1 Analysis of variance

The mathematical model describing the experiment can be formulated as

$$Y_{ijk} = \mu + \beta_i + \tau_j + \gamma_k + \varepsilon_{ijk}. \tag{6.25}$$

The analysis of the model given in equation (6.25) can be formally performed in three steps. The first step in the analysis is to consider a symmetrical balanced incomplete block design as in Chapter 5, ignoring the effect of the position in the block. This is possible because every position occurs exactly once in each block and each treatment, so positions are orthogonal to blocks and treatments. The second step is to compute the sum of squares for the positions. The sum of squares for the error in Youden square design is equal to the SS_{error} term coming from the balanced incomplete design minus the sum of squares for positions, which is computed using equation (6.9). The third step is to do an F-test for treatments.

EXAMPLE 6.6

The Texas Transportation Institute is conducting an experiment to investigate the mileage delivered by four brands of gasoline using four drivers and three automobiles. The design chosen is a Youden square where gasoline brands correspond to treatments (A, B, C, D) blocks correspond to drivers (I, II, III, IV) and automobiles correspond to positions (1, 2, 3).

In the Youden square in Table 6.22 the data are coded by subtracting

Table 6.22

		Automobile			
		1	*2*	*3*	$T_{i..}$
Driver	I	A(−1)	B(+1)	C(+2)	+2
	II	B(+2)	C(+1)	D(−2)	+1
	III	C(−3)	D(+2)	A(0)	−1
	IV	D(−1)	A(0)	B(−1)	−2
	$T_{..k}$	−3	+4	−1	$T_{...} = 0$

average gas mileage from each value. Furthermore, this is a balanced incomplete design with $t = b = 4$, $k = r = 3$ and $\lambda = 2$.

The treatment totals ($T_{.j.}$) for brands A, B, C, D are -1, $+2, 0, -1$, respectively. Also $\sum_i \sum_k Y_{ijk}^2 = 30$, $G = 0^2(12)^{-1} = 0$ and $SS_{\text{total}} = 30 - 0(12)^{-1}$. The three steps of the analysis are discussed next.

Step 1

$$SS_{\text{block}} = (\tfrac{1}{3})[(+2)^2 + (+1)^2 + (-1)^2 + (-2)^2] - G$$
$$= \tfrac{10}{3} - 0 = 3.33.$$

The adjusted treatment totals for the given data are

$$Q_A = 3(-1) - (2 - 1 - 2) = -2$$
$$Q_B = 3(+2) - (2 + 1 - 2) = +5$$
$$Q_C = 3(0) - (2 + 1 - 1) = -2$$
$$Q_D = 3(-1) - (1 - 1 - 2) = -1.$$

Using equation (5.44), the following adjusted treatment sum of squares is obtained:

$$SS_{\text{treatment(adjusted)}} = \sum \frac{Q_j^2}{(k\lambda t)}$$
$$= \frac{(-2)^2 + (+5)^2 + (-2)^2 + (-1)^2}{24}$$
$$= 1.42.$$

Step 2

$$SS_{\text{position}} = \frac{(-3)^2 + (4)^2 + (-1)^2}{4} = 6.5$$

$$SS_{\text{error}} = SS_{\text{total}} - SS_{\text{block}} - SS_{\text{treatment(adjusted)}} - SS_{\text{position}}$$
$$= 30.0 - 3.33 - 1.42 - 6.5 = 18.75.$$

Table 6.23 Analysis of variance for Example 6.6, $\alpha = 0.05$

Source	*df*	*SS*	*MS*	*F*	F_c	*Significant?*
Gasoline (adjust)	3	1.42	0.47	0.08	9.28	No
Drivers	3	3.33				
Automobiles	2	6.50	3.25	0.52	9.55	No
Error	3	18.75	6.25			
Total	11	30.00				

Step 3

The ANOVA results are shown in Table 6.23. Thus, neither gasoline nor automobile has a significant effect on gasoline mileage. In practical terms, gasoline can be bought from the lowest price supplier without significantly reducing the mileage. As an exercise, the reader can verify that drivers have no significant effect upon mileage by repeating the analysis (as in Chapter 5), by calculating driver sum of squares adjusted for gasoline and completing the ANOVA.

6.3.2 Construction of Youden squares

A cyclic Latin square is a Latin square in which the first letter of a row is moved to the end of the next row, simultaneously moving all other letters one position to the left. Starting with ABCD in the first row, the second row becomes BCDA, the third row becomes CDAB and the fourth row is DABC (Table 6.24).

To construct a Youden square with $t = 4$ and $k = 3$, any three of the four columns of a cyclic Latin square will provide the required design. A cyclic 7×7 Latin square is shown in Table 6.25. A Youden square with $t = 7$ and $k = 3$ may be constructed from this square by selecting any two adjacent columns and a third column which is one step from being adjacent. The first, second and fourth columns meet these requirements, as do the fourth, sixth and seventh columns (Tables 6.26 and 6.27). Treatments are assigned at

Table 6.24

A	B	C	D
B	C	D	A
C	D	A	B
D	A	B	C

Table 6.25

1	2	3	4	5	6	7
2	3	4	5	6	7	1
3	4	5	6	7	1	2
4	5	6	7	1	2	3
5	6	7	1	2	3	4
6	7	1	2	3	4	5
7	1	2	3	4	5	6

Table 6.26

1	2	4
2	3	5
3	4	6
4	5	7
5	6	1
6	7	2
7	1	3

Table 6.27

4	6	7
5	7	1
6	1	2
7	2	3
1	3	4
2	4	5
3	5	6

random to the numbers, as are blocks to rows and columns within the blocks.

A listing of available Youden squares and the structure of the corresponding plans can be found in Cochran and Cox, pp. 520–44 [8]. Youden [7] has extensively used the concept of these squares in the design of greenhouse experiments.

To bring about complete replications running across the blocks, further orientation of the units within the blocks was considered. This is possible when the number of blocks is equal to the number of the items to be tested and the number of the replications is also equal to the number of units in a block. The columns (and rows) of the squares selected should be free from any systematic pattern. Since the Youden squares can be considered as Latin squares, with two or more rows or columns missing, the rejected rows can be readily utilized and this provides some flexibility in the number of replications.

As pointed out in section 6.3, a Youden square is a balanced, incomplete block design, which can be produced by dividing a Latin square so that the balance is not destroyed.

A Youden square arrangement allows the design analyst to study and exploit the possible advantages of either known or suspected similarities among the various factors.

EXAMPLE 6.7

A Youden square design was used to test five different fabrics A, B, C, D, E in five runs of a Martindale machine. The weight losses in milligrams are given in Table 6.28 where 1, 2, 3 and 4 represent four different positions in the machine.

For this balanced incomplete design, $t = b = 5$, $k = r = 4$, and $\lambda = 3$. The treatment totals ($T_{.j.}$) for brands A, B, C, D, E are 1036, 1034, 1077, 1034, 1059, respectively. Also

$$\sum_i \sum_k Y_{ijk}^2 = 1\,381\,358 \qquad G = \frac{5240^2}{20} = 1\,372\,880$$

Table 6.28

	Positions			
Runs	*1*	*2*	*3*	*4*
1	A268	B249	C280	D271
2	E251	A233	B231	C291
3	D265	E249	A254	B314
4	C256	D250	E270	A281
5	B240	C250	D248	E289

and

$$SS_{\text{total}} = 1\,381\,358 - 1\,372\,880 = 8478.$$

The three steps of the analysis are discussed next.

Step 1

$$\begin{aligned} SS_{\text{block}} &= (\tfrac{1}{4})[(1068)^2 + (1006)^2 + (1082)^2 + (1057)^2 + (1027)^2] - G \\ &= 1\,373\,840.5 - 1\,372\,880 = 960.5. \end{aligned}$$

The adjusted treatment totals are:

$$\begin{aligned} Q_{\text{A}} &= 4(1036) - (1068 + 1006 + 1082 + 1057) = -69 \\ Q_{\text{B}} &= 4(1034) - (1068 + 1006 + 1082 + 1027) = -47 \\ Q_{\text{C}} &= 4(1077) - (1068 + 1006 + 1057 + 1027) = +150 \\ Q_{\text{D}} &= 4(1034) - (1068 + 1082 + 1057 + 1027) = -98 \\ Q_{\text{E}} &= 4(1059) - (1006 + 1082 + 1057 + 1027) = +64. \end{aligned}$$

Adjusted sum of squares for treatments:

$$\begin{aligned} SS_{\text{treatment(adjusted)}} &= \sum \frac{Q_j^2}{(k\lambda t)} \\ &= \frac{(-69)^2 + (-47)^2 + (+150)^2 + (-98)^2 + (+64)^2}{60} \\ &= 719.5. \end{aligned}$$

Step 2

$$SS_{\text{position}} = \frac{(1280)^2 + (1231)^2 + (1283)^2 + (1446)^2}{5} - 1\,372\,880 = 5273.2$$

$$\begin{aligned} SS_{\text{error}} &= SS_{\text{total}} - SS_{\text{block}} - SS_{\text{treatment(adjusted)}} - SS_{\text{position}} \\ &= 8478 - 960.5 - 719.5 - 5273.2 = 1524.8. \end{aligned}$$

Table 6.29 Analysis of variance for Example 6.7, $\alpha = 0.05$

Source	*df*	*SS*	*MS*	*F*	F_c	*Significant?*
Treatment (Adj.)	4	719.5	179.87	0.94	3.84	No
Blocks	4	960.5				
Positions	3	5273.2	1757.73	9.22	4.07	Yes
Error	8	1524.8	190.6			
Total	19	8478				

Step 3

The ANOVA results are shown in Table 6.29. Using a 5% level of significance, $F_{4,8} = 3.84$ and $F_{3,8} = 4.07$. Thus, it can be concluded that the treatments are not significant while the positions are.

EXERCISES

1. The strength of a synthetic fiber is to be tested to determine the effect of three treatments. Some process variation is expected and the time of day is suspected as a variable. Outline each step in the design, the model, its assumptions, a design table and the general procedure.
2. The filtration rate of a liquid product of uniform properties was tested for five filter materials. The pressure and the filter shape were also varied over a range of five conditions. The test results are shown in Table 6.30.

   (a) Verify if the design is orthogonal.
   (b) Conduct an analysis of the effect of pressure, shape and filter material, using $\alpha = 0.05$.

3. For the situation discussed in Exercise 2, assume that the given data correspond to the rearrangement of treatments shown in Table 6.31.

Table 6.30

	Position (shape)				
psig	*1*	*2*	*3*	*4*	*5*
A	f1(5.7)	f2(6.2)	f3(7.2)	f4(3.1)	f5(5.6)
B	f2(6.2)	f1(5.1)	f2(4.7)	f3(6.9)	f4(5.4)
C	f3(3.9)	f5(4.3)	f1(5.2)	f2(5.1)	f3(5.9)
D	f4(4.7)	f4(5.1)	f5(3.9)	f1(6.3)	f2(5.1)
E	f5(5.1)	f3(4.0)	f4(5.9)	f5(4.9)	f1(6.2)

Table 6.31

	Position (shape)				
psig	*1*	*2*	*3*	*4*	*5*
A	f1(5.7)	f3(6.2)	f2(7.2)	f4(3.1)	f5(5.1)
B	f5(5.1)	f2(5.1)	f1(4.7)	f3(5.1)	f4(5.4)
C	f4(3.9)	f1(4.3)	f5(5.2)	f2(6.9)	f3(5.9)
D	f3(4.7)	f5(5.1)	f4(3.9)	f1(6.3)	f2(6.2)
E	f2(6.2)	f4(4.0)	f3(5.9)	f5(4.9)	f1(5.6)

(a) Do an ANOVA significance test using $\alpha = 0.05$.
(b) Ignore shape 5 and analyze the experimental data with only shapes 1–4 as a Youden square.

4. Explain the fundamental differences among Latin squares, Graeco-Latin squares and Youden squares.
5. When are two or more Latin squares said to be orthogonal? Give examples to illustrate your definition with Latin squares of dimension 4 by 4. What is the use of orthogonal squares in experimental design? Again, give examples of typical designs based on orthogonal squares.
6. Prepare a Latin square layout for a varietal trial involving seven varieties of treatments.
7. Briefly, illustrate how to use the general regression significance test on a Latin square design.
8. Four treatments A, B, C, D are investigated; consider the hypothetical data in Table 6.32 for a Youden square, and perform the corresponding ANOVA significance test.
9. Do the ANOVA analysis on the results given in Table 6.33.
10. Do the one-way ANOVA analysis if the variation between means – corresponding to order within blocks – is not significant in Exercise 9. Nothing will be gained by the second classification.
11. A chemical company needs to measure the length of carbon–hydrogen bond in four components. There are four laboratories which can provide

Table 6.32

	Positions		
Blocks	A(23)	B(18)	C(19)
	D(30)	C(21)	A(25)
	B(16)	A(22)	D(27)
	C(24)	D(29)	B(19)

Table 6.33

	Position within blocks				
Blocks	*1*	*2*	*3*	*4*	*5*
1	C (7.6)	E(11.8)	A(17.6)	D (8.8)	B(17.9)
2	B(21.4)	D(12.9)	E(12.4)	C(15.0)	A(20.6)
3	A(16.0)	C (9.7)	D (7.4)	B(18.4)	E(16.6)
4	E(16.0)	B(18.3)	C(23.6)	A(27.4)	D(25.2)
5	D(23.3)	A(30.5)	B(25.8)	E(24.5)	C(26.6)

the required sets of measurements using four techniques. However, due to work load limitations, each laboratory can do only one measurement per day.

(a) Propose an experimental design to achieve the objective of the company in four days.
(b) For your proposed design, define the response, the treatments and any other restrictions on randomization.
(c) List all the assumptions needed for the ANOVA significance test.

12. A taxi company wishes to determine which of four brands of tires gives the greatest mileage before wearing out. One tire of each brand is installed on a single taxi. In order to eliminate car variation, only one taxi will be used in the experiment. Tires are regularly rotated, as the positions of the tires on the car could affect wear rates. The taxi has three drivers, one for each eight-hour shift, who each drive a similar number of miles. For the duration of the test, the taxi will not be assigned to any other drivers.
(a) What is the best type of experimental design for this test situation? Why?
(b) For your proposed design, define the response, the treatments and any other restrictions on randomization.
(c) Explain how you would perform the ANOVA significance test.
(d) Propose other alternative designs, possibly using more cars.

13. A manufacturer ran an experiment to determine the effect of bracket shape, machine type, pellet manufacturer and temperature on the breaking strength of injection-molded brackets. The brackets are produced on four different machines and the plastic pellets are manufactured by four suppliers. As die temperature can vary during a production run, test brackets were made with dies at four temperatures ranging from cool to hot. The corresponding results for the breaking strength in kilograms are given by A, B, C, D; the four temperatures by α, β, γ, δ; the four pellet manufacturers by I, II, III, IV; and the four machines by 1, 2, 3, 4 in Table 6.34 where the shapes are represented.

Table 6.34

	Positions			
	I	*II*	*III*	*IV*
1	Dγ(3.8)	Aβ(3.9)	Cα(4.0)	Bδ(3.4)
2	Bα(4.2)	Cδ(3.9)	Aγ(4.0)	Dβ(5.0)
3	Cβ(4.2)	Bγ(4.2)	Dδ(3.9)	Aα(4.4)
4	Aδ(4.1)	Dα(3.7)	Bβ(3.8)	Cγ(4.0)

14. An industrial engineer is seeking the quickest installation method for speedometer assemblies for a redesigned automobile assembly line. There are four different methods, involving minor design changes in the assembly. Some of the methods require the worker to assume awkward positions on average once per minute in each automobile as it comes down the line, so the engineer suspects that different workers will be able to perform at different rates. He/she therefore tests the method with four workers, and decides to measure their performance over four two-hour time periods throughout the day. The experimental results are shown in Table 6.35.

    (a) Identify the type of experimental design.
    (b) State the mathematical model.
    (c) Perform the ANOVA analysis and discuss the results.

Table 6.35

	Method	*Worker*	*Time (min)*
7.30–9.30	1	4	0.8
	2	1	0.7
	3	2	1.8
	4	3	1.1
9.30–11.30	1	3	0.7
	2	4	0.8
	3	1	1.0
	4	2	1.4
12.00–2.00	1	1	0.5
	2	2	1.0
	3	3	1.1
	4	4	0.9
2.00–4.00	1	2	1.0
	2	3	1.2
	3	4	1.4
	4	1	1.0

Table 6.36

	Run			
Positions	*I*	*II*	*III*	*IV*
1	A 26.1	D 20.0	C 32.4	B 20.7
	27.0	19.0	31.0	19.7
2	B 22.3	A 26.9	D 21.5	C 35.0
	23.1	25.0	22.0	33.8
3	C 29.0	B 22.4	A 26.3	D 22.8
	27.0	22.4	27.5	20.0
4	D 30.6	C 26.8	B 25.2	A 26.4
	31.0	28.9	25.0	27.1

15. The following experiment was designed to find out to what extent a particular type of fabric gave homogeneous results over its surface for a standard wear test. In a single run the test machine could accommodate four samples of fabric, at positions 1, 2, 3 and 4. On a large sheet of fabric four relatively small areas A, B, C and D were marked out at random at different places over the surface. From each area 8 samples were taken, and 16 samples (4 from each area) were compared in the machine. The results shown in Table 6.36, given in milligrams of wear, were obtained for 32 samples, after replicating the experiment twice. Conduct an appropriate ANOVA test to decide if there is statistical evidence to show that the four types of fabric have, or do not have, equal abrasion resistance.

REFERENCES

1. Fisher and Yates (1953) *Statistical Tables for Biological, Agricultural and Medical Research*, 4th edn, Oliver and Boyd, Edinburgh.
2. Federer, Walter T. (1955) *Experimental Design – Theory and Applications*, Macmillan Co., New York.
3. Petersen, H. C. (1977) *Rock Island Railroad Study*, Chicago, Rock Island and Pacific Railroad, Des Moines, Iowa.
4. Chakravarti, I. M., Leha, R. G. and Roy, J. (1967) *Handbook of Methods of Applied Statistics*, Vol. II, John Wiley & Sons, Inc., New York.
5. Davies, Owen L. *et al.* (1956) *The Design and Analysis of Industrial Experiments*, 2nd edn, Imperial Chemical Industries Ltd, London.
6. Youden, W. J. (1937) Use of incomplete block replications in estimating tobacco-mosaic virus. *Contr. Boyce Thompson Institute*, **9**, 41–8.
7. Youden, W. J. (1940) Experimental designs to increase accuracy of greenhouse studies. *Contr. Boyce Thompson Institute*, **11**, 219–28.

8. Cochran, W. G. and Cox, G. M. (1957) *Experimental Designs*, 2nd edn, John Wiley & Sons, Inc., New York.

FURTHER READING

Hicks, Charles R. (1982) *Fundamental Concepts in the Design of Experiments*, 3rd Edn, Holt, Rinehart and Winston, New York.

Mandl, Robert (1986) Orthogonal Latin squares: an application of experimental design to compiler testing. *Communications of the ACM*, Oct. 1985, **28**.

Montgomery, D. C. (1991) *Design and Analysis of Experiments*, 3rd edn, John Wiley & Sons, New York.

Factorial experiments

7

In Chapter 3 a totally randomized design was used to test the significance of a simple factor. The data for this experiment were arranged according to a one-way classification table with its columns corresponding to the levels of the factor. As more restrictions on randomization were added, the need arose for regrouping the data according to new ways of classification and developing new experimental designs, as was shown in Chapters 5 and 6.

The purpose of a factorial experiment is to study the effects of applying simultaneously all possible level combinations of a given number of factors. For example, in an experiment on sugar beet cultivation two important factors affecting the yield are the amount of ammonium sulfate per acre and the time of sowing. For the first factor the following levels may be chosen: (a) none, (b) 2 cwt acre^{-1} (250 kg ha^{-1}), (c) 3 cwt acre^{-1} (375 kg ha^{-1}); similarly, for the second factor two different sowing times can be chosen: (a) early May, (b) late May. The number of experimental conditions associated with this experiment is equal to $3\times 2=6$.

The methodology developed in this chapter assumes that the yields associated with the six experimental conditions are collected at random. Depending on the resources available in each particular situation, the experimental conditions may or may not be replicated. In the first case an ANOVA test can be conducted only for each main effect, since the total number of degrees of freedom is equal to 5 and the main effects of both factors will use 3 degrees of freedom (2 for the dose of ammonium sulfate and one for the date of sowing). This would leave only 2 degrees of freedom for the error. On the other hand, if the experimental conditions are each replicated twice, then the total number of degrees of freedom would be equal to 11, now leaving 8 degrees of freedom for the interaction effect and the experimental error. Of these 8 degrees of freedom the interaction effect test needs 2 degrees of freedom ($2\times 1=2$). Therefore, 5 degrees of freedom are left for the experimental error. These basic ideas are now more precisely considered in section 7.1.

7.1 INTRODUCTION

In the designs discussed in Chapters 3–6, there was a one-to-one correspondence between the treatments of the experiment and the levels of the single factor. In Chapter 7 we are interested in a more general design with several factors at different numbers of levels. Since in this new factorial design, a treatment corresponds to an experimental condition or level combination, the experiment is usually referred to as a **factorial experiment**.

The total number of treatments or experimental conditions depends on both the number of factors and the number of levels of each factor. If N and L_i are the number of factors and the number of levels of factor i, respectively, the number of experimental conditions (also referred to as treatments or cells) is equal to $L_1 L_2 \cdots L_N$. For this reason, the experiment is generally referred to as an $L_1 \times L_2 \times \cdots \times L_N$ factorial experiment.

The experimental data associated with a factorial design with N factors can generally be classified according to an N-way table with each cell corresponding to a treatment or experimental condition. If each cell has n observations, the total number of observations required for the analysis of the effects involved in the experiment is equal to $M = nL_1 L_2 \cdots L_N$. As will be seen later, we need more than one observation per cell in order to investigate the possible effects of interactions among the factors of the experiment.

It is also assumed that the order of experimentation is randomized. This can be achieved by selecting the cells (treatments or experimental conditions) at random every time an observation is going to be made. If the cell is already 'filled up', we disregard this result, choose another cell and continue in this fashion until all observations are collected.

In a one-factor-at-a-time type of experiment use is made of treatments that correspond to 'standard levels' of each factor. In this type of experiment we use treatments which differ from the control in the level of only one factor at a time. In a factorial experiment, alternatively, all experimental conditions are considered. The advantages of using factorial experiments are:

(a) They are more efficient than one-factor-at-a-time experiments.
(b) All data are used in the computation of effects.
(c) Interaction effects can be investigated.

In this chapter a collection of cells is defined as a **region**. A simple numerical example is now considered to illustrate the meaning of cells and regions. Suppose that there are three factors X_1, X_2 and X_3. The number of levels of the first factor is 2; for the second factor it is 2; and for the third factor this number is 3. Therefore, the total number of level combinations or cells is equal to $2 \times 2 \times 3 = 12$. These 12 cells can be arranged in standard order as shown in Table 7.1, where the levels for factors X_1 and X_2 are indicated by the numbers 1 and 2, and the levels for factor X_3 by the numbers 1, 2, 3.

As an example, from Table 7.1 it can be seen that the regions consisting of cells belonging to levels 1, 2, 3 of factor X_3 are $\{1, 2, 3, 4\}$, $\{5, 6, 7, 8\}$ and $\{9, 10, 11, 12\}$, respectively. Furthermore, the regions consisting of cells

Table 7.1 Cells in standard order for a $2 \times 2 \times 3$ experiment

Cell	X_1	X_2	X_3
1	1	1	1
2	2	1	1
3	1	2	1
4	2	2	1
5	1	1	2
6	2	1	2
7	1	2	2
8	2	2	2
9	1	1	3
10	2	1	3
11	1	2	3
12	2	2	3

belonging to the six combinations of the levels of factors X_1 and X_3 are $\{1, 3\}$, $\{2, 4\}$, $\{5, 7\}$, $\{6, 8\}$, $\{9, 11\}$ and $\{10, 12\}$.

The importance of identifying the cells in a region will be demonstrated later in section 7.2. As an initial step in this direction, let us assume that we are interested in the main effect of a given factor. If the experimental data are rearranged as a one-way classification table using the levels of the factor under consideration as categories, the sum of squares of this factor can be estimated following the method given in Chapter 3. In order to compute the sum of squared totals, however, we need to know which observations are included in each total. Since the observations must come only from the cells corresponding to each level of the factor under consideration, we must be able to identify in advance the region which corresponds to each level. Proceeding in this fashion it is possible to calculate sums of squares associated with the region belonging to single levels for any factors, and level combinations for the interactions of all relevant groups of factors, as shown in Example 7.1.

7.2 ANALYSIS OF VARIANCE

Let $Y_{i(c)}$ be the ith observation in cell c. If the cells are to be investigated, a suitable model which can be used for randomized order of experimentation is given by

$$Y_{i(c)} = \mu + \tau_{(c)} + \varepsilon_{i(c)} \tag{7.1}$$

where μ is the common effect in cell c, $\tau_{(c)}$ is the effect of the treatment corresponding to cell c, and $\varepsilon_{i(c)}$ is the random error. Here $i = 1, \ldots, n$ and $c = 1, 2, \ldots, \Pi L_i$.

Considering the model formulated in equation (7.1) it is possible to develop estimates for the variation of the experiment. These estimates are the MS of treatments and error. Since the numbers of degrees of freedom are known, the problem is really the identification of the sums of squares:

$$SS_{\text{total}} = SS^* - G$$

$$SS_{\text{treatment}} = \sum_c \frac{T_{(c)}{}^2}{n} - G$$

$$SS_{\text{error}} = SS_{\text{total}} - SS_{\text{treatment}}$$

where SS^* is the sum of all squared observations and G is the correction term $\sum_c T_{(c)}^2 M^{-1}$. Now suppose that the effect of treatments is not of direct interest, but that the effects of factors and interactions are the objective of the experiment. Therefore, we must break down the treatments into factors and interactions, and use the following mathematical model to describe the experiment:

$$Y_{i(c)} = \mu + \sum_{\substack{\text{main}\\ \text{effects}}} \text{sources} + \sum_{\text{interactions}} \text{sources} + \varepsilon_{i(c)}.$$

For this model, the 'fundamental equation' of ANOVA can be written as

$$SS_{\text{total}} = \sum_{\substack{\text{main}\\ \text{effects}}} SS_{\text{sources}} + \sum_{\text{interactions}} SS_{\text{sources}} + SS_{\text{error}}.$$

As an exercise, the reader can develop the ANOVA equation for the case of two factors.

It is noted that the number of degrees of freedom associated with the sum of squares of the error is equal to the total number of degrees of freedom minus the number of degrees of freedom of the treatments, that is

$$\begin{aligned} df_{\text{error}} &= (nL_1 L_2 \cdots L_N - 1) - (L_1 L_2 \cdots L_N - 1) \\ &= (L_1 L_2 \cdots L_N)(n - 1). \end{aligned}$$

Therefore, n must be larger than unity if we want to test the significance of all sources of variation (main effects and interactions).

Two procedures are now discussed. Procedure A can be followed to find the sum of squares of a main effect (factor effect). Procedure B can be followed to find the sum of squares of an interaction effect.

7.2.1 Procedure A: analysis of main effects

(a) Let R_{ij} be the region consisting of observations in cells corresponding to the jth level of factor $i, j = 1, 2, \ldots, L_i$.

(b) Let n_{ij} be the number of observations of factor i at level j, $j = 1, 2, \ldots, L_i$.
(c) Let T_{ij} be the total of observations in R_{ij}, $j = 1, 2, \ldots, L_i$.
(d) The sum of squares of factor i is therefore given by equation (7.2):

$$SS_i = \sum_{(R_{ij})} \frac{T_{ij}^2}{n_{ij}} - G. \tag{7.2}$$

7.2.2 Procedure B: analysis of interaction effects

(a) Considering only the factors in the given interaction, identify the different level combinations for these factors. Redefine R_{ij} as the region consisting of observations in cells associated with the jth level combination of the ith interaction. It is assumed that the interactions are listed in increasing number of terms. That is, all the second-order interactions first, then the third-order interactions, and so on. Usually not more than three terms are included in the interactions.
(b) Follow steps (b) and (c) of procedure A, replacing 'factor' by 'interaction' and 'level' by 'level combination'.
(c) From the *SS* value obtained using equation (7.2) of procedure A subtract all the sums of squares of the main effects and lower-order interaction effects that can be constructed from the terms of the given interaction. The result is the sum of squares of the interaction.

EXAMPLE 7.1

The effect of back rake angle, rotary speed and thrust on the performance of a center vacuum bit design can be investigated through the use of a

Table 7.2 Data for a 2^3 design for drill bit study

				Penetration rate (in min^{-1})		
Experimental conditions	*R Rake angle*	*S Rotary speed*	*L Thurst*	*Observation 1*	*Observation 2*	*Total*
1	–	–	–	4.00	1.09	5.09
2	+	–	–	4.80	3.43	8.23
3	–	+	–	12.54	5.00	17.54
4	+	+	–	5.62	5.14	10.76
5	–	–	+	6.75	8.18	14.93
6	+	–	+	16.67	22.17	38.84
7	–	+	–	6.25	20.00	26.25
8	+	+	+	26.25	20.77	47.02

factorial experiment. In Chapter 1 we consider multiple performance criteria including penetration rate, specific energy and torque (section 1.6). The purpose of this example is to illustrate how the procedures developed in Chapter 7 can be applied to do the ANOVA for penetration rate reported in Table 1.2 [1–3]. The experimental data used in this analysis are shown in Table 7.2; as can be verified, these data were previously summarized in Table 1.1.

Solution

For the experimental results given in Table 7.2, the sum of the $M = 16$ squared observations $SS^* = 2733.64$ and the grand total of these observations is $T = 168.66$. Therefore

$$\begin{aligned} SS_{\text{total}} &= SS^* - \frac{T^2}{M} \\ &= 2733.64 - \frac{(168.66)^2}{16} \\ &= 955.75. \end{aligned}$$

As an illustration of procedure A the sum of squares associated with factor R (rake angle) will be considered. It is assumed that $i = 1$ for factor R, $i = 2$ for factor S (speed) and $i = 3$ for factor L (thrust). For factor R there are two regions associated with levels $-(j = 1)$ and $+(j = 2)$. A summary of the steps taken is as follows:

(a) $R_{11} = \{1, 3, 5, 7\}$
$R_{12} = \{2, 4, 6, 8\}$
(b) $n_{11} = n_{12} = 8$.
(c) $T_{11} = 5.09 + 17.54 + 14.93 + 26.25 = 63.81$
$T_{12} = 8.23 + 10.76 + 38.84 + 47.02 = 104.85$.
(d) $SS_R = \left(\frac{1}{8}\right)[(63.81)^2 + (104.85)^2] - \frac{(168.66)^2}{16} = 105.27.$

It is noted that the number of independent squares in the computation of SS_R is $2 - 1$. Therefore, the degree of freedom associated with SS_R is 1 and $MS_R = 105.27$.

Following procedure A again, the sums of squares of rotary speed (S) and thrust (L) can be found to be equal to

$$SS_S = 74.30$$
$$SS_L = 456.04$$

with one degree of freedom for each sum of squares.

As an illustration of procedure B, we will find the sum of squares of the interaction between R and S. Assume that the interactions are listed in this order: RS, RL, SL, RSL. Then we can refer to RS as the first interaction ($i=1$), RL as the second interaction ($i=2$), and so on. A summary of the results follows:

(a) $R_{11}=\{1,5\}$, $R_{12}=\{2,6\}$, $R_{13}=\{3,7\}$, $R_{14}=\{4,8\}$.
(b) $n_{11}=n_{12}=n_{13}=n_{14}=4$
$T_{11}=5.09+14.93=20.02$
$T_{12}=8.23+38.84=47.07$
$T_{13}=17.54+26.25=43.79$
$T_{14}=10.76+47.02=57.78$.
(c) From procedure A, we obtain SS_{RS} (procedure A) as follows:

$$SS_{RS}\,(\text{procedure A})=\tfrac{1}{4}[(20.02)^2+(47.07)^2+(43.79)^2+(57.78)^2]-\frac{(168.66)^2}{16}$$
$$=190.23.$$

Since SS_{RS} (procedure A) $=SS_{RS}+SS_R+SS_S$, the sum of squares associated with the interaction between R and S is given by

$$SS_{RS}=190.23-105.27-74.30=10.66.$$

It is noted that the number of independent squares in the computation of SS_{RS} (procedure A) is $4-1=3$. Since each SS_R and SS_S has one degree of freedom, the number of degrees of freedom associated with SS_{RS} is $3-2=1$.

Following procedure B again, the sums of squares of the interaction between R and L and the interaction between S and L can be found to be equal to

$$SS_{RL}=145.93$$
$$SS_{SL}=1.28$$

each with one degree of freedom.

Finally, consider the three-factor interaction RSL. Procedure B can be used as follows:

(a) Refer to this interaction as the fourth interaction ($i=4$). For this interaction there are eight level combinations. Thus $R_{4j}=\{j\}$.
(b) $n_{4j}=2$, $j=1,\ldots,8$.
$T_{41}=5.09$, $T_{42}=8.23$, $T_{43}=17.54$, $T_{44}=10.76$, $T_{45}=14.93$, $T_{46}=38.84$, $T_{47}=26.25$, $T_{48}=47.02$.
(c) Here we obtain

$$SS_{RSL}\,(\text{procedure A})=\tfrac{1}{2}[(5.09)^2+(8.23)^2+\cdots+(47.02)^2]-\frac{(168.66)^2}{16}$$
$$=796.34.$$

Table 7.3 ANOVA table for penetration rate

Source	*df*	*SS*	*MS*	*F*	*Significant*
R	1	105.27	105.27	5.28	
S	1	74.30	74.30	3.73	
RS	1	10.66	10.66	0.54	
L	1	456.04	456.04	23.04	*
RL	1	145.93	145.93	7.32	*
SL	1	1.28	1.28	0.06	
RSL	1	2.87	2.87	0.14	
Error	8	159.41	19.93		
Total	15	955.75			

Therefore

$$SS_{RSL} = SS_{RSL}\,(\text{procedure A}) - SS_R - SS_S - SS_L - SS_{RS} - SS_{RL} - SS_{SL}$$
$$= 2.87.$$

There are seven degrees of freedom associated with SS_{RSL} (procedure A); since each of the six independent sums of squares $SS_R, SS_L, \ldots, SS_{SL}$ included in this term has one degree of freedom, SS_{RSL} must have one degree of freedom. Finally, the error sum of squares can be obtained as

$$SS_{\text{error}} = SS_{\text{total}} - SS_R - SS_S - SS_L - SS_{RS} - SS_{RL} - SS_{SL} - SS_{RSL}$$
$$= 159.41.$$

The number of degrees of freedom associated with SS_{error} is equal to eight. Therefore, $MS_{\text{error}} = 159.41(8)^{-1} = 19.93$. The ANOVA results are summarized in Table 7.3 for a level of significance equal to 0.05. As can be verified, these results are exactly the same results obtained in Table 1.2. Significant effects are identified by an asterisk in the last column of Table 7.3.

7.3 NORMAL EQUATIONS APPROACH

The procedure in this section is the same as the procedure presented in section 5.6 for the randomized block design with interactions. As an exercise, the student is asked to solve the above problem using the method of normal equations.

EXERCISES

1. A grain dryer is going to be tested to determine the effects of voltage in the electric circuit and fan speed on the drying time. Two voltages and

three speeds were chosen for a completely randomized test. After replicating each experimental condition twice, the results shown in Table 7.4 were obtained for the drying time (in minutes).

(a) Conduct an ANOVA study to test the effect of voltage, speed, and their interaction.

(b) Use a graphical approach to support your conclusions.

2. Consider the experimental results given in section 1.6. Use the ANOVA methodology to study the effect of rake angle, rotary speed, thrust, and their interactions, on specific energy and torque.

3. An oxidation process was investigated to determine the effect of the nitric acid addition time, blending time and the residue from the previous batch. A two-level factorial design was completely randomized and the results shown in Table 7.5 were obtained for the eight experimental conditions.

(a) Estimate the main effects and the interaction effects.

(b) Conduct the ANOVA and interpret your findings.

4. The object of an experiment is to study thrust forces in drilling at different speeds, feeds, and in different materials. Five speeds, three feeds and two materials were used and two samples were tested for each experimental condition. The order of experimentation was completely at random. The results shown in Table 7.6 were obtained for thrust forces.

(a) Write a mathematical model for the statistical analysis of the given data.

Table 7.4

	Revolutions per minute		
Volts	*600*	*900*	*1500*
110	48	28	7
	58	32	16
220	61	15	7
	53	11	9

Table 7.5

	No Residue HNO_3 Time (h)		*Residue addition Time (h)*	
Blending time (h)	*1*	*3*	*1*	*3*
1/4	43.0	42.2	39.2	45.1
2	41.2	41.5	42.6	41.7

Table 7.6

	Feed	Speed 100	220	475	715	870
Material 1	0.004	322	308	308	266	280
		310	285	260	250	260
	0.008	532	476	448	448	476
		530	510	495	475	510
	0.014	840	812	743	812	896
		700	700	650	810	810
Material 2	0.004	392	336	322	308	336
		370	330	285	275	275
	0.008	686	533	518	672	699
		565	530	530	550	590
	0.014	1010	979	1010	1093	2020
		925	870	950	1090	1090

(b) Code the data by subtracting 200 from each entry, and perform the corresponding analysis.

5. Solve Exercise 4 using the least-squares significance test.
6. The results given in Table 7.7 were obtained from an experiment to test the main effects and interaction. The two factors are the content of nickel and the content of manganese. The experimental response is the breaking strength (in ft-lb) of a certain product. Conduct the corresponding ANOVA tests.
7. The data in Table 7.8 came from a completely randomized experiment to compare the reflective properties of four kinds of paint. Only 19 test specimens were available, so that one treatment had to be replicated only four times. The test specimens were allotted to the paints at random, and the experimental response was obtained by using an optical instrument.

Table 7.7

Nickel	Manganese (%) 1.0	2.0	3.0	4.0
0.0	50.9	51.2	53.2	50.6
	51.3	50.0	53.5	51.0
1.5	50.3	50.7	54.2	52.0
	50.0	50.7	55.0	51.5
3.0	46.5	47.2	49.8	48.3
	47.0	47.0	50.0	49.0

Table 7.8

P1	*P2*	*P3*	*P4*
195	45	195	120
150	40	230	55
205	195	115	50
110	65	235	80
160	145	225	
820	490	1000	305

Peform a significance test (use a 5% level of significance) using the general least-squares procedure. Write the normal equations, and solve them under a condition that will make the solution a legitimate set of estimators. Indicate which parameters are estimated by the corresponding estimators.

8. Consider the data in Table 7.9, which appear in a paper by Box and Cox [4]. These are survival times of groups of four animals randomly allocated to three poisons (I, II, III) and four treatments (A, B, C, D). The experiment was part of an investigation to combat the effects of certain toxic agents. Perform an appropriate analysis of variance and summarize your results in an ANOVA table. Code the experimental data by multiplying each observation by 100.

Table 7.9

	A	*B*	*C*	*D*
I	0.31	0.82	0.43	0.45
	0.45	1.10	0.45	0.71
	0.46	0.88	0.63	0.66
	0.43	0.72	0.76	0.62
II	0.36	0.92	0.44	0.56
	0.29	0.61	0.35	1.02
	0.40	0.49	0.31	0.71
	0.23	1.24	0.40	0.38
III	0.22	0.30	0.23	0.30
	0.21	0.37	0.25	0.36
	0.18	0.38	0.24	0.31
	0.23	0.29	0.22	0.33

REFERENCES

1. Atkins, B. C. (1982) Drilling applications successes using STRATAPAX blank bits in mining and construction. Paper presented at the Australian Drilling Association Symposium, January 18–20, 1982.
2. Hough, C. L., Obioha, M. O. and Garcia-Diaz, A. (1983) Performance, rake angle and wear test analysis of advance technology coal mine roof bolt drill bits. *Contractor report SAND83-7006*, November, Sandia National Laboratories Albuquerque New Mexico.
3. Hough, C. L. (1986) The effect of back rake angle on the performance of small diameter polycrystalline diamond rock bits: ANOVA test. *ASME Journal of Energy, Resources Technology*, **198**, 305–9, December.
5. Box, G. E. P and Cox, D. R. (1964) An analysis of transformations. *J. Roy. Stat. Soc.*, **B26**, 211.

Two-level factorial experiments

8

In Chapter 7 a methodology was developed for deciding if the main effects and interaction effects are significant in a system whose experimental response is affected by multiple factors. In order to test the significance of the various effects, a complete factorial experiment was designed and run with an equal number of replications for each experimental condition. One limitation of the general factorial experiment discussed in Chapter 7 is that no single estimate was actually obtained to quantify the effect due to a main factor or an interaction. Instead, only a statistical test was conducted to verify whether the effects are negligible or not.

In this new chapter a special case of the multi-factor experiment is considered: the particular simplification of the factorial experiment is that only two levels are allowed for each factor; these two levels may indicate two possible numerical values for quantitative factors or two possible non-numerical choices for qualitative factors. The special attractive feature of two-level factorial experiments is that single quantitative estimates can be obtained for each effect to supplement the results of the significant tests. Furthermore, confidence intervals can be found for the true effects from the effect estimates themselves.

A vast number of experiments conducted in the fields of chemistry, physics and engineering are designed to study the effects of a group of two-level factors as well as their interactions. The two levels of each factor correspond to numerical values in the case of quantitative factors and to two choices in the case of qualitative factors. These experiments are intended to determine the main and interaction effects on a particular experimental response, such as yield of a chemical process, amount of a product, performance of a machine, relative strength of rubber, fuel consumption of a process, service life of a tool, and so on. In the experiment on sugar beet cultivation introduced in Chapter 7, we can have a two-level factorial experiment if we try two different doses of ammonium sulfate (0 and 2 cwt acre^{-1} (250 kg ha^{-1})) and two different dates of sowing (early May and late May). In this case there are $2 \times 2 = 4$ possible combinations or treatments. One of the treatments can be used as a control, usually the combination of standard levels.

It will be shown in this chapter that the relevant effects can be estimated from the experimental results following a simple process that involves only additions and subtractions.

This chapter is divided into six sections; the first section gives basic notation and definitions; the second shows a geometric representation of two-level designs with two and three factors; the third develops the analytical methodology to estimate main and interaction effects; the fourth presents the ANOVA methodology for the 2^n experiment and the procedure for finding confidence intervals; the fifth describes an algorithm due to Yates [1] for estimating effects in 2^n factorial designs; the last section illustrates the use of 2^n factorial experiments in the study of weld strength [2, 3].

8.1 DEFINITIONS AND NOTATION

Consider a factorial experiment involving n factors $X_1, X_2, \ldots, X_n$, each having two levels indicated by + and −. The number of experimental conditions or treatments for which data need to be collected is equal to 2^n; for this reason, two-level factorial experiments are usually referred to as **2^n factorial designs**. Although the order of actual running should, of course, be randomized, the 2^n experimental conditions are usually rearranged in the standard or normal order indicated by Table 8.1. This table is known as the **design matrix** of a 2^n factorial experiment.

The entries of the design matrix can be interpreted as coded values of −1 or +1. Each row of the design matrix corresponds to an experimental condition. In each experimental condition a value of −1 indicates that the corresponding factor is set at its 'low level' if it is a quantitative factor; similarly, a coded value of +1 indicates that the factor is at its 'high level'. In the case of qualitative factors the coded values −1 and +1 represent the choices available for each factor.

Table 8.1 Design matrix of a 2^n factorial experiment

Treatment	X_1	X_2	X_3	⋯	X_n
1	−	−	−	⋯	−
2	+	−	−	⋯	−
3	−	+	−	⋯	−
4	+	+	−	⋯	−
⋮	⋮	⋮	⋮	⋯	⋮
2^n	+	+	+	⋯	+

Let r be the number of replications for each of the 2^n treatments or experimental conditions. It is assumed that:

(a) The $r2^n$ observations are collected at random.
(b) The observations corresponding to the cth experimental condition came from a normal population with mean $\mu_{(c)}$ and variance σ^2.
(c) All 2^n normal distributions are independent.

Let $T_{(c)}$ be the total of the r observations collected for the cth experimental condition; therefore, an estimate of $\mu_{(c)}$ is given by $\bar{Y}_{(c)} = T_{(c)}r^{-1}$.

Definition: A linear combination

$$L = \sum_c a_{(c)} \bar{Y}_{(c)}$$

is defined as a contrast if $\sum_c a_{(c)} = 0$. Two contrasts $L_1 = \sum_c a_{(c)} \bar{Y}_{(c)}$ and $L_2 = \sum_c b_{(c)} \bar{Y}_{(c)}$ are said to be orthogonal if $\sum_c a_{(c)} b_{(c)} = 0$.

Definition: A contrast $L = \sum_c a_{(c)} \bar{Y}_{(c)}$ belongs to the main effect of factor X_i if its coefficients $a_{(c)}$ are independent of the levels of factors other than X_i. As a result of this, the coefficients of all treatment means in which factor X_i occurs at a given level are the same.

Definition: A contrast $L = \sum_c a_{(c)} \bar{Y}_{(c)}$ belongs to the interaction between factors X_i and X_j if (a) the coefficients $a_{(c)}$ are independent of the levels of factors other than X_i and X_j; (b) the contrast is orthogonal to all contrasts belonging to the main effects of X_i and X_j.

Definition: A contrast $L = \sum_c a_{(c)} \bar{Y}_{(c)}$ belongs to the mth factors interaction $X_1 X_2 \cdots X_m$ if (a) the coefficients $a_{(c)}$ are independent of the levels of factors other than $X_1, X_2, \ldots, X_m$; and (b) the contrast is orthogonal to all contrasts belonging to the main effects of factors $X_1, X_2, \ldots, X_m$, as well as to all contrasts belonging to the k-factor interactions of subgroups of k factors, for $k = 2, \ldots, m-1$.

The above definitions will now be illustrated by considering a 2^3 factorial design; the total number of effects that can be tested by using the design is equal to seven: three main effects, three two-factor interactions and one three-factor interaction. Table 8.2 shows the design matrix and the coefficients of the contrasts belonging to all the effects associated with a 2^3 design. In this table the coefficients for the interactions have been generated using the algebraic relationships $(+)(+) = (-)(-) = (+)$ and $(-)(+) = (+)(-) = (-)$. It is formally understood that $+$ and $-$ actually represent the values $+1$ and -1 respectively.

Using the basic definitions given in this section, it can be verified that the following contrasts L_1, L_2, L_3 and L_4 belong to the effects of X_1, X_2, X_1X_3

Table 8.2 Design matrix and coefficients for orthogonal contrasts belonging to main and interaction effects of a 2^3 experiment

Treatment	X_1	X_2	X_3	X_1X_2	X_1X_3	X_2X_3	$X_1X_2X_3$
1	−	−	−	+	+	+	−
2	+	−	−	−	−	+	+
3	−	+	−	−	+	−	+
4	+	+	−	+	−	−	−
5	−	−	+	+	−	−	+
6	+	−	+	−	+	−	−
7	−	+	+	−	−	+	−
8	+	+	+	+	+	+	+

and $X_1X_2X_3$, respectively:

$$L_1 = -\bar{Y}_{(1)} + \bar{Y}_{(2)} - \bar{Y}_{(3)} + \bar{Y}_{(4)} - \bar{Y}_{(5)} + \bar{Y}_{(6)} - \bar{Y}_{(7)} + \bar{Y}_{(8)}$$
$$L_2 = -\bar{Y}_{(1)} - \bar{Y}_{(2)} + \bar{Y}_{(3)} + \bar{Y}_{(4)} - \bar{Y}_{(5)} - \bar{Y}_{(6)} + \bar{Y}_{(7)} + \bar{Y}_{(8)}$$
$$L_3 = +\bar{Y}_{(1)} - \bar{Y}_{(2)} + \bar{Y}_{(3)} - \bar{Y}_{(4)} - \bar{Y}_{(5)} + \bar{Y}_{(6)} - \bar{Y}_{(7)} + \bar{Y}_{(8)}$$
$$L_4 = -\bar{Y}_{(1)} + \bar{Y}_{(2)} + \bar{Y}_{(3)} - \bar{Y}_{(4)} + \bar{Y}_{(5)} - \bar{Y}_{(6)} - \bar{Y}_{(7)} + \bar{Y}_{(8)}.$$

8.2 GEOMETRIC REPRESENTATION OF 2^2 AND 2^3 DESIGNS

A 2^2 factorial design will be considered first. The four experimental conditions of this design are listed in standard order in Table 8.3, where it is assumed that the total of each experimental condition corresponds to r random observations.

As can be observed in Table 8.3, the total of the r observations corresponding to the experimental condition defined by the coded level $X_1 = -1$ and $X_2 = -1$ is represented by the symbol (1); the total associated with the experimental condition defined by the coded levels $X_1 = +1$ and $X_2 = -1$ is represented by the symbol a; and so on. Although this notation looks confusing it is used in a good number of textbooks because it is actually

Table 8.3 Design matrix for a 2^2 factorial experiment

Condition	X_1	X_2	*Total*
1	−	−	(1)
2	+	−	a
3	−	+	b
4	+	+	ab

quite simple and functional. If the letters a and b are associated with the + level of factors X_1 and X_2, respectively, then the symbols a, b and ab can be used to represent the totals corresponding to the experimental conditions where only $X_1 = 1, X_2 = 1$ and $X_1 = X_2 = 1$. In this notation, the symbol (1) is used to represent the total of the experimental condition where all variables are set at the − level (that is, $X_1 = -1$ and $X_2 = -1$).

If we consider a 2^3 factorial experiment, according to the above notation, the symbols a, b and c would be associated with factors X_1, X_2 and X_3, respectively. If the symbols representing the totals corresponding to the experimental conditions of the 2^2 design shown in Table 8.3 (in standard order) are 'multiplied' by the symbol c associated with X_3, the following sequence of symbols is generated: c, ac, bc, abc. Table 8.4 shows the design matrix of the 2^3 factorial experiment with the eight experimental conditions arranged in standard order. It is noted that the total symbols included in

Table 8.4 Design matrix for 2^3 factorial experiment

Condition	X_1	X_2	X_3	*Total*
1	−	−	−	(1)
2	+	−	−	a
3	−	+	−	b
4	+	+	−	ab
5	−	−	+	c
6	+	−	+	ac
7	−	+	+	bc
8	+	+	+	abc

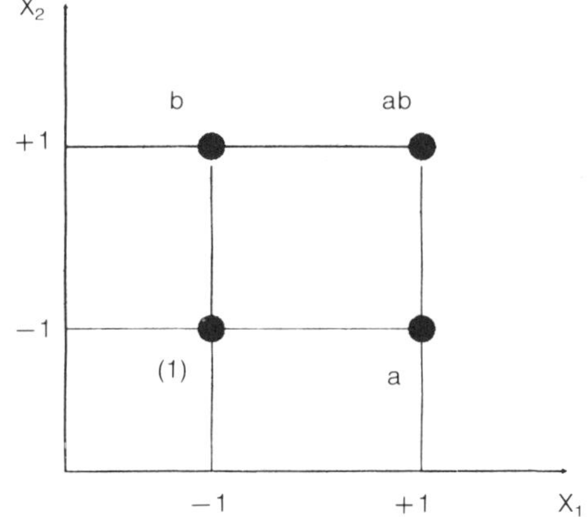

Figure 8.1 Geometric representation of a 2^2 design.

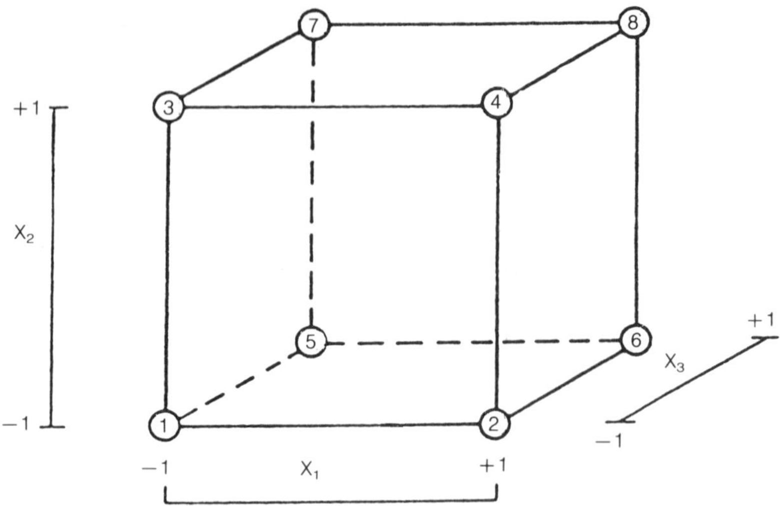

Figure 8.2 Geometric representaion of a 2^3 design.

the column for total in Table 8.4 correspond to the sequence just previously generated.

To finish with the present notation, in a 2^5 factorial experiment, for instance, the symbol *ace* would represent the total of *r* random observations corresponding to the experimental condition defined by the coded levels $X_1 = +1$, $X_2 = -1$, $X_3 = +1$, $X_4 = -1$ and $X_5 = +1$.

The four experimental conditions of the 2^2 design can be geometrically represented by the corners of the square shown in Figure 8.1. Similarly, the eight experimental conditions of the 2^3 design can be represented by the corners of the cube shown in Figure 8.2.

8.3 EFFECT ESTIMATION METHODOLOGY FOR 2^n FACTORIAL EXPERIMENTS

The purpose of this section is threefold:

(a) to define the effect of a factor or an interaction in the context of a 2^n factorial design;
(b) to develop an ANOVA test to verify if the effect of a factor or interaction is or is not significant;
(c) to show how to find confidence intervals for the effects of a 2^n design.

Two bibliographical references that can be recommended to the student for supplementary reading concerning the material covered in this section are the excellent books by Box *et al.* [4] and Davies *et al.* [5].

8.3.1 Analysis of main effects

Let us consider a two-level factorial design with r replications for each experimental condition. The total number of observations required to run the complete experiment is equal to $r2^n$, where $\frac{1}{2}r2^n$ observations are associated with the $-$ level of X_i and $\frac{1}{2}r2^n$ observations are associated with the $+$ level of the same factor. Let T_i^+ be the sum of the $r2^{n-1}$ observations corresponding to the 2^{n-1} experimental conditions having $X_i = +1$, and let T_i^- be the sum of the $r2^{n-1}$ observations corresponding to the 2^{n-1} experimental conditions having $X_i = -1$. The main effect of factor X_i is estimated as the difference between the two averages

$$\bar{Y}_i^+ = \frac{T_i^+}{r2^{n-1}} \quad \text{and} \quad \bar{Y}_i^- = \frac{T_i^-}{r2^{n-1}}$$

that is

$$E_i = \frac{1}{r2^{n-1}}(T_i^+ - T_i^-). \tag{8.1}$$

Using the terminology introduced in Chapter 7, the region R_{i1} contains the cells or experimental conditions associated with the coded level $X_i = +1$ and the region R_{i2} contains the cells associated with the coded level $X_i = -1$. It is now noted that

$$T_i^+ = \sum_{c \in R_{i1}} T_{(c)}$$

and

$$T_i^- = \sum_{c \in R_{i2}} T_{(c)}.$$

Therefore, equation (8.1) can be rewritten as

$$E_i = \frac{1}{2^{n-1}} \sum_{c \in R_{i1} \cup R_{i2}} a_{(c)} \frac{T_{(c)}}{r}$$

or

$$E_i = \frac{1}{2^{n-1}} \sum_{c \in R_{i1} \cup R_{i2}} a_{(c)} \bar{Y}_{(c)} \tag{8.2}$$

where $a_{(c)} = 1$ for $c \in R_{i1}$, $a_{(c)} = -1$ for $c \in R_{i2}$ and $\sum a_{(c)} = 0$.

Using the basic definitions given in section 8.1, equation (8.2) can be finally expressed as

$$E_i = \frac{L_i}{2^{n-1}} \tag{8.3}$$

where

$$L_i = \sum_{c \in R_{i1} \cup R_{i2}} a_{(c)} \bar{Y}_{(c)} \tag{8.4}$$

is a contrast belonging to the main effect of factor X_i.

As an illustration of the above results, in the case of a 2^2 design, the effects of factors X_1 and X_2 can be estimated by

$$E_1 = \frac{L_1}{2}$$

$$E_2 = \frac{L_2}{2}$$

where L_1 and L_2 are obtained from Table 8.3:

$$L_1 = \frac{1}{r}(-(1) + a - b + ab)$$

$$L_2 = \frac{1}{r}(-(1) - a + b + ab).$$

Equivalently,

$$E_1 = \frac{a + ab}{2r} - \frac{(1) + b}{2r} \tag{8.5}$$

and

$$E_2 = \frac{b + ab}{2r} - \frac{(1) + a}{2r}. \tag{8.6}$$

It is noted that equation (8.5) actually estimates the effect of factor X_i as the difference of two averages: the first average corresponds to the two experimental conditions having $X_1 = +1$, and the second corresponds to the two conditions associated with $X_1 = -1$. Using the graphical representation given in Figure 8.1, it is concluded that E_1 is the difference between the average of the $r2^{n-1}$ observations corresponding to the right vertical side and the average of the $r2^{n-1}$ observations corresponding to the left vertical side. A similar analysis based on equation (8.6) concludes that E_2 is the difference between the top horizontal side average of the square of Figure 8.1 and the bottom horizontal side average. These remarks are illustrated in Figure 8.3.

As a second illustration of the results obtained from equations (8.3) and (8.4), consider a 2^3 design; from Table 8.4, it is possible to identify the following orthogonal contrasts belonging to the effects of X_1, X_2 and X_3, respectively:

$$L_1 = \frac{1}{r}(-(1) + a - b + ab - c + ac - bc + abc)$$

$$L_2 = \frac{1}{r}(-(1) - a + b + ab - c - ac + bc + abc)$$

$$L_3 = \frac{1}{r}(-(1) - a - b - ab + c + ac + bc + abc).$$

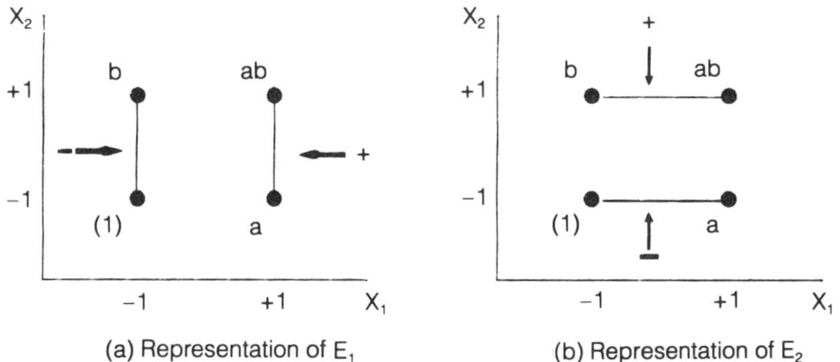

Figure 8.3 Graphical interpretation of main effects in a 2^2 design: (a) representation of E_1, (b) representation of E_2.

From equation (8.3) the following estimates can now be obtained for the effects of X_1, X_2 and X_3:

$$E_1 = \frac{a + ab + ac + abc}{4r} - \frac{(1) + b + c + bc}{4r} \tag{8.7}$$

$$E_2 = \frac{b + ab + bc + abc}{4r} - \frac{(1) + a + b + ab}{4r} \tag{8.8}$$

$$E_3 = \frac{c + ac + bc + abc}{4r} - \frac{(1) + a + b + ab}{4r}. \tag{8.9}$$

Again, in the context of the graphical representation shown in Figure 8.2, the results obtained from equations (8.7)–(8.9) correspond to differences between averages associated with parallel sides of the cube generated by the eight experimental conditions of the 2^3 design. Figure 8.4 illustrates this interpretation.

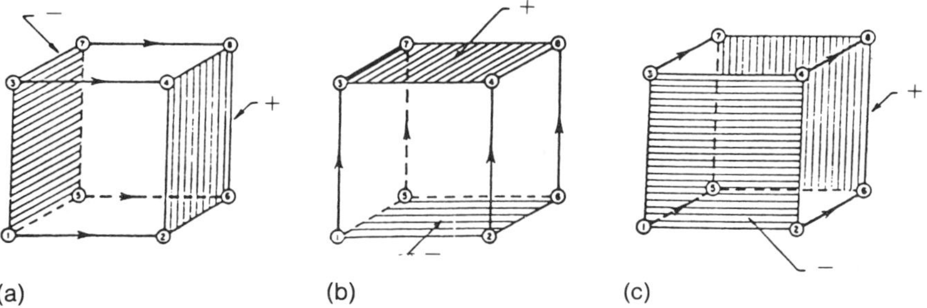

Figure 8.4 Graphical interpretation of main effects in a 2^3 design: (a) representation of E_1, (b) representation of E_2, (c) representation of E_3.

8.3.2 Analysis of interaction effects

The general procedure developed in this section is based on the following considerations. If we set any factor at either of its two levels, main effect estimates can be computed for all other factors. Actually, two main effect estimates can be computed for each other factor, one corresponding to each of the two levels of the factor initially selected to be set at a particular level. These two main effects can be regarded as conditional main effect estimates since in each case a chosen factor is preset. If the two conditional main effects for a factor are about the same, then this would indicate that there is no interaction between this factor and the factor initially preset. If the two conditional effect estimates, on the other hand, are very different, this would indicate that the interaction effect is important.

Proceeding in this similar line of thinking, we can now also compute two-factor interaction effects for all other factors different from that one initially selected to be preset at a given level. Again, for each pair of factors two conditional two-factor interaction effects can be estimated. If the two conditional estimates are similar, then we say that the corresponding three-factor interaction effect is not important. But if the two conditional two-factor interaction effects are very different, this would indicate that the three-factor interaction effect is important. We can generalize this discussion to relate any interaction effect to the difference between two conditional interaction effects of lower order.

Let us consider any two factors X_i and X_j each having two coded levels -1 and $+1$. Also, let $L_j = \sum_c b_{(c)} \bar{Y}_{(c)}$ be a contrast belonging to the main effect of factor X_j. The conditional effect of factor X_j given that factor $X_i = +1$ can be estimated from the 2^{n-1} runs in contrast, L_j corresponding to the + level for X_i; as defined in Section 8.3.1, these runs are in set R_{i1}. The estimate is obtained from equations (8.3) and (8.4) after substituting $n-1$ for n:

$$E_j(X_i^+) = \frac{\sum_{c\in R_{i1}} b_{(c)} \bar{Y}_{(c)}}{2^{n-2}} \tag{8.10}$$

Similarly, the conditional effect of factor X_j given that factor $X_i = -1$ can be estimated from the 2^{n-1} runs in contrast, L_j having X_i set at its $-$ level:

$$E_j(X_i^-) = \frac{\sum_{c\in R_{i2}} b_{(c)} \bar{Y}_{(c)}}{2^{n-2}} \tag{8.11}$$

The effect of the interaction between factors X_i and X_j is defined as the **average difference** between the conditional effects estimated in equations (8.10) and (8.11). The interaction effect E_{ij} can be estimated, therefore, as

$$E_{ij} = \frac{E_j(X_i^+) - E_j(X_i^-)}{2} \tag{8.12}$$

Substituting equations (8.10) and (8.11) into equation (8.12), E_{ij} can be written as

$$E_{ij} = \frac{\sum_{c \in R_{i1}} b_{(c)} \bar{Y}_{(c)} - \sum_{c \in R_{i2}} b_{(c)} \bar{Y}_{(c)}}{2^{n-1}} \tag{8.13}$$

Now let us note that R_{i1} is actually the union of the following two sets of experimental conditions:

$$S_{ij}^{++} = \{c \mid X_i = +1, X_j = +1\}$$
$$S_{ij}^{+-} = \{c \mid X_i = +1, X_j = -1\}$$

Additionally, R_{i2} is the union of the following two sets:

$$S_{ij}^{-+} = \{c \mid X_i = -1, X_j = +1\}$$
$$S_{ij}^{--} = \{c \mid X_i = -1, X_j = -1\}$$

Therefore, equation (8.13) can be rewritten as

$$E_{ij} = \frac{1}{2^{n-1}}\left(\sum_{c \in S_{ij}^{++}} b_{(c)} \bar{Y}_{(c)} + \sum_{c \in S_{ij}^{+-}} b_{(c)} \bar{Y}_{(c)} - \sum_{c \in S_{ij}^{-+}} b_{(c)} \bar{Y}_{(c)} - \sum_{c \in S_{ij}^{--}} b_{(c)} \bar{Y}_{(c)}\right)$$
$$= \frac{1}{2^{n-1}}\left(\sum_{c \in S_{ij}^{++}} \bar{Y}_{(c)} - \sum_{c \in S_{ij}^{+-}} \bar{Y}_{(c)} - \sum_{c \in S_{ij}^{-+}} \bar{Y}_{(c)} + \sum_{c \in S_{ij}^{--}} \bar{Y}_{(c)}\right).$$

The term between parentheses in the above relationship is now more closely examined. Let L_i be a contrast belonging to the main effect of factor X_i, and also let L_i be orthogonal to L_j. Using the coefficients obtained from the design matrix given in Table 8.1, L_i and L_j can be expressed as

$$L_i = \sum_c a_{(c)} \bar{Y}_{(c)}$$

$$L_j = \sum_c b_{(c)} \bar{Y}_{(c)}$$

where, as already known, $a_{(c)} = -1$, $b_{(c)} = -1$ or $+1$, $\sum_c a_{(c)} = 0$, $\sum_c b_{(c)} = 0$ and $\sum_c a_{(c)} b_{(c)} = 0$. It is easily noted that

$$a_{(c)} b_{(c)} = +1 \quad \text{for } c \in S_{ij}^{++}$$
$$a_{(c)} b_{(c)} = -1 \quad \text{for } c \in S_{ij}^{+-}$$
$$a_{(c)} b_{(c)} = -1 \quad \text{for } c \in S_{ij}^{-+}$$
$$a_{(c)} b_{(c)} = +1 \quad \text{for } c \in S_{ij}^{--}$$

As a result of the above observations, the contrast

$$L_{ij} = \sum_{c \in S_{ij}^{++}} \bar{Y}_{(c)} - \sum_{c \in S_{ij}^{+-}} \bar{Y}_{(c)} - \sum_{c \in S_{ij}^{-+}} \bar{Y}_{(c)} + \sum_{c \in S_{ij}^{--}} \bar{Y}_{(c)}$$

belongs to the interaction effect due to factors X_i and X_j. For this reason, we can finally write for the estimate of the interaction effect between X_i and X_j:

$$E_{ij} = \frac{L_{ij}}{2^{n-1}} \tag{8.14}$$

which is similar to equation (8.3). The three-factor interaction associated with factors X_i, X_j and X_k can be analyzed following a procedure that is basically identical to that followed to develop equation (8.14). The estimate of a three-factor interaction effect E_{ijk} can be defined as the average difference between the conditional interaction effects $E_{ij}(X_k^+)$ and $E_{ij}(X_k^-)$:

$$E_{ijk} = \frac{E_{ij}(X_k^+) - E_{ij}(X_k^-)}{2}. \tag{8.15}$$

Equation (8.15) is similar to equation (8.12); the reader can verify that if L_{ijk} is an orthogonal contrast belonging to the interaction effect of factors X_i, X_j and X_k, the estimate E_{ijk} is given by

$$E_{ijk} = \frac{L_{ijk}}{2^{n-1}} \tag{8.16}$$

which, again, resembles equations (8.3) and (8.14). The results for the interaction effects can be easily illustrated on 2^2 and 2^3 factorial experiments. First, let us consider a two-level 2^2 experiment; the design matrix, the coefficients for the contrasts belonging to the interaction effect and the totals of the r observations for each experimental condition are summarized in Table 8.5.

From Table 8.5 the following results are readily available:

$$L_{12} = \frac{(1) - a - b + ab}{r}.$$

Therefore, using equation (8.14):

$$E_{12} = \frac{(1) + ab}{2r} - \frac{a + b}{2r}$$

Table 8.5 Design matrix and additional information for a 2^2 experiment

Condition	X_1	X_2	*Total*	
1	−	−	+	(1)
2	+	−	−	*a*
3	−	+	−	*b*
4	+	+	+	*ab*

which, as illustrated in Figure 8.5, is the difference between the averages corresponding to the two diagonals of the square representing the experimental conditions of the 2^2 design.

A second illustration of the results that can be obtained from equation (8.14) and (8.15) is provided by the analysis of a 2^3 factorial experiment. As an

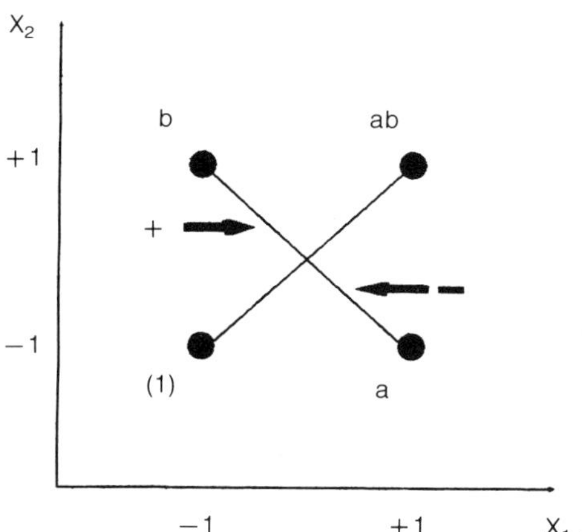

Figure 8.5 Interpretation of the two-factor interaction effect in a 2^2 experiment.

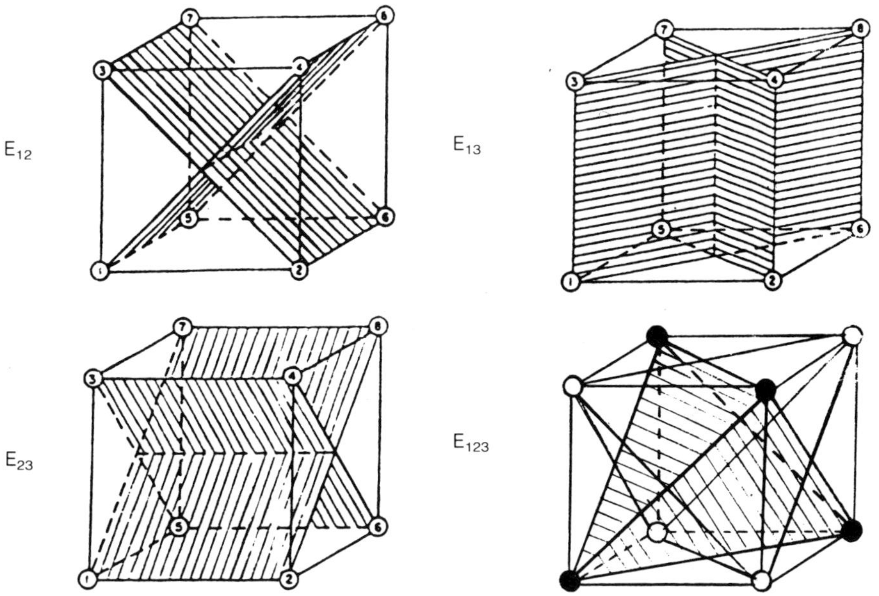

Figure 8.6 Interpretation of interaction effects in a 2^3 experiment.

exercise, the reader may verify the interpretation given in Figure 8.6 for two-factor interactions and for the three-factor interaction. It is shown in this figure that two-factor interaction effects correspond to average differences associated with diagonal planes, and that the three-factor interaction effect corresponds to the difference between two averages associated with pyramids. In general any effect or order m is estimated as the difference between two averages associated with two hyperplanes of dimension $m-1$.

EXAMPLE 8.1

This example presents an application of the methodology developed in this chapter to study the effects of temperature, adult density (number of mating pairs per unit area) and female size on the reproduction of southern pine beetle populations in east Texas. As indicated by Wagner *et al.* [6], these populations are typically small during the winter months, increasing during the spring and reaching their peak in June. As a result of low temperature, southern pine beetle populations during the winter can be found only in isolated trees. As the end of winter approaches, and during the early spring, the number of adults begins to increase as temperature, density and size conditions all favor reproduction. Thus the potential for destructive infestations increases as the weather conditions continue to be more and more favorable during early spring. Statistical analyses thus play an important role in the understanding of the independences of those factors on the southern pine beetle.

The following data correspond to a 2^3 factorial experiment designed to investigate the effects of the total number of eggs oviposited per female southern pine beetle. Three factors are considered important in this analysis:

(a) the density or number of mating pairs introduced into the bark sample;
(b) the temperature;
(c) the time of year.

This third factor is important since the size of the insect changes during the year. Each of these factors has two levels as indicated in Table 8.6.

Table 8.6

	X_1 *Density* (dm^{-2})	X_2 *Temperature* (°C)	X_3 *Time of year*
− level	2	15	February
+ level	4	25	June

Table 8.7

Run	X_1	X_2	X_3	*Replications* 1	2	3
1	−	−	−	55	48	65
2	+	−	−	22	28	32
3	−	+	−	45	25	35
4	+	+	−	37	30	28
5	−	−	+	33	38	30
6	+	−	+	20	23	28
7	−	+	+	15	32	22
8	+	+	+	11	13	7

Table 8.8

Condition	X_1	X_2	X_3	X_1X_2	X_1X_3	X_2X_3	$X_1X_2X_3$	*Total*
1	−1	−1	−1	+1	+1	+1	−1	168
2	+1	−1	−1	−1	−1	+1	+1	82
3	−1	+1	−1	−1	+1	−1	+1	105
4	+1	+1	−1	+1	−1	−1	−1	95
5	−1	−1	+1	+1	−1	−1	+1	101
6	+1	−1	+1	−1	+1	−1	−1	71
7	−1	+1	+1	−1	−1	+1	−1	69
8	+1	+1	+1	+1	+1	+1	+1	31

Each experimental condition was replicated 3 times and the 24 observations were collected at random. The results are summarized in Table 8.7, where the experimental conditions are listed in standard order. The coefficients of the seven orthogonal contrasts that can be generated from the design matrix of this 2^3 experiment are shown in Table 8.8.

Main effects

Using equation (8.3) the main effects are estimated as follows:

$$E_1 = \frac{L_1}{4} = \tfrac{1}{12}(-168 + 82 - 105 + 95 - 101 + 71 - 69 + 31) = -13.67$$

$$E_2 = \frac{L_2}{4} = \tfrac{1}{12}(-168 - 82 + 105 + 95 - 101 - 71 + 69 + 31) = -10.17$$

$$E_3 = \frac{L_3}{4} = \tfrac{1}{12}(-168 - 82 - 105 - 95 + 101 + 71 + 69 + 31) = -14.83.$$

Two-factor interactions

Using equation (8.14) the two-factor interaction effects are estimated as follows:

$$E_{12} = \frac{L_{12}}{4} = \tfrac{1}{12}(168 - 82 - 105 + 95 + 101 - 71 - 69 + 31) = 5.67$$

$$E_{13} = \frac{L_{13}}{4} = \tfrac{1}{12}(168 - 82 + 105 - 95 - 101 + 71 - 69 + 31) = 2.33$$

$$E_{23} = \frac{L_{23}}{4} = \tfrac{1}{12}(168 + 82 - 105 - 95 - 101 - 71 + 69 + 31) = -1.83.$$

Three-factor interaction

Using equation (8.16) the three-factor interaction effect is estimated by

$$E_{123} = \frac{L_{123}}{4} = \frac{1}{12}(-168 + 82 + 105 - 95 + 101 - 71 - 69 + 31) = -7.00.$$

All main effects appear to be significant. Since these effects are negative, a change for each factor from its low level to its high level would reduce the value of the experimental response (number of eggs). Also, one two-factor interaction effect and the three-factor interaction effect seem to be significant. A more conclusive study of the significance of both main effects and interaction effects is provided by the ANOVA methodology. Section 8.4 summarizes the specific results for 2^n factorial experiments.

8.4 ANOVA METHODOLOGY FOR 2^n DESIGNS

Once a particular effect is estimated it would be desirable to test if the effect is statistically significant or not. This section uses the ANOVA methodology to test the hypothesis that the effect due to factor or the interaction of a group of factors (usually two or three) is not significant.

Let E be the estimate of the effect to be tested and let L be the orthogonal contrast belonging to this effect that can be generated from the design matrix. As proved in equations (8.3), (8.14) and (8.16),

$$E = \frac{L}{2^{n-1}}$$

where

$$L = \sum_{c=1}^{m} a_{(c)} \bar{Y}_{(c)}.$$

In the above contrast definition, $m = 2^n$ and $a_{(c)} = +1$ or -1. From equation (4.11), it is concluded that

$$SS_L = L^2\left(\sum_c \frac{a_{(c)}^2}{r}\right)^{-1}$$

or

$$SS_L = \frac{rL^2}{2^n}. \tag{8.17}$$

With n factors at two levels, the total number of experimental conditions is equal to 2^n. The number of treatment means is, of course, equal to 2^n. With these 2^n treatment means a set of $2^n - 1$ mutually orthogonal contrasts $\sum a_{(c)} Y_{(c)}$ may be defined, in a manner that each contrast will represent either a main effect or an interaction effect. For a given n, we have

n main effects

$\binom{n}{2}$ two-factor interactions

$\binom{n}{3}$ three-factor interactions

$\vdots$

$\binom{n}{n}$ n-factor interactions.

Each effect has one degree of freedom. The total number of degrees of freedom is given by $r2^{n-1}$. Therefore, the number of degrees of freedom of the error is equal to

$$r2^n - 1 - 2^n + 1 = 2^n(r-1)$$

which implies that the number of replications must be $r \geqslant 2$ if the interaction effects are going to be investigated.

The sum of squares due to the random error can be computed as

$$SS_{\text{error}} = SS_{\text{total}} - \sum_i SS_i - \sum_i \sum_j SS_{ij} - \ldots, SS_{12\ldots n}$$

where SS_i is the sum of squares due to factor X_i, SS_{ij} is the sum of squares due to the interaction of factors X_i and X_j, and so on. The total sum of squares is equal to

$$SS_{\text{total}} = \sum_{c=1}^{m} \sum_{k=1}^{r} Y_{ck}^2 - \frac{T^2}{N}$$

where Y_{ck} is the kth observation taken for the cth experimental condition, $m = 2^n$, T is the grand total of all observations and $N = r2^n$.

In order to test if an effect (main or interaction) is significant, the value of

$$F = \frac{SS_L}{MS_{\text{error}}}$$

should be compared to the critical value $F_{1-\alpha}(1, \nu)$ where $\nu = 2^n(r-1)$.

Confidence intervals

Let E be one of the $2^n - 1$ estimates that can be computed from the $r2^n$ random observations collected for a 2^n factorial experiment. We know that

$$E = \frac{L}{2^{n-1}}$$

where L is a contrast belonging to the effect estimated by E:

$$L = \sum_c a_{(c)} \bar{Y}_{(c)}.$$

From the above relationship, the expected value of the estimate E is found to be equal to

$$\mu_E = \frac{\sum_c a_{(c)} \mu_{(c)}}{2^{n-1}}.$$

Also

$$\sigma_E^2 = \frac{\sigma_L^2}{2^{2n-2}}$$

$$= \sum_c \frac{\sigma^2}{r} (2^{2n-2})^{-1}$$

$$= \frac{2^n \sigma^2}{r 2^{2n-2}}.$$

Simplifying,

$$\sigma_E^2 = \frac{4\sigma^2}{r2^n} \tag{8.18}$$

where σ^2 is the variance of the error. Since σ^2 is generally unknown, it can be estimated by $S^2 = MS_{\text{error}}$; therefore, using equation (8.18), an estimate of σ_E^2 is provided by

$$\hat{\sigma}_E^2 = \frac{4S^2}{r2^n}. \tag{8.19}$$

From the normality assumption, it is concluded that

$$\frac{E - \mu_E}{\hat{\sigma}_E} \sim t_v$$

where $v = (r-1)2^n$. A $100(1-\alpha)\%$ confidence interval for μ_E can be obtained from the above results:

$$E \pm t_{1-\alpha/2,v} \frac{2S}{(r2^n)^{1/2}}. \tag{8.20}$$

EXAMPLE 8.2

The purpose of this example is to illustrate the ANOVA methodology on the experiment described in Example 8.1. Using equation (8.17) we obtain the following results: $SS_1 = 1120.67$,. $SS_2 = 620.17$, $SS_3 = 1320.17$, $SS_{12} = 192.67$, $SS_{13} = 32.67$, $SS_{23} = 20.17$, $SS_{123} = 294.00$. The sum of squares due to the total, treatments and error are computed as follows:

$$SS_{\text{total}} = \sum_{c=1}^{8} \sum_{k=1}^{3} Y_{ck}^2 - \frac{T^2}{N} = 25\,992 - \frac{(722)^2}{24} = 4271.83$$

$$SS_{\text{treatment}} = \sum_{i=1}^{3} SS_i + \sum_{i=1}^{3} \sum_{j=1}^{3} SS_{ij} + SS_{123} = 3600.52$$

$$SS_{\text{error}} = SS_{\text{total}} - SS_{\text{treatment}} = 671.31.$$

Table 8.9 summarizes the ANOVA results using a level of significance $\alpha = 0.05$; the critical value of the F-statistic to be used is equal to $F_{0.95}(1, 16) = 4.49$.

As a result of the above analysis, the main effects are all found to be significant; the only two-factor interaction effect that is significant is due to

Table 8.9

Source	*df*	*SS*	*MS*	F_0	*Significant?*
X_1	1	1120.67	1120.67	26.71	Yes
X_2	1	620.17	620.17	14.78	Yes
X_3	1	1320.17	1320.17	31.46	Yes
X_1X_2	1	192.67	192.67	4.59	Yes
X_1X_3	1	32.67	32.67	0.78	No
X_2X_3	1	20.17	20.17	0.48	No
$X_1X_2X_3$	1	294.00	294.00	7.01	Yes
Error	16	671.31	671.31		
Total	23				

density and temperature; finally, the interaction of the three factors under investigation is also significant.

8.5 YATES ALGORITHM

An alternative systematic method for the computation of effects and the sums of squares due to them is given by Yates [1]. This is illustrated in Table 8.10 for a 2^3 experiment. This technique is also easily generalized for $n \geqslant 4$. For $n \geqslant 4$, instead of three operations X, Y, Z there have to be n similar operations.

The same data obtained from the experiment of Example 8.1 will be analyzed using the alternative computational procedure developed by Yates. Following the calculations indicated in Table 8.10, we obtain the results summarized in Tables 8.11 and 8.12. In Table 8.11, the effect estimate corresponding to any row (except the first) is equal to the value in the third column divided by 12 (or $r2^{n-1}$). Additionally, the corresponding sum of squares is equal to the same value squared and divided by 24 (or $r2^n$). For the first row we obtain the average $\bar{Y}$ and the correction term T^2/N dividing by $r2^n$.

Table 8.10 Yates algorithm for a 2^3 design

Treatment	*Total*	*(1)*	*(2)*	*(3) = rL*
$- - -(1)$	T_1	$X_1 = T_1 + T_2$	$Y_1 = X_1 + X_2$	$Z_1 = Y_1 + Y_2$
$+ - - a$	T_2	$X_2 = T_3 + T_4$	$Y_2 = X_3 + X_4$	$Z_2 = Y_3 + Y_4$
$- + - b$	T_3	$X_3 = T_5 + T_6$	$Y_3 = X_5 + X_6$	$Z_3 = Y_5 + Y_6$
$+ + - ab$	T_4	$X_4 = T_7 + T_8$	$Y_4 = X_7 + X_8$	$Z_4 = Y_7 + Y_8$
$- - + c$	T_5	$X_5 = T_2 - T_1$	$Y_5 = X_2 - X_1$	$Z_5 = Y_2 - Y_1$
$+ - + ac$	T_6	$X_6 = T_4 - T_3$	$Y_6 = X_4 - X_3$	$Z_6 = Y_4 - Y_3$
$- + + bc$	T_7	$X_7 = T_6 - T_5$	$Y_7 = X_6 - X_5$	$Z_7 = Y_6 - Y_5$
$+ + + abc$	T_8	$X_8 = T_8 + T_7$	$Y_8 = X_8 - X_7$	$Z_8 = Y_8 - Y_7$

Table 8.11 Yates algorithm for Example 8.1

Run	*Total*	*(1)*	*(2)*	*(3)*
1	168	250	450	722
2	82	200	272	−164
3	105	172	−96	−122
4	95	100	−68	68
5	101	−86	−50	−178
6	71	−10	−72	28
7	69	−30	76	−22
8	31	−38	−8	−84

Table 8.12 Effects and sums of squares in Example 8.2

X_1	X_2	X_3	*Effect*	*Sum of squares*
−	−	−	$\bar{Y} = 30.08$	$T^2/N = 21\,720.17$
+	−	−	$E_1 = -13.67$	$SS_1 = 1\,120.67$
−	+	−	$E_2 = -10.17$	$SS_2 = 620.17$
+	+	−	$E_{12} = 5.67$	$SS_{12} = 192.67$
−	−	+	$E_3 = -14.83$	$SS_3 = 1\,320.17$
+	−	+	$E_{13} = 2.33$	$SS_{13} = 32.67$
−	+	+	$E_{23} = -1.83$	$SS_{23} = 20.17$
+	+	+	$E_{123} = 7.00$	$SS_{123} = 294.00$

8.6 AN APPLICATION: A STUDY OF WELD STRENGTH

This section summarizes an application documented by Wang and DeVries in reference [2]. The material included in this summary consists of a brief definition of the problem, identification of meaningful levels for each factor, data for the experimental conditions associated with the design, estimation of main and interaction effects, confidence intervals, ANOVA results and a discussion of the results.

(a) Problem definition

It is desired to quantify the effects of the ambient temperature (T), the wind velocity (V) and the bar size (S) on the strength of welded rail steel joints.

(b) Factor levels

For the three variables included in the study, a 2^3 full factorial design was run considering the levels given in Table 8.13.

(c) Experimental data

A set of coded variables can be used to represent the eight experimental conditions of this design. A coded variable takes on the value +1 at the

Table 8.13

Factor	*Low level (−)*	*High level (+)*
T(°F)	0	70
V (mile h^{-1})	0	20
S (in)	0.364	1.00

high level of the corresponding original factor, and the value -1 at the low level of the same factor. Let X_1, X_2, X_3 be the coded variables corresponding to T, V and S, respectively. These coded variables are defined as

$$X_1 = \frac{T-35}{35}$$

$$X_2 = \frac{V-10}{10}$$

$$X_3 = \frac{S-0.682}{0.318}.$$

Each of the eight experimental conditions was replicated twice, and the corresponding results are summarized in Table 8.14.

(d) Effect estimates and sums of squares

From the experimental data given in Table 8.15, the results obtained using Yates algorithm are shown in that table.

Table 8.14 Tensile strength (10^3 psi)

Order	X_1	X_2	X_3	*Run 1*	*Run 2*
1	−	−	−	86.0	89.0
2	+	−	−	88.6	86.0
3	−	+	−	73.6	82.0
4	+	+	−	82.0	92.0
5	−	−	+	79.7	78.5
6	+	−	+	99.7	95.4
7	−	+	+	80.7	76.5
8	+	+	+	89.7	85.7

Table 8.15 Yates algorithm

Standard order	*Total*	*(1)*	*(2)*	*(3)*	*Effects (10^3 psi)*	*Sums of squares*
1	175.0	349.6	679.2	1365.1	$\bar{Y} = 85.319$	$T^2/N = 1\,161\,468.626$
2	174.6	329.6	685.9	73.1	$E_1 = 9.138$	$SS_1 = 333.976$
3	155.6	353.3	18.0	−40.7	$E_2 = 5.088$	$SS_2 = 103.531$
4	174.0	332.6	55.1	0.1	$E_{12} = 0.013$	$SS_{12} = 0.001$
5	158.2	−0.4	−20.0	6.7	$E_3 = 0.898$	$SS_3 = 2.806$
6	195.1	18.4	−20.7	37.1	$E_{13} = 4.638$	$SS_{13} = 86.026$
7	157.2	36.9	18.8	−0.7	$E_{23} = 0.088$	$SS_{23} = 0.031$
8	175.4	18.2	−18.7	−37.5	$E_{123} = 4.688$	$SS_{123} = 87.891$

(e) Confidence intervals

$$SS_{\text{total}} = 117\,202.830 - 116\,468.626 = 734.204$$
$$SS_{\text{treatment}} = 614.259$$
$$SS_{\text{error}} = 119.945$$
$$S^2 = MS_{\text{error}} = 14.993.$$

From equation (8.19)

$$\hat{\sigma}_E^2 = \frac{4S^2}{r2^n} = 3.748$$

or, equivalently,

$$\hat{\sigma}_E = 1.936 \quad (10^3 \text{ psi}).$$

Therefore, a confidence interval for any effect is given by

$$E \pm 1.936t_c$$

where, for $\alpha = 0.05$ and $v = (r-1)2^n = 8$, $t_c = 2.306$. Hence, the 95% confidence interval for the effect estimated by the value E is given by the range $E \pm (1.936)(2.306)$, or equivalently, $E \pm 4.464$ psi (0.314 kg cm^{-2}).

EXERCISES

1. Prove that the variance of the effects in a replicated 2^n factorial experiment is given by

$$V[\text{effect}] = \left(\frac{4}{rN}\right)S^2$$

   where N is the number of experimental conditions, r the number of replications and S^2 the pooled estimate of run variance.
2. Calculate the main effects and interactions on the basis of the data in Table 8.16 obtained from three replications of an experiment. Use the method of orthogonal contrasts.
3. Calculate the standard errors of the main effects and interactions in the experiment of Exercise 2. Find confidence intervals for all the main effects and interactions.
4. Use Yates' algorithm to calculate and test the significance of effects in Exercise 2. Identify the corresponding contrasts and solve the same problem by establishing a relationship between the effects and the contrasts.
5. Solve Exercise 2 using a graphical approach.
6. Solve Exercise 2 using the general ANOVA approach.
7. The yield of a sucrose derivative from a new process using corn waste

material was investigated to assess the effect of four factors. Each factor had two levels, as indicated in Table 8.17. After testing each level combination once, in random order, the results given in Table 8.18 were obtained.

(a) Set up a standard design table including the observed yields.
(b) Determine the main effects and interaction effects using Yates' algorithm.

Table 8.16

Depth of planting (in)	*Watering (number of times per day)*	*Type of lima bean considered*	*Yield* 1	2	3
1.5	2	Large	5	5	4
0.5	1	Baby	6	7	6
0.5	2	Baby	10	9	8
1.5	1	Large	3	3	1
1.5	1	Baby	4	5	5
0.5	2	Large	8	7	7
0.5	1	Large	4	5	4
1.5	2	Baby	7	7	6

Table 8.17

Factor	Low level (−)	High level (+)
Acid concentration (N)	2	2.7
Reaction time (min)	10	20
Acid quantity (l)	0.10	0.15
Temperature (°C)	50	75

Table 8.18 Yield of sucrose derivative

		Temperature (°C)			
		50		*75*	
Acid concentration (N)	*Reaction time (min)*	*Acid quantity* 0.1	*0.15*	*Acid quantity* 0.1	*0.15*
2	10	5.07	5.2	5.66	5.61
	20	5.41	5.01	5.62	5.39
2.7	10	5.01	5.06	5.59	5.27
	20	5.41	5.23	5.02	5.13

Table 8.19

Run	*T(°C)*	*C(%)*	K	*y*
1	160	20	A	60
2	180	20	A	72
3	160	40	A	54
4	180	40	A	68
5	160	20	B	52
6	180	20	B	83
7	160	40	B	45
8	180	40	B	80

(c) Perform an ANOVA and find an estimate of the error variance.
(d) Find 95% confidence intervals for the most significant effects.

8. Show that the standard deviation of the estimate of a main effect or interaction is twice the standard deviation of the estimate of the common effect.
9. Clearly explain the differences among the following estimates: S^2, SS_L and $4S^2(rN)^{-1}$.
10. A completely randomized experiment with two temperatures (factor A), two lengths of washing time (factor B) and five replications per treatment gave the following results:
    (a) average for all experimental conditions having both factors at their low levels: 6.2;
    (b) average for all experimental conditions having factor A at its high level and factor B at its low level: 14.1;
    (c) average for all experimental conditions having factor B at its high level and factor A at its low level: 5.4;
    (d) average for all experimental conditions having both factors at their high levels: 7.8;

    The variable of interest was the amount of impurities (in grams) remaining in the material washed. Estimate main and two-factor interaction effects. Interpret your results.
11. The data in Table 8.19 come from a pilot plant investigation of a process. The factors under consideration are two quantitative variables (temperature and concentration) and a single qualitative variable (catalyst). The response is the chemical yield in grams. Calculate all effects.

REFERENCES

1. Yates, F. (1937) The design and analysis of factorial experiments. *Imperial Bureau of Soil Science Bulletin 35*, Hafner Publishing Company.

2. Wang, K. K. and DeVries, M. F. (1969) Investigation of manufacturing processes by statistical experimental design techniques. *University of Wisconsin, Engr. Expt. Sta., Reprint No. 1369*, March.
3. Wu, S. M. (1964) Analysis of rail steel bar welds by two-level factorial design. *Welding Journal Research Supplement*, April.
4. Box, G. E. P., Hunter, W. G. and Hunter, J. S. (1978) *Statistics for Experimenters*, John Wiley & Sons, New York.
5. Davies, O. L. *et al.* (1971) *The Design and Analysis of Industrial Experiments*, Hafner Publishing Company, New York.
6. Wagner, T. L., Feldman, R. M., Gagne, J. A., Cover, J. D., Coulson, R. N. and Schoolfield, R. M. (1981) Factors affecting gallery construction, oviposition, and reemergence of *Dendroctonus frontalis* in the laboratory. *Annals of the Entomological Society of America*, **74**(3), 255–73.

Blocking in two-level factorial experiments

9

In this chapter we will combine the principles of blocking from Chapter 5 with those of 2^n factorial experiments discussed in Chapter 8. As defined in Chapter 5, blocks are groups of treatments created to increase the homogeneity of the experimental material. In other words, blocks correspond to groups of rows (cells) of the design matrix of an experiment. This chapter is divided into five sections. The general effect of blocking in factorial designs is considered in section 9.1. Different blocking arrangements for different block sizes are studied in section 9.2. Section 9.3 contains an introduction to block designs with partial confounding. An algorithm for recovery of interblock information for partial confounding is considered in section 9.3. Finally, some recommended blocking arrangements are listed in section 9.4.

9.1 INTRODUCTION

When several factors are investigated simultaneously a factorial experiment is desirable; it allows both main and interaction effects to be estimated. Chapter 8 considered an ideal situation for factorial experiments in the sense that there were no restrictions on the homogeneity of the experimental material and then all runs could be carried out in one block. However, in practice the number of factor-level combinations may become so large that it would be impossible to carry out all the runs within a single block. The cost, the batch size of material, the limitations of the equipment used, variations in operator skills, and time available for experimentation are typical reasons for the need to consider blocking. Since in this case the blocks would introduce their own source of variation into the experiment, special techniques are required to separate block effects from main and interaction effects.

The fundamental concept of blocking will be introduced through the following hypothetical situation illustrated in Figure 9.1. A manufacturing process requires that a base material be mixed with three additives in a blender. It is desired to investigate the effect of each additive by means of a 2^3 factorial experiment. However, if the capacity of the blender is less than

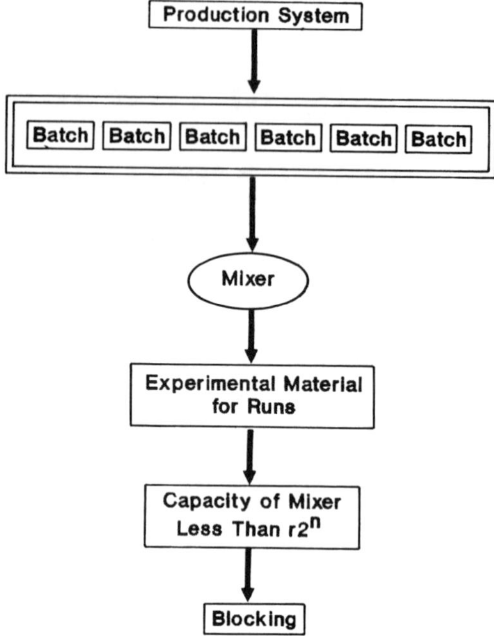

Figure 9.1 Blocking as a local control strategy.

eight runs, more than one blend will be needed. Since the quality of the product is affected by the blend, different blends may lead to different results. In this case, an appropriate experiment would involve blocking, with all runs with the same blend considered as a block.

Assuming that the capacity of the blender is large enough to produce homogeneous experimental material for four runs, the blender will be used twice; that is, two blocks will be needed. However, which of the eight runs of the 2^3 factorial experiment will be together in each of the two blocks? The total number of possible ways to accommodate eight runs into two blocks of four runs each is exactly equal to 70. Each of these possibilities or blocking arrangements will have its own advantages and disadvantages, depending on what effects become confounded with block effects. Two or more effects are said to be confounded when they are inseparable and hence cannot be estimated individually. A key concept underlying the design of blocking arrangements is that no important main or interaction effect should be confounded with block effects.

In order to illustrate the meaning of confounding suppose that in the manufacturing process under consideration the two blocks correspond to the treatment arrangements shown in Table 9.1, where the treatments are identified by the level combinations and the treatment yields are given in

Table 9.1 Two blocks of size $k = 4$

Block	X_1	X_2	X_3	*Yield*
	$-$	$-$	$-$	(1)
	+	+	$-$	ab
1	+	$-$	+	ac
	$-$	+	+	bc
	+	$-$	$-$	$a + x$
	$-$	+	$-$	$b + x$
2	$-$	$-$	+	$c + x$
	+	+	+	$abc + x$

standard order. In Table 9.1 it is assumed that the block effect is a fixed unknown value x that is part of each treatment yield in block 2. The three-factor interaction effect can be estimated as the value of $\frac{1}{4}L(X_1, X_2, X_3)$. That is,

$$\begin{aligned} E_{123} &= \tfrac{1}{4}[-(1) + (a + x) + (b + x) - ab + (c + x) - ac - bc + (abc + x)] \\ &= \tfrac{1}{4}[-(1) + a + b - ab + c - ac - bc + abc] + x. \end{aligned}$$

As can be seen in the above expression, the numerical value of E_{123} will estimate the three-factor interaction effect only if the block effect is negligible ($x = 0$). Otherwise, no separate estimates for the three-factor interaction effect and the block effect can be obtained. Since it is necessary to estimate the block effect, the three-factor interaction effect can be assumed to be negligible. It should be noted that the present blocking arrangement allows the main effects to be unconfounded. For instance,

$$\begin{aligned} E_1 &= \tfrac{1}{4}[-(1) + (a + x) - (b + x) + ab - (c + x) + ac - bc + (abc + x)] \\ &= \tfrac{1}{4}[-(1) + a - b + ab - c + ac - bc + abc] \end{aligned}$$

which, as can be seen, is independent of the block effect. As an exercise, the student should verify that other effects are not confounded with the block effect in this arrangement of treatments. In order to avoid the loss of information on essential effects, the grouping of treatments into blocks must be carefully considered. A general technique for blocking and identifying those effects confounded with block effects will be discussed in the following section.

9.2 BLOCK SIZE AND BLOCKING VARIABLES

The size of a block depends on the number of factors, replications and blocks. The block size in two-level factorial experiments can be obtained from the following formula:

$$k = \frac{r2^n}{b} \tag{9.1}$$

where k is the block size or number of runs in each block, b the number of blocks (including all replicates), r the number of replicates for each experimental condition and n the number of factors. From equation (9.1) the number of blocks can be obtained as

$$b = \frac{r2^n}{k}. \tag{9.2}$$

Any combination of integer values for r, n and k resulting in an integer value for b is feasible. For instance, if $r=1$ and $n=3$, we would need four blocks of size $k=2$. If $r=4$, $n=3$ and $k=2$, the number of blocks required is 16.

A blocking variable is a main effect or an interaction effect that can be used to identify the experimental conditions belonging to each block. Let v be the number of blocking variables per replicate. Obviously, the total number of level combinations corresponding to these v two-level variables is equal to 2^v. This number should be equal to the number of blocks per replicate, br^{-1}:

$$2^v = \frac{b}{r}. \tag{9.3}$$

Runs	X_1	X_2	X_3
1	-	-	-
2	+	-	-
3	-	+	-
4	+	+	-
5	-	-	+
6	+	-	+
7	-	+	+
8	+	+	+

Block	Runs	
Block 1	Run 2	■
	Run 7	■
Block2	Run 3	■
	Run 6	■
Block 3	Run 4	■
	Run 5	■
Block 4	Run 1	■
	Run 8	■

Case 1: No replications per run

Runs	X_1	X_2	X_3
1	-	-	-
2	+	-	-
3	-	+	-
4	+	+	-
5	-	-	+
6	+	-	+
7	-	+	+
8	+	+	+

Block	Runs	
Block 1	Run 2	■ ■ ■ ■
	Run 7	■ ■ ■ ■
Block2	Run 3	■ ■ ■ ■
	Run 6	■ ■ ■ ■
Block 3	Run 4	■ ■ ■ ■
	Run 5	■ ■ ■ ■
Block 4	Run 1	■ ■ ■ ■
	Run 8	■ ■ ■ ■

Case 2: Four replications per run

Figure 9.2 Illustrations of blocking.

As an illustration, consider a case with $k=2$, $n=3$ and $r=1$: from equation (9.2), the number of blocks is given by $b=(r2^n)k^{-1}=8(2)^{-1}=4$, and from equation (9.3) the number of level combinations associated with the blocking variables is equal to $2^v=br^{-1}=4(1)^{-1}=4$. Therefore, the number of blocking variables needed is $v=2$. As a second illustration, we can consider $n=3$, $k=2$ and $r=4$. In this case, $2^v=br^{-1}=16(4)^{-1}$, and hence $v=2$. In both of these illustrations we have four main blocks, although in the second illustration each main block is replicated four times. If, for example, block 1 is defined as the group having experimental conditions 2 and 7 (the numbers correspond to the standard order), block 2 as having conditions 3 and 6, block 3 as having conditions 4 and 5 and block 4 as having conditions 1 and 8, the two cases considered can be represented as in Figure 9.2. The fundamental purpose of this chapter is to study procedures that will allow us to determine which experimental conditions are placed in each block.

9.2.1 Use of blocking variables

The procedure of using blocking variables to group the experimental conditions of a 2^n design is simple. The 2^n experimental conditions are arranged in standard order and the signs $(+, -)$ of the 2^n-1 orthogonal contrasts are obtained as indicated in Chapter 8. The next step is to choose v of the 2^n-1 contrasts (or columns), consider them as the blocking variables of the design, identify the 2^v level combinations associated with the blocking variables, and finally assign to the same block all experimental conditions (runs) corresponding to each level combination.

The above procedure can be illustrated using the manufacturing example of section 9.1. In this example, $n=3$, $r=1$ and $k=4$. Therefore, $b=r2^nk^{-1}=2$ and $2^v=br^{-1}=2$. That is, $v=1$. Let B be the required single blocking variable. From the design matrix it is possible to generate a column of $+$ and $-$ signs for the contrast belonging to the interaction of the three factors under consideration. The corresponding results are shown in Table 9.2. Note that these results reproduce the blocking arrangement shown in Table 9.1.

Table 9.2 Identification of blocks using $B=X_1X_2X_3$

Run	X_1	X_2	X_3	$B=X_1X_2X_3$	*Yield*	*Block*
1	$-$	$-$	$-$	$-$	(1)	1
2	$+$	$-$	$-$	$+$	*a*	2
3	$-$	$+$	$-$	$+$	*b*	2
4	$+$	$+$	$-$	$-$	*ab*	1
5	$-$	$-$	$+$	$+$	*c*	2
6	$+$	$-$	$+$	$-$	*ac*	1
7	$-$	$+$	$+$	$-$	*bc*	1
8	$+$	$+$	$+$	$+$	*abc*	2

9.2.2 Use of defining contrasts in blocking

The grouping of experimental conditions into blocks can also be carried out by the use of a defining contrast, a concept related to that of blocking variables. Let B be a blocking variable defined as $B = X_1^{a_1}, X_2^{a_2}, \ldots, X_n^{a_n}$, where a_i is equal to either 0 or 1 for $i = 1, 2, \ldots, n$. The defining contrast corresponding to the blocking variable B is defined as

$$L = a_1x_1 + a_2x_2 + \cdots + a_nx_n.$$

As an example, the defining contrast corresponding to the blocking variable $B = X_1X_3$ is given by

$$L = (1)X_1 + (0)X_2 + (1)X_3 = X_1 + X_3$$

since $a_1 = 1$, $a_2 = 0$ and $a_3 = 1$. Considering again the manufacturing application of section 9.1, the numerical value of L(mod 2) can be computed for each of the eight experimental conditions, as shown in Table 9.3, where $X_i = 0$ when X_i is set at the low level ($-$) and $X_i = 1$ when it is set at the high level ($+$).

In Table 9.3 all runs having L(mod 2) equal to zero are put in block 1 and all runs having L(mod 2) equal to one are put in block 2. Note that these results are exactly the same as those shown in Table 9.2.

As a second example, consider a 2^3 factorial design with one replicate and four blocks. As can be verified, this design requires the use of two blocking variables. Let these variables be arbitrarily chosen as $B_1 = X_1X_3$ and $B_2 = X_1X_2X_3$. In this case there are two defining contrasts given by

$$L_1 = X_1 + X_2 + X_3$$

and

$$L_2 = X_1 + X_3$$

for B_1 and B_2, respectively. The numerical values of L_1 (mod 2) and L_2 (mod 2) are shown in Table 9.4. As can be seen in this table, all experimental conditions having $L_1 = 0$, $L_2 = 0$ are in block 1, all having $L_1 = 0$, $L_2 = 1$

Table 9.3 Identification of blocks using $L = X_1 + X_2 + X_3$

Experimental condition	X_1	X_2	X_3	*L(mod 2)*	*Block*
1	0	0	0	0	1
2	1	0	0	1	2
3	0	1	0	1	2
4	1	1	0	0	1
5	0	0	1	1	2
6	1	0	1	0	1
7	0	1	1	0	1
8	1	1	1	1	2

Table 9.4 Blocking using two defining contrasts

Experimental condition	Design matrix X_1	X_2	X_3	Treatment yield	L_1 (mod 2)	L_2 (mod 2)	Block
1	0	0	0	(1)	0	0	1
2	1	0	0	*a*	1	1	4
3	0	1	0	*b*	1	0	3
4	1	1	0	*ab*	0	1	2
5	0	0	1	*c*	1	1	4
6	1	0	1	*ac*	0	0	1
7	0	1	1	*bc*	0	1	2
8	1	1	1	*abc*	1	0	3

are in block 2, all having $L_1 = 1$, $L_2 = 0$ are in block 3 and all having $L_1 = 1$, $L_2 = 1$ are in block 4.

In the above example, there are four blocks generated by two blocking variables $B_1 = X_1X_3$ and $B_2 = X_1X_2X_3$. As already known, the number of degrees of freedom associated with four blocks is equal to three. However, the two contrasts corresponding to B_1 and B_2 only contribute with one degree of freedom each. The third degree of freedom actually corresponds to the interaction of B_1 and B_2. This interaction can be found as follows:

$$\begin{aligned} B_1B_2 &= (X_1X_3)(X_1X_2X_3) \\ &= (X_1X_1)(X_2)(X_3X_3) \\ &= X_2 \end{aligned}$$

since $X_1X_1 = X_3X_3 = I$, where I is an identity vector. Since X_2 is a main effect, which usually bears important information, the blocking arrangement using $B_1 = X_1X_3$ and $B_2 = X_1X_2X_3$ is not desirable, unless it is assumed that X_2 is not a significant factor. A more appropriate blocking strategy would be to define $B_1 = X_1X_2$ and $B_2 = X_1X_3$ since in this case $B_1B_2 = X_2X_3$. For a more detailed treatment of this topic, the student can consult reference [1].

EXAMPLE 9.1

The data in Table 9.5 are used to illustrate the ANOVA for the replicated block design introduced in Figure 9.2. Here $B_1 = X_1X_2$, $B_2 = X_1X_3$ and $B_1B_2 = X_2X_3$. Since the blocking variables are the same for the four replications, this design is said to be a design with 'total confounding', as opposed to a design with 'partial confounding' where each replication may have different blocking variables.

Using the results developed in Chapter 8, it is easy to verify that $E_1 = 5.75$, $E_2 = -1.25$, $E_3 = 0.375$, $E_{B_1} = 0.375$, $E_{B_2} = 2.5$, $E_{B_1B_2} = 0$ and $E_{123} = 0.125$.

Table 9.5 Design with total confounding

Run	$X_1X_2X_3$	B_1B_2	*Replications*	*Total*	*Block*
1	$---$	$++$	15 20 10 15	60	4
2	$+--$	$--$	19 20 21 12	72	1
3	$-+-$	$-+$	17 18 9 10	54	2
4	$++-$	$+-$	15 17 14 22	68	3
5	$--+$	$+-$	13 15 12 12	52	3
6	$+-+$	$-+$	20 20 25 18	83	2
7	$-++$	$--$	10 12 14 9	45	1
8	$+++$	$++$	20 15 25 20	80	4

The sums of squares for these estimates, as well as that for the total, are as follows:

$$SS_1 = \frac{rL^2}{2^n} = 8(E^2) = 8(5.75)^2 = 264.5$$

$$SS_2 = 12.50$$

$$SS_3 = 1.125$$

$$SS_{\text{block}} = SS_{B_1} + SS_{B_2} + SS_{B_1B_2} = 51.125$$

$$SS_{123} = 0.125$$

$$SS_{\text{total}} = 8886 - \frac{514^2}{32} = 629.875.$$

Note that SS_{block} can also be computed as follows, using the block totals (117, 137, 120, 140):

$$SS_{\text{block}} = \frac{117^2}{8} + \frac{137^2}{8} + \frac{120^2}{8} + \frac{140^2}{8} - \frac{514^2}{32} = 51.125$$

$$SS_{\text{replication}} = \frac{129^2 + 137^2 + 130^2 + 118^2}{8} - \frac{514^2}{32} = 23.125.$$

Instead of considering 4 main blocks, we can consider 16 blocks with totals shown in Table 9.6. Considering the data in Table 9.6, the following sums

Table 9.6

	Block 1	*Block 2*	*Block 3*	*Block 4*
Replication 1	35	29	37	28
Replication 2	35	32	38	32
Replication 3	35	35	34	26
Replication 4	35	21	28	34

of squares can be easily computed:

$$SS_{16\text{block}} = \frac{35^2 + \cdots + 34^2}{2} - \frac{514^2}{32} = 155.875$$

$$SS_{16\text{block}} = SS_{4\text{block}} + SS_{\text{replication}} + SS_{\text{rep*block}}$$

$$SS_{\text{rep*block}} = 155.875 - 51.125 - 23.125 = 81.625.$$

The fundamental ANOVA equation for this example can be written as follows:

$$SS_{\text{total}} = SS_1 + SS_2 + SS_3 + SS_{123} + SS_{4\text{block}} + SS_{\text{rep}} + SS_{\text{rep*block}} + SS_{\text{error}}.$$

From this equation it is possible to verify that the error sum of squares is equal to $SS_{\text{error}} = 195.75$ with 12 degrees of freedom.

9.3 PARTIAL CONFOUNDING

Consider a 2^3 experimental design with four blocks. If this experiment were going to be replicated four times, an alternative would be to have the same blocking arrangement for all the four replications. That is, each replication would have the same two experimental conditions in each block. If the two-factor interactions $B_1 = X_1X_2$ and $B_2 = X_1X_3$ are used as the block variables, the four replications would contain the experimental condition shown in Table 9.7.

Another alternative would be to use a different set of block variables in each replication. This design is known as a **partially confounded design**. As an example of such a scheme let us consider the following block variables: $B_1 = X_1X_2$ and $B_2 = X_1X_3$ for the first replication, $B_1 = X_1X_2$ and $B_2 = X_1X_2X_3$ for the second, $B_1 = X_1X_3$ and $B_2 = X_1X_2X_3$ for the third, and

Table 9.7 Design with total confounding

	Replications			
	1	*2*	*3*	*4*
Block 1	(1) *abc*	(1) *abc*	(1) *abc*	(1) *abc*
Block 2	*a* *bc*	*a* *bc*	*a* *bc*	*a* *bc*
Block 3	*ab* *c*	*ab* *c*	*ab* *c*	*ab* *c*
Block 4	*b* *ac*	*b* *ac*	*b* *ac*	*b* *ac*

Table 9.8 Design with partial confounding

	Replications			
	1	*2*	*3*	*4*
Block 1	(1) *abc*	*c* *abc*	*b* *abc*	*a* *abc*
Block 2	*a* *bc*	*ac* *bc*	*ab* *bc*	*ab* *ac*
Block 3	*ab* *c*	(1) *ab*	(1) *ac*	(1) *bc*
Block 4	*b* *ac*	*a* *b*	*a* *c*	*b* *c*

$B_1 = X_2X_3$ and $B_2 = X_1X_2X_3$ for the fourth. The corresponding experimental conditions for each block in each replication are shown in Table 9.8.

In the blocking arrangement given in Table 9.8, different interactions have been confounded in different replications. The effect of the interaction X_1X_2 is confounded in replications 1 and 2, but it can still be estimated from replications 3 and 4. The interactions X_1X_3 and X_2X_3 can be estimated similarly from two of four replications. Table 9.9 shows which effects are confounded (c) or unconfounded (u) in each replication. In this table n_u is the number of replications in which an effect is unconfounded and n_c the number of replications where it is confounded. Evidently an effect with a high n_u value can be estimated with more accuracy than another with a low n_u value.

The topic of recovery of interblock information [2, 3] will now be briefly described. The fundamental idea is that the estimate of an effect can be improved by including its interblock information. This is accomplished by combining the effect estimates E_u and E_c, based on the average of the n_u unconfounded estimates and n_c confounded estimates, respectively. Yates [1, 4] developed the basic interblock recovery procedure. Box *et al.* expanded upon Yates' work with the following weighting equation being the primary

Table 9.9 Illustration of partial confounding

Replications	X_1	X_2	X_3	X_1X_2	X_1X_3	X_2X_3	$X_1X_2X_3$
1	u	u	u	c	c	c	u
2	u	u	c	c	u	u	c
3	u	c	u	u	c	u	c
4	c	u	u	u	u	c	c
n_u	3	3	3	2	2	2	1
n_c	1	1	1	2	2	2	3

result:

$$E = \frac{W_u E_u + W_c E_c}{W_u + W_c} \tag{9.4}$$

where $W_u = n_u(S_u^2)^{-1}$ and $W_c = n_c(S_c^2)^{-1}$. S_u^2 and S_c^2 are the separate variance estimates for unconfounded and confounded effects, respectively.

As an illustration, let us consider again the four separate 2^3 factorial designs summarized in Table 9.8. The main effect for factor X_1 is unconfounded in replications 1, 2 and 3, and confounded in replication 4. Therefore, to estimate the main effect for factor X_1, we can use the data from replications 1, 2 and 3. To estimate the main effect of factor X_2, we can use replications 1, 2 and 4, and for estimating the effect of factor X_3 we can use replications 1, 3 and 4. Similarly, an estimate of the two-factor interaction X_1X_2 can be obtained by combining the data from replications 3 and 4. In fact, all the factorial effects can be estimated by combining the information from the replications in which they are not confounded with blocks. Proceeding in this fashion, we obtain the estimates represented by the symbol E_u in equation (9.4).

We can use the four blocks in replication 4 to obtain yet another estimate of the main effect of X_1. The main effect of factor X_1 can be estimated from a simple contrast between these four blocks. Likewise, the blocks of replications 3 and 2 will provide estimates of the main effects of factors X_2 and X_3, respectively. The block-based estimates of the interaction effects may be similarly obtained. These are the estimates represented by the symbol E_c in equation (9.4).

For any effect (main or interaction) two estimates E_u and E_c are combined as indicated in equation (9.4) by weighting them inversely with respect to their variances, $S_u^2 n_u^{-1}$ and $S_c^2 n_c^{-1}$. These variances are obtained from the ANOVA table (see Table 9.11). An example problem is illustrated below.

EXAMPLE 9.2

The data shown in Table 9.10 are used to illustrate the recovery of interblock information of a 2^3 factorial design with partial confounding. The four replications being considered are those previously indicated in Table 9.8. All steps of the computational process needed to estimate the effects are first illustrated in detail for the main effect of factor X_1.

Step 1

Calculate E_u and SS_u. This is accomplished by combining the information from replications 1, 2 and 3 in which factor X_1 is unconfounded:

$$L_u = \frac{-(15+20+10)+(19+20+21)-\cdots-(10+12+14)+(20+15+25)}{3} = 22$$

Table 9.10

Run	$X_1X_2X_3$	*Replications*
1	− − −	15 20 10 15
2	+ − −	19 20 21 12
3	− + −	17 18 9 10
4	+ + −	15 17 14 22
5	− − +	13 15 12 12
6	+ − +	20 20 25 18
7	− + +	10 12 14 9
8	+ + +	20 15 25 20

$$E_u = \frac{L_u}{2^{3-1}} = 5.5 \qquad SS_u = \frac{3L_u^2}{2^3} = 181.5.$$

Step 2

Calculate E_c and SS_c. This is accomplished by combining the information from replication 4:

$$L_c = \frac{-15 + 12 - 10 + 22 - 12 + 18 - 9 + 20}{1} = 26$$

$$E_u = \frac{L_u}{2^{3-1}} = 5.5 \qquad SS_u = \frac{3L_u^2}{2^3} = 181.5.$$

Step 3

Calculate variance estimates. These can be obtained from the ANOVA table (Table 9.11). From the ANOVA table, an estimate of the variance for each confounded effect is equal to $S_c^2 = 3.442$. Similarly, an estimate of the variance for each unconfounded effect is equal to $S_u^2 = 21.561$.

Step 4

Calculate weighted effect estimates. For confounded and unconfounded effect estimates the corresponding weights are defined as $W_c = n_c(S_c^2)^{-1}$ and $W_u = n_u(S_u^2)^{-1}$. As an illustration for X_1, $W_c = 0.291$ and $W_u = 0.139$. Finally, the estimate for each effect is calculated by combining their confounded and unconfounded estimates, E_c and E_u, and weighting them inversely with respect to their variances. As an illustration for factor X_1, this is shown using equation (9.4):

$$E_1 = \frac{(0.139)(5.5) + (0.291)(6.5)}{0.139 + 0.291} = 6.177.$$

The result for all effect estimates using partial confounding in this analysis are provided in Table 9.12.

Table 9.11

Source	*df*	*SS*	*MS*
Blocks	15	190.875	12.725
Replications	3	23.125	7.708
Confounded:			
E_1	1	84.500	84.500
E_2	1	4.500	4.500
E_3	1	21.125	21.125
E_{12}	1	2.250	2.250
E_{13}	1	33.0625	33.0625
E_{23}	1	3.0625	3.0625
E_{123}	1	2.0417	2.0417
Error	5	17.2083	3.442
Unconfounded:			
E_1	1	181.500	181.500
E_2	1	8.167	8.167
E_3	1	15.042	15.042
E_{12}	1	9.000	9.000
E_{13}	1	18.060	18.060
E_{23}	1	3.060	3.060
E_{123}	1	10.125	10.125
Error	9	194.046	21.561
Total	31	629.875	

Table 9.12

Effects	E_c	E_u	n_c	n_u	W_c	W_u	E
1	6.500	5.500	1	3	0.291	0.139	6.177
2	−1.500	−1.167	1	3	0.291	0.139	−1.392
3	−3.250	1.583	1	3	0.291	0.139	−1.688
12	−0.750	1.500	2	2	0.581	0.093	−0.440
13	2.875	2.125	2	2	0.581	0.093	2.772
23	−0.875	0.875	2	2	0.581	0.093	−0.634
123	−0.583	2.250	3	1	0.872	0.046	−0.441

9.4 SOME RECOMMENDED BLOCKING STRATEGIES

Table 9.13 taken from reference [2] illustrates some desirable blocking arrangements for two-level designs. In Table 9.13 a slight change in the notation followed in this book has been made. For instance, $B = 123$ is used instead of $B = X_1X_2X_3$.

Table 9.13 Blocking arrangements for 2^k factorial designs

No. Vars	*Block size*	*Block generator*	*Interaction confounded with blocks*
3	4	$B_1 = 123$	123
	2	$B_1 = 12, B = 13$	12, 13, 23
4	8	$B_1 = 1234$	1234
	4	$B_1 = 124, B_2 = 134$	124, 134, 23
	2	$B_1 = 12, B_2 = 23, B_3 = 34$	12, 23, 34, 13, 1234, 24, 14
5	16	$B_1 = 12345$	12345
	8	$B_1 = 123, B_2 = 345$	123, 234, 1245
	4	$B_1 = 125, B_2 = 235, B_3 = 345$	125, 235, 345, 13, 1234, 24, 145
	2	$B_1 = 12, B_2 = 13, B_3 = 34, B_4 = 45$	12, 13, 34, 45, 23, 1234, 1245, 14, 1345 35, 24, 2345, 1235, 15, 25 i.e. all 2*fi* and 4*fi**
6	32	$B_1 = 123456$	123456
	16	$B_1 = 1236, B_2 = 3456$	1236, 3456, 1245
	8	$B_1 = 135, B_2 = 1256, B_3 = 1234$	135, 1256, 1234, 236, 245, 3456, 146
	4	$B_1 = 126, B_2 = 136, B_3 = 346, B_4 = 456$	126, 136, 346, 456, 23, 1234, 1245 14, 1345, 35, 246, 23456, 12356, 156, 25
	2	$B_1 = 12, B_2 = 23, B_3 = 34, B_4 = 45, B_5 = 56$	all 2*fi*, 4*fi* and 6*fi*
7	64	$B_1 = 1234567$	1234567
	32	$B_1 = 12367, B_2 = 34567$	12367, 34567, 1245
	16	$B_1 = 123, B_2 = 456, B_3 = 167$	123, 456, 167, 123456, 2367, 1457,
	8	$B_1 = 1234, B_2 = 567, B_3 = 345, B_4 = 147$	23457, 1234, 567, 345, 147, 1234567, 125, 237, 3467, 1456, 1357, 1267, 2356, 136, 246
	4	$B_1 = 127, B_2 = 237, B_3 = 347, B_4 = 457, B_5 = 567$	127, 237, 347, 457, 567, 13, 1234, 1245, 1256, 24, 2345, 2356, 3456, 35, 1234567, 46, 147, 13457, 13567, 12467, 12357, 257, 24567, 23467, 367, 15, 1456, 1346, 1236, 26, 167
	2	$B_1 = 12, B_2 = 23, B_3 = 34, B_4 = 45, B_5 = 56, B_6 = 67$	all 2*fi*, 4*fi* and 6*fi*

*'*fi*' is an abbreviation for 'factor interaction'; thus, for example 2*fi* means two-factor interaction

As an illustration of the results given in Table 9.13, the case $n = 5$, $k = 4$ and $r = 1$ will be considered. From equation (9.2), $b = 2^5(4)^{-1} = 8$, and from equation (9.3), $2^v = 8$. Therefore, $v = 3$. Thus block variables B_1, B_2, B_3 will generate four interactions B_1B_2, B_1B_3, B_2B_3, $B_1B_2B_3$. If we define (as in Table 9.13)

$$B_1 = X_1X_2X_5$$
$$B_2 = X_2X_3X_5$$
$$B_3 = X_3X_4X_5$$

the following confounding interactions will be generated:

$$B_1B_2 = X_1X_3$$
$$B_1B_3 = X_1X_2X_3X_4$$
$$B_2B_3 = X_2X_4$$
$$B_1B_2X_3 = X_1X_4X_5.$$

EXERCISES

1. Consider a factorial experiment with five factors each at two levels. For blocks of size equal to 16, 8 and 2, identify a good set of blocking variables and investigate in each case the interactions confounded with block-related effects.
2. Using the geometric interpretation of two-level factorial experiments, illustrate how main effects can be cleared in the case of three factors and two blocks.
3. Consider the data for a 2^3 factorial design in Table 9.14. Conduct an ANOVA for the case where the experimental data are arranged into blocks of size four. Use the same block generators for both replications.
4. Consider a factorial design with five factors, each at two levels. Suppose that the conditions for running the experiments were such that only two observations could be collected at a time. Propose a meaningful blocking strategy and test the significance of the block effect. Use the following experimental data in your analysis, assuming that the runs are given in standard order:

Runs 1–8:	2.152	2.057	2.111	2.037	2.267	2.210	2.301	2.236
Runs 9–16:	2.170	2.033	3.164	1.978	2.301	2.215	2.332	2.072
Runs 17–24:	2.025	2.025	1.944	1.991	2.053	1.944	2.220	1.898
Runs 25–32:	2.004	2.057	2.146	1.857	2.114	1.919	2.161	2.041

Table 9.14

X_1	X_2	X_3	*Observations*	
–	–	–	10	11
+	–	–	20	24
–	+	–	10	12
+	+	–	12	13
–	–	+	10	12
+	–	+	20	20
–	+	+	10	9
+	+	+	22	19

Table 9.15

Run	X_1	X_2	X_3	*Activity level*
1	−	−	−	10 11
2	+	−	−	20 24
3	−	+	−	10 12
4	+	+	−	12 13
5	−	−	+	10 12
6	+	−	+	20 20
7	−	+	+	10 9
8	+	+	+	22 19

5. A drug company is currently developing a new vitamin. The research chemists of the company are studying the influence that three variables may have on the new drug. In order to measure the effect of the three variables the following design was run using mice and choosing the level of activity as the response of the experiment. The corresponding results in standard order are shown in Table 9.15, where each treatment was replicated twice.

   In your analysis, assume that the 16 mice used were divided into two groups according to age, and that younger mice were used for runs 2, 3, 5 and 8. Do a complete analysis of this factorial experiment.
6. Consider a design with four variables in four blocks each containing four runs. Show how an ANOVA test can be performed. Specify your assumptions. Illustrate your complete analysis using the following data, corresponding to two replications of the design and block variables $B_1 = X_1X_2X_4$ and $B_1 = X_2X_3X_4$:

Block 1:	12	14	16	19	Block 1:	23	33	43	55
Block 2:	20	24	22	30	Block 2:	30	40	35	45
Block 3:	56	59	48	40	Block 3:	55	67	77	88
Block 4:	16	15	17	28	Block 4:	12	22	15	19

REFERENCES

1. Yates, F. (1939) The recovery of inter-block information in variety trials arranged in three-dimensional lattices. *Ann. Eugen.*, **9**, 136–56.
2. Box, G. E. P., Hunter, W. G. and Hunter, J. S. (1978) *Statistics for Experimenters*, John Wiley & Sons, New York.
3. Cochran, W. G. and Cox, G. M. (1978) *Experimental Designs*, John Wiley & Sons, New York.
4. Yates, F. (1940) The recovery of inter-block information in balanced incomplete block designs. *Ann. Eugen.*, **10**, 317–25.

Special topics in the analysis of unreplicated two-level factorial experiments

10

Factorial designs are widely used in experiments involving several factors where it is necessary to study the joint effects of these factors on a specified experimental response. The large number of treatment combinations which results in experiments employing many factors may be very difficult or impossible to sample. To reduce the size of the experiment only one replication may be used. This, however, results in zero degrees of freedom for the error and no direct estimate for it is available. The purpose of this chapter is to present and discuss two standard methodologies to investigate the effects when replication is not present. The discussion will be limited to factors at exactly two levels. The methodologies are: (a) use of a reference distribution, and (b) normal probability plots.

10.1 USE OF A REFERENCE DISTRIBUTION

A reference distribution provides a means of determining if a result is explained by mere chance variation or whether it is explained by a significant effect. To make this decision the investigator must in some way produce a relevant reference set that represents a characteristic set of outcomes which could occur entirely without effect. The actual result may then by compared with this reference set. If it is found to be exceptional, the result is declared to be explained by a significant effect.

In unreplicated experiments, the results are assessed using a reference distribution scaled by an error estimate obtained from higher-order interactions. The higher-order interactions will provide an upper limit for the value of the error. Since, in general, these interactions will be small in comparison with the error, they may be used to provide an estimate of experimental error. Cochran and Cox [1] suggest that the interactions constituting the estimate of error be chosen before the results have been inspected.

This procedure is, of course, not free from criticism. The practice of examining the higher-order interactions and using those that are small as errors leads to serious underestimation of the true error variance. On the other hand, if some of the interactions happen to be large, the error mean square that is used will overestimate the true error variance, and the fact that the interactions are large will not be discovered. When a series of experiments of the same general type are being conducted, some safeguard is obtained both by examining the higher-order interactions in experiments that are replicated and by watching the two-factor interactions in experiments with single replication. If most two-factor interactions are small, it seems very unlikely that interactions of a higher order will be large. If many two-factor interactions are found to be large, this suggests that some three-factor interactions may be substantial.

Once the effects which are not significant are identified and an estimate for the error standard deviation is computed, the main effect and interactions are plotted together with a reference distribution which is scaled by a factor equal to the estimated standard deviation. The effects that are significant are then distinguishable from those that are insignificant (random noise).

According to John and Quenouille [2] it may be generally preferable to test a large number of factors without replication rather than a smaller number with replication since the former results in a greater amount of generality and wider applicability of the conclusions.

Most of the material introduced in this chapter will be illustrated on an application of experimental design due to Close [3]. In this application it is intended to identify the most significant effects of factors or interactions affecting the performance of a typical solar water-heater system.

The basic relations governing the behavior of a solar water-heater system have been formulated for solution by a digital computer. Calculations performed for a typical design suggest that a controller that causes the circulating pump to operate whenever the collector temperature rises above the minimum storage temperature should be used; that the flow rate of the circulating water is relatively unimportant; and that days of intermittent insolation can be simulated by simple sine functions. Although there were many factors affecting the response of the experiment, four factors were considered to be the most desirable for a factorial design and analysis.

The four factors selected for the study are listed in Table 10.1 where both lower and upper levels are also indicated for each factor. The experimental results for the 2^4 factorial experiment to be investigated are given in Table 10.2. A simplified representation of the solar water-heater system is shown in Figure 10.1.

In Table 10.2, $-$ represents a total daily insolation of 100 Btu $\text{ft}^{-2}\ \text{h}^{-1}$ and

Table 10.1 Factors and level for each factor

Factor	*Low level (−)*	*High level (+)*
Total daily insolation (Btu ft^{-2} h^{-1})	100	200
Tank storage capacity (Btu ft^{-1})	5	50
Measure of insolation intermittency	0	40
Water mass flow rate (Btu F^{-1} h^{-1})	5	50

Btu ft^{-2} h^{-1} = 3.15457 Wm^{-2}
Btu ft^{-1} = 0.82684 kcal m^{-1}
Btu $F^{-1}h^{-1}$ = 0.52740 W°C^{-1}

Table 10.2 Experimental results in standard order

Treatment	X_1	X_2	X_3	X_4	*Efficiency (%)*
1	−	−	−	−	82.0
2	+	−	−	−	83.7
3	−	+	−	−	61.7
4	+	+	−	−	100.0
5	−	−	+	−	82.1
6	+	−	+	−	84.1
7	−	+	+	−	67.7
8	+	+	+	−	100.0
9	−	−	−	+	82.0
10	+	−	−	+	86.3
11	−	+	−	+	66.0
12	+	+	−	+	100.0
13	−	−	+	+	82.2
14	+	−	+	+	89.8
15	−	+	+	+	68.6
16	+	+	+	+	100.0

+ a total daily insolation of 200 Btu ft^{-2} h^{-1} for factor 1; similarly, − represents a tank storage capacity of 5 Btu ft^{-1} and + a capacity of 50 Btu ft^{-1} for factor 2; also − represents 0 days of insolation intermittency and + 40 days of insolation intermittency for factor 3; finally, − represents a flow rate equal to 5 Btu F^{-1} h^{-1} and + a rate of 50 Btu F^{-1} h^{-1} for factor 4.

Table 10.3 shows the coefficients of all orthogonal contrasts belonging to the main effects and interaction effects associated with the four factors under consideration. Here the value −1 is represented by a minus sign (−) and the value +1 by a plus sign (+). This table also shows the response value for each experimental condition. Using the information given in Table 10.3, any effect can be estimated by dividing the corresponding contrast value by 8; the average or common effect (which corresponds to a column of +1s),

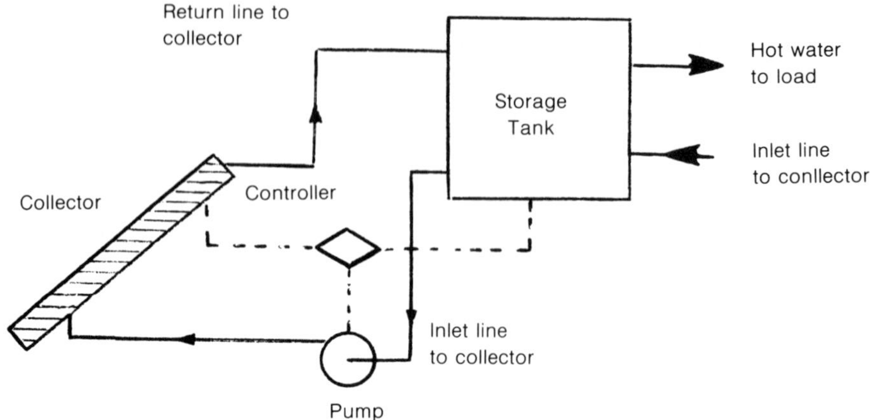

Figure 10.1 Diagram of a typical solar water-heater system.

Table 10.3 Orthogonal contrasts for calculating effect estimates

1	*2*	*3*	*4*	*12*	*13*	*14*	*23*	*24*	*34*	*123*	*124*	*134*	*234*	*1234*	*Efficiency(%)*
−	−	−	−	+	+	+	+	+	+	−	−	−	−	+	82.0
+	−	−	−	−	−	−	+	+	+	+	+	+	−	−	83.7
−	+	−	−	−	+	+	−	−	+	+	+	−	+	−	61.7
+	+	−	−	+	−	−	−	−	+	−	−	+	+	+	100.0
−	−	+	−	+	−	+	−	+	−	+	−	+	+	−	82.1
+	−	+	−	−	+	−	−	+	−	−	+	−	+	+	84.1
−	+	+	−	−	−	+	−	−	−	−	+	+	−	+	67.7
+	+	+	−	+	+	−	+	−	−	+	−	−	−	−	100.0
−	−	−	+	+	+	−	+	−	−	−	+	+	+	−	82.0
+	−	−	+	−	−	+	+	−	−	+	−	−	+	+	86.3
−	+	−	+	−	+	−	−	+	−	+	−	+	−	+	66.0
+	+	−	+	+	−	+	−	+	−	−	+	−	−	−	100.0
−	−	+	+	+	−	−	−	−	+	+	+	−	−	+	82.2
+	−	+	+	−	+	+	−	−	+	−	−	+	−	−	89.8
−	+	+	+	−	−	−	+	+	+	−	−	−	+	−	68.6
+	+	+	+	+	+	+	+	+	+	+	+	+	+	+	100.0

however, is estimated when the value of the contrast belonging to this effect is divided by 16.

The effects of the factors being studied can also be estimated using Yates algorithm, which was described in Chapter 8; for reasons that will be clear as our development continues, the Yates algorithm will be more appropriately referred to as the **forward Yates algorithm**. Table 10.4 summarizes the results obtained from this procedure considering the data shown in Tables 10.2 or 10.3.

Table 10.4 Forward Yates algorithm

Run	*Treatment*	*Y*	*(1)*	*(2)*	*(3)*	*(4)*	*Effect*	*Identity*
1	− − − −	82.0	165.7	327.4	661.3	1336.2	83.510	Mean
2	+ − − −	83.7	161.7	333.9	674.9	151.6	18.950	X_1
3	− + − −	61.7	166.2	334.3	74.3	−8.2	−1.030	X_2
4	+ + − −	100.0	167.7	340.6	77.3	120.4	15.050	X_1X_2
5	− − + −	82.1	168.3	40.0	−2.5	12.8	1.600	X_3
6	+ − + −	84.1	166.0	34.3	−5.7	−5.0	−0.625	X_1X_3
7	− + + −	67.7	172.0	38.3	66.9	4.4	0.550	X_2X_3
8	+ + + −	100.0	168.6	39.0	53.5	−12.2	−1.525	$X_1X_2X_3$
9	− − − +	82.0	1.7	−4.0	6.5	13.6	1.700	X_4
10	+ − − +	86.3	38.3	1.5	6.3	3.0	0.375	X_1X_4
11	− + − +	66.0	2.0	−2.3	−5.7	−3.2	−0.400	X_2X_4
12	+ + − +	100.0	32.3	−3.4	0.7	−13.4	−1.675	$X_1X_2X_4$
13	− − + +	82.2	4.3	36.6	5.5	−0.2	−0.025	X_3X_4
14	+ − + +	89.8	34.0	30.3	−1.1	6.4	0.800	$X_1X_3X_4$
15	− + + +	68.6	7.6	29.7	−6.3	−6.6	−0.830	$X_2X_3X_4$
16	+ + + +	100.0	31.4	23.8	−5.9	0.4	0.050	$X_1X_2X_3X_4$

Since no direct estimate of σ^2 can be obtained from the 16 runs of Table 10.4, an estimate can be found if we assume that all three-factor interactions and the four-factor interaction are negligible; in this case, it is assumed that each of the selected interaction terms has an expected value equal to zero, and, therefore, it contributes with one degree of freedom to the sum of squares associated with those interactions. Table 10.5 summarizes the computation of an estimate of the error variance having 5 degrees of freedom; in this table an estimate of $SS_{\text{error}} = 6.4626$ was obtained and then divided by the number of degrees of freedom (i.e. 5) to get the vaue $MS_{\text{error}} =$ 1.292 52.

Using the estimate $\hat{\sigma}^2 = 1.292\,52$, it is concluded that for any effect estimator E the standard deviation is equal to $\hat{\sigma}_E^2 = (1.292\,52)^{1/2} = 1.1369$; for the common effect, the standard deviation of the estimate is equal to half this value, that is, $\hat{\sigma}_{\bar{Y}} = 0.57$. Table 10.6 shows both effect and standard deviation

Table 10.5 Error variance estimation

Interaction	*Effect*	*Effect*^{2}
$X_1X_2X_3$	−1.525	2.3256
$X_2X_3X_4$	−1.675	2.8056
$X_1X_3X_4$	0.800	0.6400
$X_2X_3X_4$	−0.830	0.6889
$X_1X_2X_3X_4$	0.050	0.0025
		6.4626

Table 10.6 Standard deviation of effect estimates

Effect	*Estimate*	*Standard deviation*
Average	83.51	0.57
1	18.95	1.14
2	−1.03	1.14
3	1.60	1.14
4	1.70	1.14
12	15.05	1.14
13	−0.625	1.14
14	−0.375	1.14
23	0.550	1.14
24	−0.400	1.14
34	−0.025	1.14
123	−1.525	1.14
124	−1.675	1.14
134	0.800	1.14
234	−0.830	1.14
1234	0.050	1.14

estimates for the common effect, the main effects and the interaction effects. In Table 10.6 the effects are identified (using the notation introduced in Table 9.13) by the numbers 1, 2, 3, 4, 12, ..., instead of the symbols $X_1, X_2, X_3, X_4, X_1X_2$, etc. It is noted that X_1 and X_1X_2 are substantially larger than all other effects.

Once the standard deviation of the effect estimates has been computed, it is possible to construct confidence intervals for each effect. For a given significance level α, the confidence interval for an effect with estimate equal to E is given by $E \pm t_{1-\alpha/2}\hat{\sigma}_E$. In the example being considered in this chapter, a 95% confidence interval for the effect estimate E_{12} is given by the range $15.05 \pm (2.571)(1.1369)$. Using two decimal figures this range is found to be interval [12.13, 17.97].

To determine which treatments are significant their effect estimates are plotted together with a reference t-distribution centered at zero and having five degrees of freedom and a scale factor equal to 1.14. This is shown in Figure 10.2, where it is easily seen that the effects of X_1 and X_1X_2 are very significant.

A brief interpretation of the results obtained from the analysis of the effects using the reference distribution shown in Figure 10.2 is given as follows:

(a) An increase in the total daily insolation from 100 to 200 Btu ft^{-2} h^{-1} would increase the efficiency of the system by about 19%.
(b) There is an appreciable interaction between total daily isolation and tank storage capacity.
(c) Individually the tank storage capacity has no important effect; however,

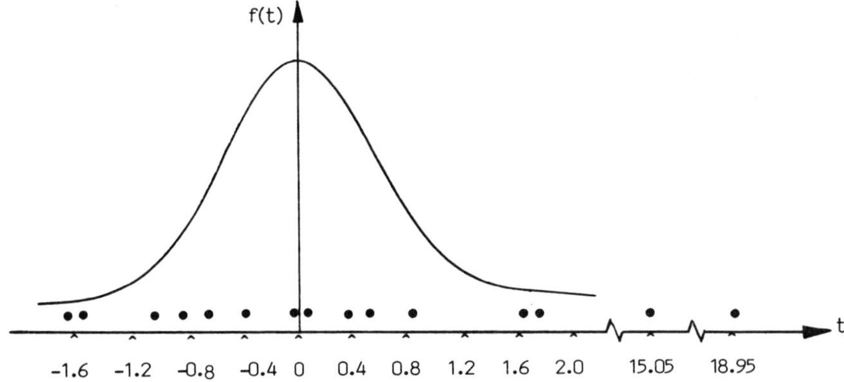

Figure 10.2 Analysis of effects using a reference distribution.

as explained above, the effect of this factor must be considered jointly with total daily insolation.

(d) The measure of insolation intermittency and the water mass flow rate have no significant effect on the performance of the system.

10.2 USE OF NORMAL PROBABILITY PLOTS

When high-order interaction effects are significant the use of a reference distribution would not be appropriate, since this procedure considers those interaction effects as acceptable components of the random error. In cases where there may be a few meaningful high-order interactions a method developed by C. Daniel [4] can be applied.

Basically, Daniel's method uses normal probability plots of the effects in order to identify those that are significant. Once these effects are identified, a regression model is developed to estimate the value of the response for each experimental condition; then it is possible to compute the residuals or difference between the observed response and the estimated value. Finally, the method plots these residuals on normal probability paper to verify if indeed the effects previously considered to be negligible are not important. Daniel's method of normal probability plots will be illustrated using the example discussed in section 10.1.

A simple cumulative distribution can be constructed by ordering the effect estimates from lowest to highest according to their magnitude; an empiric cumulative distribution of the effects can be obtained when each effect is plotted against the percentage of effects that are less than or equal to that effect. That is, the ith ordered effect estimate in a sample of m values would be plotted against the percent probability $P_i = 100im^{-1}$; if each point is moved into a percent probability point, P_i is redefined as $P_i = 100(i - \frac{1}{2})m^{-1}$, where $i = 1, 2, \ldots, m$ and $m = 2^n - 1$.

Fitting a theoretical distribution to the sample of m effect estimates can easily be carried out graphically by plotting their empiric cumulative distribution on normal probability paper. The data and probability scales on the probability paper are so constructed that the cumulative distribution of a normal variable will plot as a straight line as in Fig. 10.3. Figure B.1 (Appendix B) shows a few sample scales for normal plots [5].

10.2.1 The reverse Yates algorithm

Before discussing in more detail the use of normal probability plots, we will pause to consider a method proposed by Daniel [4] to compute the residuals associated with a regression model that includes only effects considered to be significant. Continuing with the analysis of the solar water-heater system described in section 10.1, there is some statistical evidence that all effects but the common effect, the main effect of total daily insolation and the interaction effect due to total daily insolation and tank storage capacity are negligible. As a result of these findings, the response of the experiments conducted to test the significance of the effects can be approximated by the model

$$Y^* = 83.51 + \left(\frac{18.95}{2}\right)X_1 + \left(\frac{15.05}{2}\right)X_1X_2$$

where the coefficients being divided by 2 come from Table 10.4. It is noted that the coefficients of X_1 and X_1X_2 are equal to one-half of the calculated

Table 10.7 Calculation of residuals

Y	Y^*	$Y - Y^*$
82.0	81.56	0.44
83.7	85.46	−1.76
61.7	66.51	−4.81
100.0	100.51	−0.51
82.1	81.56	0.54
84.1	85.46	−1.36
67.7	66.51	1.19
100.0	100.51	−0.51
82.0	81.56	0.44
86.3	85.46	0.84
66.0	66.51	−0.51
100.0	100.51	−0.51
82.2	81.56	0.64
89.8	85.46	4.34
68.6	66.51	2.09
100.0	100.51	−0.51

effects, since a change from a coded value of -1 to a coded value of $+1$ involves a change size of two units.

The residual associated with any estimate Y^* is equal to the difference $Y - Y^*$, where Y is the actual observed experimental response; since the total number of experimental conditions is equal to 2^n, there will be 2^n residuals. For the application under study, the total number of residuals is 16. Table 10.7 summarizes the values of Y, Y^* and $Y - Y^*$ for these 16 residuals.

The residuals $Y - Y^*$ can alternatively and more easily be computed following the reverse Yates algorithm [4]. To perform the algorithm we start with the nth column of the regular forward Yates algorithm, setting all negligible effects equal to zero in this column, and putting the elements in a reversed standard order. Then we execute the regular sequence of additions and subtractions to generate n columns and divide the elements of the last column by 2^n. This reverse Yates computation actually produces fitted values of the response for each of the 2^n experimental conditions. By subtraction, the actual residuals are computed, as indicated in Table 10.8 for the solar water-heater study.

In order to verify if the residuals are normally distributed, they can be plotted on normal probability paper and check if their empiric cumulative distribution can be fitted to a straight line. Once the residuals are ordered according to their magnitude, the percent probability of the ith residual is obtained from the relationship $P_i = 100(i - \frac{1}{2})m^{-1}$, where $i = 1, 2, \ldots, m$ and $m = 2$. The analysis of effects and residuals will be conducted in the remaining portion of this chapter.

Table 10.8 Reverse Yates algorithm

Run	*(0)*	*(1)*	*(2)*	*(3)*	*(4)*	Y^*	Y	$Y - Y^*$
16	0.0	0.0	0.0	0.0	1608.2	100.51	100.0	−0.51
15	0.0	0.0	0.0	1608.2	1064.2	66.51	68.6	2.09
14	0.0	0.0	0.0	0.0	1367.4	85.46	89.8	4.34
13	0.0	0.0	1608.2	1064.2	1305.0	81.56	82.2	0.64
12	0.0	0.0	0.0	0.0	1608.2	100.51	100.0	−0.51
11	0.0	0.0	0.0	1367.4	1064.2	66.51	66.0	−0.51
10	0.0	120.4	0.0	0.0	1367.4	85.46	86.3	0.84
9	0.0	1487.8	1064.2	1305.0	1305.0	81.56	82.0	0.44
8	0.0	0.0	0.0	0.0	1608.2	100.51	100.0	−0.51
7	0.0	0.0	0.0	1608.2	1064.2	66.51	67.7	1.19
6	0.0	0.0	0.0	0.0	1367.4	85.46	84.1	−1.36
5	0.0	0.0	1367.4	1064.2	1305.0	81.56	82.1	0.54
4	120.4	0.0	0.0	0.0	1608.2	100.51	100.0	−0.51
3	0.0	0.0	0.0	1367.4	1064.2	66.51	61.7	−4.81
2	151.6	−120.4	0.0	0.0	1367.4	85.46	83.7	−1.76
1	1336.2	1184.6	1305.0	1305.0	1305.0	81.56	82.0	0.44

Table 10.9 Empiric cumulative distribution of effect

Order i	1	2	3	4	5	6	7
Effect value	−1.675	−1.515	−1.03	−0.83	−0.625	−0.4	−0.025
Identity	124	123	2	234	13	24	34
$P_i = 100\left(\frac{i-\frac{1}{2}}{15}\right)$	3.3	10	16.7	23.3	30	36.7	43.3

Order i	8	9	10	11	12	13	14	15
Effect value	0.05	0.375	0.55	0.8	1.6	1.7	15.05	18.95
Identity	1234	14	23	134	3	4	12	1
$P_i = 100\left(\frac{i-\frac{1}{2}}{15}\right)$	50	56.7	63.3	70	76.7	83.3	90	96.7

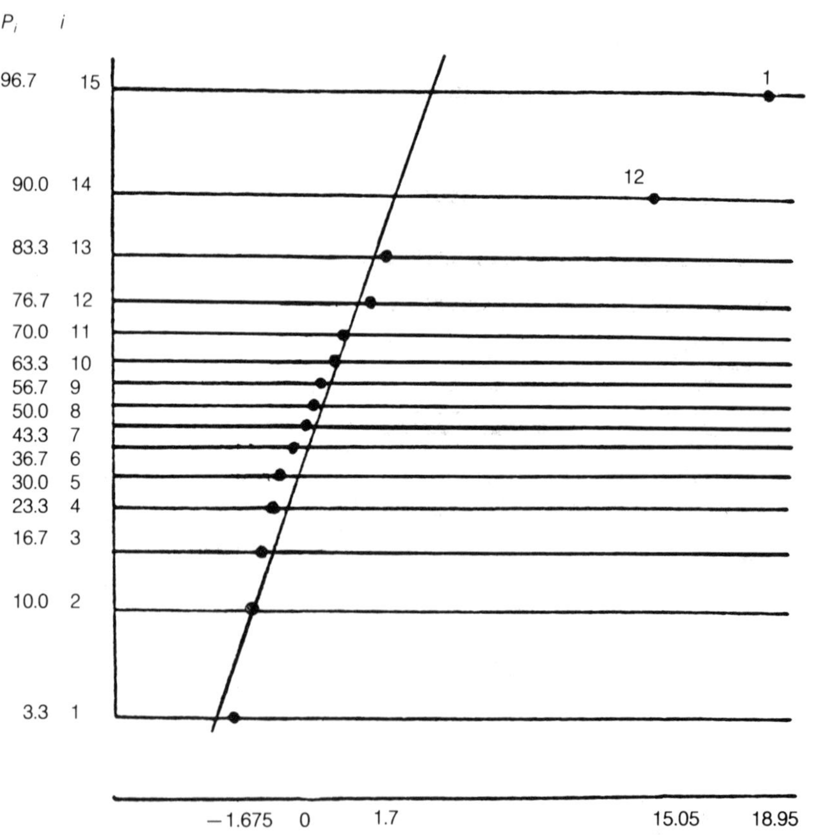

Figure 10.3 Normal probability plots for effects.

10.2.2 Analysis of effects

Suppose that the results given in Table 10.3 were not affected by the changes in the levels of the factors considered in this study. In that case, the 15 effects would represent 15 contrasts between pairs of averages containing 8 observations each, and being roughly normal with mean equal to zero. They would, therefore, be plotted as a straight line on normal probability paper.

To verify if the 15 effects fit as a straight line on normal probability paper (see Figure B.1, Appendix B) the empiric cumulative distribution of the effects must first be obtained, as indicated in Table 10.9.

Figure 10.3 shows the normal probability plots of the effect estimates and their empiric cumulative distribution summarized in Table 10.9. As can be seen in Figure 10.3, 13 estimates fit a straight line reasonably well; only 2 estimates associated with X_1 and X_1X_2 are clearly not on the straight line. Therefore, it is concluded that these two effects are significant.

It is noted that the method of probability plots in this case yields the same results obtained when a reference distribution was used in section 10.1; this, of course, is coincidental, since in the normal probability plot analysis no need exists to assume beforehand that higher-order interactions were not significant.

10.2.3 Analysis of residuals

The empiric cumulative distribution of the residuals given in Tables 10.7 and 10.8 is shown in Table 10.10. This distribution corresponds to residuals

Table 10.10 Empiric cumulative distribution of residuals

Order i	1	2	3	4	5	6	7	8
Residual	−4.81	−1.76	−1.36	−0.51	−0.51	−0.51	−0.51	−0.51
Run No.*	3	2	6	–	–	–	–	–
$P_i = 100\left(\frac{i-\frac{1}{2}}{16}\right)$	3.13	9.38	15.63	21.88	28.13	34.38	40.63	46.80

Order i	9	10	11	12	13	14	15	16
Residual	0.44	0.44	0.54	0.64	0.84	1.19	2.09	4.34
Run No*	–	–	5	13	10	7	15	14
$P_i = 100\left(\frac{i-\frac{1}{2}}{16}\right)$	53.13	59.58	65.67	1.88	78.13	84.3	90.6	96.38

*This is the order of the experimental condition in standard order. Only those runs with no ties in their residuals are indicated. For a discussion of rounding in Yates' algorithm the reader is advised to consult the work by Daniel [4].

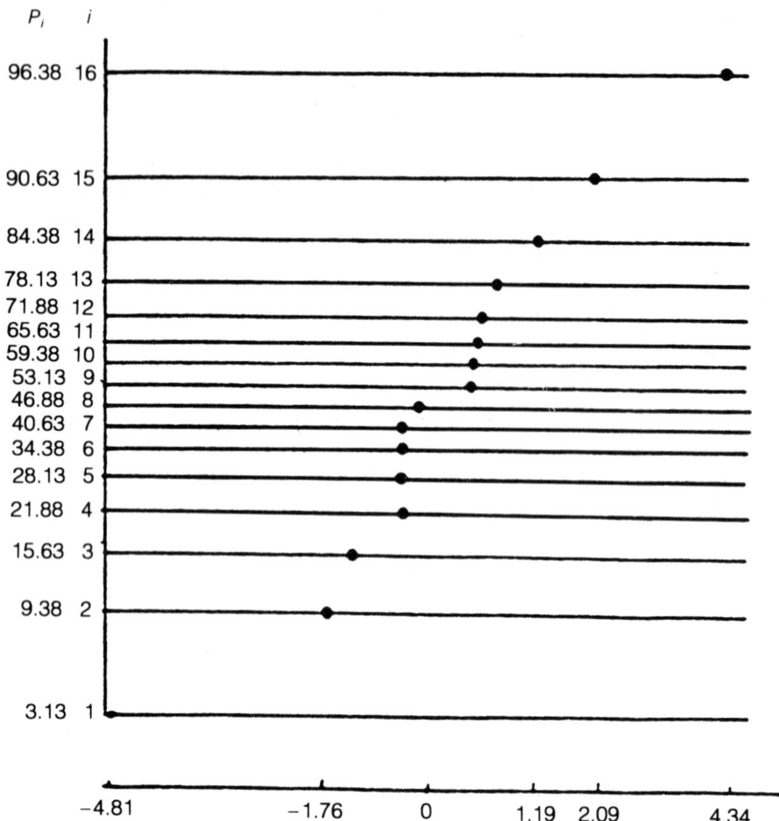

Figure 10.4 Normal probability plots for residuals.

computed on the basis of a model that includes only those effects that appear to be significant. As shown by the use of a reference distribution or probability plots, only X_1 and X_1X_2 should be used in the fitting model.

The normal probability plots of the empiric distribution of residuals is shown in Figure 10.4. Unlike the probability plots of the effects, all points from the residual distribution lie close to a straight line, confirming the conjecture that effects other than those of X_1 and X_1X_2 can be considered to be negligible.

The reader is recommended to consult the excellent book by Box, Hunter and Hunter [5] for more enlightening applications of reference distributions and normal probability plots in the study of unreplicated two-level factorial experiments.

EXERCISES

1. The toxicity of airborne materials in the workplace is of particular concern to the workers and plant management because of the danger to health and loss of production. A dry grinding operation was suspected to be the source of high levels of lead oxide in the area because a leaded steel part was commonly in process. The local industrial engineer and industrial hygiene engineer determined that the four variables listed in Table 10.11 were important in controlling the process to minimize the lead dust generation. Two levels were selected for each variable and an appropriate statistical design was developed.
   (a) Determine the effects from the data in Table 10.12.
   (b) Plot these effects on normal probability paper and report your findings.
   (c) Use reverse Yates algorithm to determine the residual values.
   (d) Plot the residuals on normal probability paper to determine if your conclusions in parts (a) and (b) are consistent.

Table 10.11

Factor	+	−
Wheel revolutions perminute	10 000	6000
Contact time (min)	2	1.3
Surface condition	Prescraped	No change
Orientation of part	Vertical	Horizontal

Table 10.12

Order of run	*1 2 3 4*	*Result*
2	− − − −	0.5
6	+ − − −	0.3
12	− + − −	0.33
4	+ + − −	0.2
1	− − + −	0.57
7	+ − + −	0.46
14	− + + −	0.49
3	+ + + −	0.16
8	− − − +	0.60
10	+ − − +	0.29
15	− + − +	0.31
11	+ + − +	0.14
16	− − + +	0.52
9	+ − + +	0.36
5	− + + +	0.33
13	+ + + +	0.22

(e) Plot the residuals against the Y-values. Report your findings.
(f) State the conclusions and their effect on production, as you would make them for the plant manager.

2. A chemical engineer studied the effects of temperature (X_1), pressure (X_2), alkalinity (X_3) and flow rate (X_4) on a pilot study regarding production of NaOH. Each factor is tested at two levels. The results are given below in standard order:

27.4, 29.9, 19.8, 19.4, 14.5, 14.8, 18.1, 18.4, 15.7, 27.3, 28.2, 16.0, 15.1, 18.9, 22.0, 18.9.

(a) Develop a complete standard design table of all effects and interactions.
(b) Calculate all effects using the forward Yates algorithm.
(c) Calculate residuals using the reverse Yates algorithm.
(d) Plot residuals on normal probability paper.
(e) Plot residuals against results.
(f) If needed, make appropriate transformations on the observed values.

3. Consider a two-level factorial design with results given below in standard order:

1, −9, 20, 12, −7, −9, 17, 10, −10, −20, 19, 13, −11, −21, 15, 8.

(a) Develop a complete standard design table of all effects and interactions.
(b) Determine the effects using Yates algorithm.
(c) Calculate the standard deviation of the effects assuming all three- and four-factor interactions are negligible.
(d) Find all significant effects using the appropriate reference distribution.

4. Use the reference distribution and probability plot methods to study the effects in the experimental situation given in Exercise 11, Chapter 8.
5. Discuss the implications of viewing the residuals as reflections of random error. For an example, the reader can consult reference [6].

REFERENCES

1. Cochran, William G. and Cox, Gertrude M. (1950) *Experimental Designs.* John Wiley & Sons, New York.
2. John, J. A. and Quenouille, M. H. (1977) *Experiments: Design and Analysis.* Charles Griffin & Company Ltd, London.
3. Close, D. J. (1967) A design approach for solar processes. *Solar Energy*, **11**, 112.
4. Daniel, Cuthbert (1976) *Applications of Statistics to Industrial Experimentation.* John Wiley & Sons, New York.
5. Box, George E., Hunter, J. Stewart and Hunter, William G. (1978) *Statistics for Experimenters*, John Wiley & Sons, New York.
6. Federer, Walter T. (1955) *Experimental Design.* The Macmillan Company, New York.

Two-level fractional factorial designs

11

In the factorial designs studied in Chapters 7 and 8, the number of experimental conditions rapidly increases as the number of factors under investigation becomes progressively greater. As a result of this reality, full factorial designs may be undesirable in some practical experimental situations; for instance, a 2^8 full factorial design with each treatment being replicated four times would involve over 1000 runs, which could easily exceed the time, money and other resources available. In other applications, the experimenter may not be interested in all the information that can be derived from a full design, particularly the higher-order interactions. In this chapter we will introduce a methodology that allows the experimenter to reduce the amount of experimentation necessary to obtain almost as much usable information without running a full factorial design. The concept involved in the corresponding analysis is known as a **fractional factorial design**.

Fractional factorial designs were first proposed by Finney [1]. In these designs only a fraction of the experimental conditions is sampled and main effects and lower-order interaction effects are estimated under the assumption that the higher-order interaction effects are negligible. The principles involved in the selection of a fractional factorial design are essentially those followed in confounding blocks with effects in full factorial designs (Chapter 9).

11.1 FUNDAMENTAL CONCEPTS

Consider a factorial experiment involving k factors $X_1, X_2, \ldots, X_k$, each having two levels. A full factorial design to study these factors would require a total of 2^k experimental conditions or runs; as indicated in Chapter 8, a full design allows us to estimate not only main effects and two-factor interactions, but also three-factor and higher-order interactions, which may not be meaningful or easy to explain. As an illustration, in the case of a two-level factorial experiment involving 7 variables, a full design will provide estimates for 7 main effects, 21 two-factor interactions, and 99 higher-order interactions; it could very well be the case that only 28 estimates are desirable (main effects and two-factor interactions) and, therefore, a total of 128 runs may not be reasonable. Actually, it will be shown in this chapter that the 28 desired estimates can be obtained with only 32 runs.

In general, instead of 2^k runs it is possible to estimate main effects and lower-order interaction effects with only 2^m runs, where $m < k$; the fraction of the total number of runs that is actually used is equal to $2^m(2^k)^{-1} = (2^{k-m})^{-1}$ and the collection of the 2^m runs is referred to as a 2^m base design.

A 2^m base design provides estimates for $2^m - 1$ effects, of which m are main effects and $2^m - m - 1$ are interactions. Since $m < k$, not all original k factors can be accommodated as main effects in this 2^m base design. However, if m original factors $X_1, X_2, \ldots, X_m$ are associated with the m main effects of the base design, and the remaining $k - m$ original factors $X_{m+1}, X_{m+2}, \ldots, X_k$ are associated with $k - m$ interactions of the base design, then all original k main effects can be estimated from 2^m runs, assuming that the interactions associated with original factors are negligible.

As an illustration of the basic concept explained in the above paragraph, let us assume that we are only interested in the main effects of seven variables. These main effects can be estimated using a 2^3 base design, which provides seven orthogonal contrasts. Three of these contrasts belong to the main effects of three factors and the remaining four contrasts to the interactions between those factors. Therefore, factors X_1, X_2 and X_3 can be associated with the contrasts belonging to the main effects, and, if their interactions are negligible, factors X_4, X_5, X_6 and X_7 can be assigned to the four contrasts belonging to the interaction effects. Table 11.1 shows the orthogonal contrasts and the effects associated with them. It is noted that contrasts belonging to the interactions are labeled $X_4 = X_1X_2, X_5 = X_1X_3$, $X_6 = X_2X_3$ and $X_7 = X_1X_2X_3$. This design is referred to as a 2^{7-4} design.

In the fractional factorial design in Table 11.1 only main effects can be estimated. If estimates for the 21 two-factors interactions are also desired a 2^5 base design would be needed. The corresponding fractional factorial design would be referred to as a 2^{7-2} design.

The relationship used to indicate which original factors are confounded with which interations of the base design is referred to as the **generator** of

Table 11.1 Contrasts associated with a 2^{7-4} design

Run	X_1	X_2	X_3	$X_4 = X_1X_2$	$X_5 = X_1X_3$	$X_6 = X_2X_3$	$X_7 = X_1X_2X_3$
1	−	−	−	+	+	+	−
2	+	−	−	−	−	+	+
3	−	+	−	−	+	−	+
4	+	+	−	+	−	−	−
5	−	−	+	+	−	−	+
6	+	−	+	−	+	−	−
7	−	+	+	−	−	+	−
8	+	+	+	+	+	+	+

the fractional factorial design. In the 2^{7-4} fractional factorial design shown in Table 11.1 there are four generators:

$$X_4 = X_1 X_2$$
$$X_5 = X_1 X_3$$
$$X_6 = X_2 X_3$$
$$X_7 = X_1 X_2 X_3.$$

If the symbol I is used to represent a column of 2^m + 's, the above generators can alternatively be expressed as

$$I = X_1 X_2 X_4$$
$$I = X_1 X_3 X_5$$
$$I = X_2 X_3 X_6$$
$$I = X_1 X_2 X_3 X_7.$$

The above generators could be labeled I_1, I_2, I_3 and I_4. This notation simplifies the presentation of a new concept referred to as a **definining relation**.

The defining relation of a fractional factorial design is a set or relationships that defines all possible confoundings or aliases that can be obtained from the generators of the design. In order to find the defining relation of a given design its generators are multiplied two at a time, three at a time and so on, until all possible combinations are exhausted.

For the 2^{7-4} fractional factorial design shown in Table 11.1, the defining relation can be obtained as follows:

$$\begin{aligned} I &= I_1 = I_2 = I_3 = I_4 \\ &= I_1 I_2 = I_1 I_3 = I_1 I_4 = I_2 I_3 = I_2 I_4 = I_3 I_4 \\ &= I_1 I_2 I_3 = I_1 I_2 I_4 = I_1 I_3 I_4 = I_2 I_3 I_4 \\ &= I_1 I_2 I_3 I_4. \end{aligned}$$

Equivalently, the defining relation given can be expressed more meaningfully in terms of the original factors:

$$\begin{aligned} I &= X_1X_2X_4 = X_1X_3X_5 = X_2X_3X_6 = X_1X_2X_3X_7 \\ &= X_2X_3X_4X_5 = X_1X_3X_4X_6 = X_3X_4X_7 = X_1X_2X_5X_6 = X_2X_5X_7 = X_1X_6X_7 \\ &= X_4X_5X_6 = X_1X_4X_5X_7 = X_2X_4X_6X_7 = X_3X_5X_6X_7 \\ &= X_1X_2X_3X_4X_5X_6X_7. \end{aligned}$$

The defining relation is used to find the **alias structure** of all the effects associated with a fractional factorial design. The alias structure of a given effect is a set of relationships that indicates which other effects are confounded with it. In order to find the alias structure of an effect all words of the defining relation are multiplied by the given effect; as an illustration, the alias structure of the effect of factor X_1 in the 2^{7-4} fractional factorial design being considered is

$$\begin{aligned}X_1 &= X_2X_4 = X_3X_5 = X_1X_2X_3X_6 = X_2X_3X_7\\ &= X_1X_2X_3X_4X_5 = X_3X_4X_6 = X_1X_3X_4X_7 = X_2X_5X_6 = X_1X_2X_5X_7 = X_6X_7\\ &= X_1X_4X_5X_6 = X_4X_5X_7 = X_1X_2X_4X_6X_7 = X_1X_3X_5X_6X_7\\ &= X_2X_3X_4X_5X_6X_7.\end{aligned}$$

Ignoring all three-factor and higher-order interactions we can reasonably conclude that

$$X_1 = X_2X_4 = X_3X_5 = X_6X_7$$

which means that the main effect of factor X_1 is confounded with three-two-factor interactions.

11.2 COMPUTATION AND ANALYSIS OF EFFECTS

A fractional factorial design for studying k factors with the aid of p generators is generally known as 2^{k-p} fractional factorial design; this design is a $(\frac{1}{2})^p$ fraction of the corresponding full design, requires 2^{k-p} runs and has a defining relation consisting of 2^p words, including I.

The effects associated with a fractional factorial design belong to a set of $2^{k-p}-1$ orthogonal contrasts and can be estimated following the same procedure developed in Chapter 8 for 2^n factorial designs. The corresponding relationships for L, E and SS_L are obtained as follows, after setting $n=k-p$ into the formulas for 2^n designs:

$$L = \sum_c \frac{a_{(c)}T_{(c)}}{r} \tag{11.1}$$

$$E = \left(\frac{1}{2}\right)^{k-p-1} L \tag{11.2}$$

$$SS_L = \frac{rL^2}{2^{k-p}}. \tag{11.3}$$

11.3 AN APPLICATION: A STUDY OF AIR CONDITIONER NOISE LEVEL

A Midwestern air conditioner manufacturer has designed a new type of belt-driven remote cooling unit considered to be more energy efficient than the units previously manufactured. When the acoustics laboratory tested a prototype unit in a 'dead' acoustic test room, it was found that the remote unit produced a 3000 Hz noise at a level of 46 dB. (A typical office has a

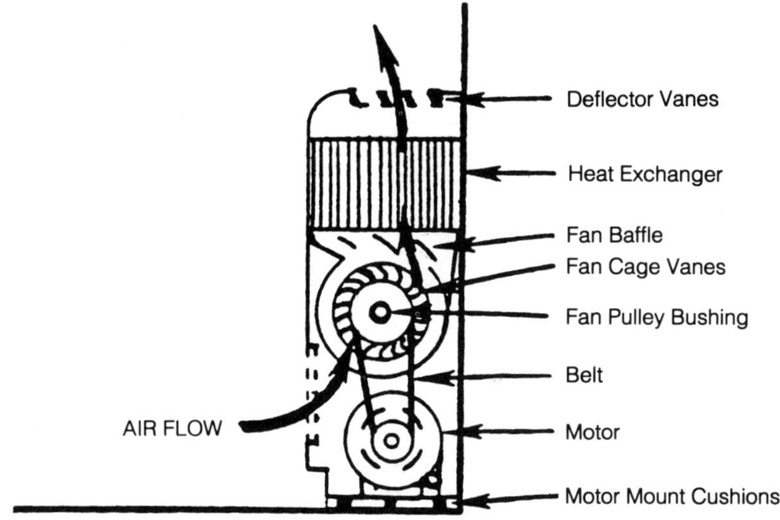

Figure 11.1 Cross section of air conditioner remote unit.

noise level of 50 dB, a living room has a level of 40 dB and a bedroom a level of 30 dB.)

Since the competition has just designed a remote unit which produces a noise level of 40 dB maximum for any frequency, company officials fear that their higher noise level in the 3000 Hz range might hurt sales, especially to hotels and motels. Further testing revealed that the remote unit was quiet in all other frequency ranges not exceeding 34 dB. A decision was made to produce enough modified prototype units to identify the factors affecting the noise level and correct the problem.

Figure 11.1 shows a cross-section of the new remote unit. It is believed, after some testing, that the excessive noise level is not related to the heat exchanger unit itself or to the thermostat/control assembly. For this reason, the test engineers believe that the noise is probably being generated in the fan/airflow portion of the unit.

Table 11.2 indicates seven factors that may affect the noise level of the air conditioner unit; two levels are given for each of the factors. Since the construction and testing of a new prototype remote unit cost over $1000, the use of full design for investigating the seven factors of Table 11.2 would be very expensive (over $100 000). On the other hand, lost sales due to the excessive noise level could amount to a few million dollars.

Since there is some evidence that the higher-order interactions are not very significant (perhaps even negligible), the testing engineers decided that instead of 128 tests they could initially conduct 8 tests, using a 2^{7-4} fractional factorial design.

The response for each experimental condition was defined as the sound

Table 11.2 Factors to be investigated

Noise level factors	*−1*	*+1*
X_1: Friction coating on the belt	Uncoated	Coated
X_2: Fan baffle for even cooling	No baffle	With baffle
X_3: Fan motor brand	Bodine	GE
X_4: Deflector vanes	Metal	Plastic
X_5: Use of a rubber bushing for pulley	Yes	No
X_6: Cushion mount for large motor	Yes	No
X_7: Fastening of squirrel cage fan vanes	Crimped	Brazed

level measured in decibels in the 'dead' room of the acoustics laboratory at the 3000 Hz frequency, using a full-spectrum sound level recording meter. Although it was doubtful that any of the tests would cause a major increase in decibels at other frequencies, the use of full-spectrum tests would identify any such occurrences.

The following generators were used for the last four factors of Table 11.2:

$$X_4 = X_1X_2 \qquad X_6 = X_2X_3$$
$$X_5 = X_1X_3 \qquad X_7 = X_1X_2X_3$$

which, as the reader can verify, are the same generators used in the 2^{7-4} fractional factorial design discussed in section 11.1.

Using the above generators eight experimental conditions involving factors X_1 through X_7 were tested and the corresponding experimental results are given in Table 11.3.

For the design being used we have already found the defining relation in section 11.1. In that section we also identified the two-factor interactions confounded with factor X_1. Proceeding in a similar fashion, the reader can verify that the alias structures for factors $X_2, X_3, \ldots, X_7$ are those given below, after assuming that three-factor and higher-order interactions are

Table 11.3 Noise levels for initial 2^{7-4} design

Test	X_1	X_2	X_3	$X_4=X_1X_2$	$X_5=X_1X_3$	$X_6=X_2X_3$	$X_7=X_1X_2X_3$	*dB*
1	−	−	−	+	+	+	−	34
2	+	−	−	−	−	+	+	46
3	−	+	−	−	+	−	+	34
4	+	+	−	+	−	−	−	47
5	−	−	+	+	−	−	+	34
6	+	−	+	−	+	−	−	46
7	−	+	+	−	−	+	−	35
8	+	+	+	+	+	+	+	45

negligible:

$$X_1 = X_2X_4 = X_3X_5 = X_6X_7 \qquad X_5 = X_1X_3 = X_2X_7 = X_4X_6$$
$$X_2 = X_1X_4 = X_3X_6 = X_5X_7 \qquad X_6 = X_1X_7 = X_2X_3 = X_4X_5$$
$$X_3 = X_1X_5 = X_2X_6 = X_4X_7 \qquad X_7 = X_1X_6 = X_2X_5 = X_3X_4.$$
$$X_4 = X_1X_2 = X_3X_7 = X_5X_6$$

As a result of the above alias structures, each main effect is confounded with three two-factor interactions. For this reason, on the basis of the data shown in Table 11.3, no clear estimates can be computed for the main effects. The effect estimates actually indicate the average effect that an aggregate of factors and interactions have on the noise level of the air conditioner remote unit.

The estimates of the seven aggregate effects can be computed using the results given in section 11.2. As an illustration, the effect of the confounded group X_1,X_2X_4, X_3X_5 and X_6X_7, which will be represented symbolically by

$$X_1 + X_2X_4 + X_3X_5 + X_6X_7,$$

can be obtained as:

(a) The contrast belonging to the first column of Table 11.3 is equal to

$$L_1 = -34 + 46 - 34 + 47 - 34 + 46 - 35 + 45 = 47.$$

(b) From equation (11.2) the estimate of the effect of $X_1 + X_2X_4 + X_3X_5 + X_6X_7$ is found to be equal to $E_1 = (\frac{1}{2})^2 L_1 = 47(\frac{1}{4}) = 11.75$.

Following the procedure previously outlined, the aggregate effects shown below are obtained:

$$\begin{aligned}
X_1 + X_2X_4 + X_3X_5 + X_6X_7 &= +11.75\,\text{dB}\\
X_2 + X_1X_4 + X_3X_6 + X_5X_7 &= +0.25\,\text{dB}\\
X_3 + X_1X_5 + X_2X_6 + X_4X_7 &= -0.25\,\text{dB}\\
X_4 + X_1X_2 + X_3X_7 + X_5X_6 &= -0.25\,\text{dB}\\
X_5 + X_1X_3 + X_2X_7 + X_4X_6 &= -0.75\,\text{dB}\\
X_6 + X_1X_7 + X_2X_3 + X_4X_5 &= -0.25\,\text{dB}\\
X_7 + X_1X_6 + X_2X_5 + X_3X_4 &= -0.75\,\text{dB}
\end{aligned}$$

According to the results given above, only the aggregate effect of X_1 and the three two-factor interactions X_2X_4, X_3X_5 and X_6X_7 seems to be significant. This, however, does not mean that only factor X_1 has a significant effect, since the net effect of each group may be the result of negative as well as positive components.

Section 11.4 describes a methodology that can be used to obtain clear estimates for the main effects. The methodology will be illustrated using the air conditioner noise study experimental data.

11.4 COMPLEMENTARY FRACTIONAL FACTORIALS: MIRROR IMAGE DESIGNS

A total of 2^p fractional designs can be generated from the relationship $I = \pm I_1 = \pm I_2 = \cdots = \pm I_p$. As an illustration, let us consider again a 2^{7-4} fractional factorial design. Since the number of generators is equal to 4, there are 16 designs that can be considered as possible alternatives. These 16 fractional designs can be generated from the relationship $I = \pm I_1 = \pm I_2 = \pm I_3 = \pm I_4$. Table 11.4 shows the signs for each set of four generators that can be used to design a 2^{7-4} fractional factorial experiment.

As can be easily verified, the 2^{7-4} fractional factorial design used in the air conditioner noise study is fraction No. 16 in Table 11.4, when $I_1 = X_1X_2X_4$, $I_2 = X_1X_3X_5$, $I_3 = X_2X_3X_6$ and $I_4 = X_1X_2X_3X_7$. As an illustration of another complementary fraction, let us consider now the following choices for the generators:

$$I = -I_1 = -I_2 = -I_3 = +I_4.$$

That is,

$$\begin{aligned} I_1 &= -X_1X_2X_4 \quad \text{or} \quad X_4 = -X_1X_2 \\ I_2 &= -X_1X_3X_5 \quad \text{or} \quad X_5 = -X_1X_3 \\ I_3 &= -X_2X_3X_6 \quad \text{or} \quad X_6 = -X_2X_3 \\ I_4 &= X_1X_2X_3X_7 \quad \text{or} \quad X_7 = X_1X_2X_3. \end{aligned}$$

Table 11.4 Complementary 2^{7-4} fractional factorial designs

Fraction No.	I_1	I_2	I_3	I_4
1	−1	−1	−1	−1
2	+1	−1	−1	−1
3	−1	+1	−1	−1
4	+1	+1	−1	−1
5	−1	−1	+1	−1
6	+1	−1	+1	−1
7	−1	+1	+1	−1
8	+1	+1	+1	−1
9	−1	−1	−1	+1
10	+1	−1	−1	+1
11	−1	+1	−1	+1
12	+1	+1	−1	+1
13	−1	−1	+1	+1
14	+1	−1	+1	+1
15	−1	+1	+1	+1
16	+1	+1	+1	+1

Table 11.5 Mirror image design

Run	X_1	X_2	X_3	$X_4=-X_1X_2$	$X_5=-X_1X_3$	$X_6=-X_2X_3$	$X_7=X_1X_2X_3$
1	+	+	+	−	−	−	+
2	−	+	+	+	+	−	−
3	+	−	+	+	−	+	−
4	−	−	+	−	+	+	+
5	+	+	−	−	+	+	−
6	−	+	−	+	−	+	+
7	+	−	−	+	+	−	+
8	−	−	−	−	−	−	−

The choice of generators given above corresponds to fraction No. 9. The set of orthogonal contrasts belonging to this design is shown in Table 11.5. It is noted in this table that all the signs of the original fractional factorial design (also referred to as the **principal fraction**) shown in Table 11.1 have been reversed. For this reason, the new fraction is known as a **mirror image** design.

Proceeding as we did in section 11.1 to identify the alias structure of X_1 using the principal fraction of a 2^{7-4} design, the following results are obtained for the mirror image design shown in Table 11.5. As we did before, it is assumed that three-factor and higher-order interactions are negligible.

$$X_1 = -X_2X_4 - X_3X_5 - X_6X_7$$
$$X_2 = -X_1X_4 - X_3X_6 - X_5X_7$$
$$X_3 = -X_1X_5 - X_2X_6 - X_4X_7$$
$$X_4 = -X_1X_2 - X_3X_7 - X_5X_6$$
$$X_5 = -X_1X_3 - X_2X_7 - X_4X_6$$
$$X_6 = -X_2X_3 - X_1X_7 - X_4X_5$$
$$X_7 = -X_3X_4 - X_2X_5 - X_1X_6.$$

Table 11.6 Noise levels for 2^{7-4} mirror image design

Test	X_1	X_2	X_3	$X_4=-X_1X_2$	$X_5=-X_1X_3$	$X_6=-X_2X_3$	$X_7=X_1X_2X_3$	*dB*
9	+	+	+	−	−	−	+	46
10	−	+	+	+	+	−	−	37
11	+	−	+	+	−	+	−	45
12	−	−	+	−	+	+	+	36
13	+	+	−	−	+	+	−	46
14	−	+	−	+	−	+	+	36
15	+	−	−	+	+	−	+	47
16	−	−	−	−	−	−	−	38

Comparing the alias structures obtained from the principal fraction and from the mirror image fraction, it is noted that each main effect is confounded with the same three two-factor interactions. It is also noted that the signs of the interactions in an alias structure are the opposite of those in the other alias structure. This implies that clear estimates for main effects can be obtained by averaging the estimates provided by the principal and the mirror image fractional factorial designs.

Considering again the air conditioner noise level study, it was decided to run eight more tests using the mirror image design. The results from these additional tests are summarized in Table 11.6. From those results given in Table 11.6, and using equations (11.1) and (11.2), the following estimates are obtained:

$$\begin{aligned}
X_1 - X_2X_4 - X_3X_5 - X_6X_7 &= +9.25\\
X_2 - X_1X_4 - X_3X_6 - X_5X_7 &= -0.25\\
X_3 - X_1X_5 - X_2X_6 - X_4X_7 &= -0.75\\
X_4 - X_1X_2 - X_3X_7 - X_5X_6 &= -0.25\\
X_5 - X_1X_3 - X_2X_7 - X_4X_6 &= +0.25\\
X_6 - X_2X_3 - X_1X_7 - X_4X_5 &= -1.25\\
X_7 - X_3X_4 - X_2X_5 - X_1X_5 &= -0.25.
\end{aligned}$$

As the principal fraction provided estimates of the main effects added to their aliases, and the mirror image fraction gave estimates of the same effects with the same aliases subtracted, it is possible to clear the main effect estimates by averaring the results from both fractions. For example, for factor X_1 the clear estimate would be equal to $\frac{1}{2}(11.75 + 9.25) = 10.50$. Proceeding in a similar fashion, the other main effects are also cleared. A summary of the results follows:

$$\begin{aligned}
E_1 &= 10.50\,\text{dB} & E_5 &= -0.25\,\text{dB}\\
E_2 &= 0.00\,\text{dB} & E_6 &= -0.75\,\text{dB}\\
E_3 &= -0.50\,\text{dB} & E_7 &= -0.50\,\text{dB}.\\
E_4 &= -0.25\,\text{dB} & &
\end{aligned}$$

The results from the principal and the mirror image fractions can also be used to provide more precise information on the two-factor interactions. The effect of each group of three two-factor interactions can be obtained by subtracting the results provided by the mirror image design from those provided by the principal fraction and then dividing by two. Proceeding as indicated, the following estimates are available:

$$\begin{aligned}
E_{24+35+67} &= \tfrac{1}{2}(11.75 - 9.25) = 1.25\,\text{dB}\\
E_{14+36+57} &= \tfrac{1}{2}(0.25 + 0.25) = 0.25\,\text{dB}\\
E_{15+26+47} &= \tfrac{1}{2}(-0.25 + 0.75) = 0.25\,\text{dB}
\end{aligned}$$

$$E_{12+37+56} = \tfrac{1}{2}(-0.25 + 0.25) = 0.00\,\text{dB}$$
$$E_{13+27+46} = \tfrac{1}{2}(-0.75 - 0.25) = -0.50\,\text{dB}$$
$$E_{17+23+45} = \tfrac{1}{2}(-0.25 + 1.25) = 0.50\,\text{dB}$$
$$E_{16+25+34} = \tfrac{1}{2}(-0.75 + 0.25) = -0.25\,\text{dB}.$$

As all aliases had been canceled, the engineers concluded that the new, energy-efficient friction coating on the belt was generating the offending 3000 Hz noise as it passed over the pulleys. Fortunately, energy consumption tests revealed that the coating did not significantly reduce energy consumed, and this design feature was eliminated from the production model of the remote unit. The total costs for all tests were under 25 000.00, which was less than one-quarter of the costs of a full factorial experiment for noise alone. Tests of production models have consistently been under 37 dB for all frequency ranges, allowing the company to advertise their new model as 'The Quiet One', which has helped to boost sales.

11.5 RESOLUTION OF FRACTIONAL FACTORIAL DESIGNS

From discussions in previous sections, it is evident that when there is a choice of interactions for introducing additional factors, some options may be considerably more desirable than others, from the point of view of the alias structures generated. A concept that aids in the selection of a fractional factorial design is known as the **resolution** of the design.

In general, a design with the resolution equal to R is a design in which no q-factor ($q < R$) effect is confounded with any other effect containing fewer than $R - q$ factors. As a result of this definition, a design with resolution equal to 3 would be such that no main effect is confounded with another main effect but with two-factor interactions, and these are aliased with other two-factor interactions. Similarly, a design with resolution equal to 4 would have no main effects confounded with main effects or two-factor interactions, although two-factor interactions are aliased with other two-factor interactions.

Finally, a design having resolution equal to 5 would have no main effects aliased with other main effects or with two-factor or three-factor interactions, while two-factor interactions are confounded with three-factor interactions.

In the case of the 2^{7-4} fractional design with generators

$$I = X_1X_2X_4 = X_1X_3X_5 = X_2X_3X_6 = X_1X_2X_3X_7$$

we have already shown that the defining relation is given by

$$\begin{aligned} I &= X_1X_2X_4 = X_1X_3X_5 = X_2X_3X_6 = X_1X_2X_3X_7 \\ &= X_2X_3X_4X_5 = X_1X_3X_4X_6 = X_3X_4X_7 = X_1X_2X_5X_6 \\ &= X_2X_5X_7 = X_1X_6X_7 \end{aligned}$$

$$=X_4X_5X_6=X_1X_4X_5X_7=X_2X_4X_6X_7=X_3X_5X_6X_7$$
$$=X_1X_2X_3X_4X_5X_6X_7.$$

We have also found the following alias structure for X_1:

$$X_1=X_2X_4=X_3X_5=X_1X_2X_3X_6=X_2X_3X_7$$
$$=X_1X_2X_3X_4X_5=X_3X_4X_6=X_1X_3X_4X_7=X_2X_5X_6$$
$$=X_1X_2X_5X_7=X_6X_7$$
$$=X_1X_4X_5X_6=X_4X_5X_7=X_1X_2X_4X_6X_7=X_1X_3X_5X_6X_7$$
$$=X_2X_3X_4X_5X_6X_7.$$

From the above result, the alias structure for the interaction X_1X_7, for instance, will be given by

$$X_1X_7=X_2X_4X_7=X_3X_5X_7=X_1X_2X_3X_6X_7=X_2X_3$$
$$=X_1X_3X_4X_5X_7=X_3X_4X_6X_7=X_1X_3X_4=X_2X_5X_6X_7$$
$$=X_1X_2X_5=X_6$$
$$=X_1X_4X_5X_6X_7=X_4X_5=X_1X_2X_4X_6=X_1X_3X_5X_6$$
$$=X_2X_3X_4X_5X_6.$$

It is noted from the two alias structures given above, that the main effect is not confounded with other main effects but with two-factor interactions, while the two-factor interaction is confounded with other two-factor interactions. In conclusion, this design has resolution equal to 3; for this reason the design is referred to as a 2_{III}^{7-4} fractional factorial design.

Finally, it can be easily seen that the shortest word in the defining relation of the design under consideration has three 'letters'; this is, in fact, a result of the general observation that the resolution of a design is equal to the length of the shortest word in the defining relation. As an illustration we now show how to generate 2_{IV}^{8-4} from a 2_{III}^{7-4}. First, we consider runs 1 through 8 for a 2^{7-4} design with generators $X_4=X_1X_2$, $X_5=X_1X_3$, $X_6=X_2X_3$ and $X_7=X_1X_2X_3$. Second, we consider the mirror image of this design for runs

Table 11.7 Generation of a 2_{IV}^{8-4} design

Run	X_1	X_2	X_3	$X_4=X_1X_2X_8$	$X_5=X_1X_3X_8$	$X_6=X_2X_3X_8$	$X_7=X_1X_2X_3$	X_8
1	−	−	−	+	+	+	−	+
2	+	−	−	−	−	+	+	+
⋮	⋮	⋮	⋮	⋮	⋮	⋮	⋮	⋮
8	+	+	+	+	+	+	+	+
9	+	+	+	−	−	−	+	−
10	−	+	+	+	+	−	−	−
⋮	⋮	⋮	⋮	⋮	⋮	⋮	⋮	⋮
16	−	−	−	−	−	−	−	−

Table 11.8 Resolution III, IV and V designs

R	$q<R$	$R-q$	$<R-q$
3	1	2	1
	2	1	–
4	1	3	1, 2
	2	2	1
	3	1	–
5	1	4	1, 2, 3
	2	3	1, 2
	3	2	1
	4	1	–

9 through 16. Third, we add a new column associated with X_8, as indicated in Table 11.7. Note that for the new design $X_4 = X_1X_2X_8$, $X_5 = X_1X_3X_8$, $X_6 = X_2X_3X_8$ and $X_7 = X_1X_2X_3$.

Table 11.8 illustrates the definition of resolution using three typical values of R. The first column shows the values of the resolution, the second gives the possible values for q, the third identifies the corresponding values for $R-q$ and the last column shows the effects that are not confounded with the effects given in the second column. As can be seen in this table, a design with resolution equal to 4 (that is, a resolution IV design) is one such that no main effects are aliased with other main effects or with two-factor interactions, but two-factor interactions may be confounded with other two-factor interactions. Similar illustrations are given in Table 11.8 for resolution III and resolution V designs.

11.6 BLOCKING IN FRACTIONAL FACTORIAL DESIGNS

Often changes in the response for an experiment may occur as a result of nuisance variables which are not directly under investigation. Changes in the environment of the experiment with time may impose extraneous influences on the measured response which could obscure or inflate the estimates of the effects of the main variables under consideration. Such effects could be in the form of raw material differences, operator or machine differences or environmental differences. In experimental situations where these differences affect the homogeneity of the data to be analyzed, the runs of the design should be grouped as to increase the sensitivity of the significance tests concerning main effects and lower-order interaction effects.

The methodology developed in Chapter 9 for full factorial designs is essentially the same used in the case of fractional designs. In this new situation, however, blocking variables and their interactions have the same status as main effects. As an illustration of the principles involved in blocking the runs of a fractional factorial design, let us consider the following example.

EXAMPLE 11.1

Consider a screening situation in which 7 factors need to be studied in only 16 tests. A certain raw material must be prepared to run the 16 experimental conditions, but a single batch can accommodate only 4 tests. For this reason, it is necessary to use a fractional factorial design with four blocks.

Since two block variables are needed to generate four blocks, the number of generators is five instead of three. The following generators will be considered:

$$I = X_1X_2X_5 \qquad I = X_1X_4B_1$$
$$I = X_1X_3X_6 \qquad I = X_2X_3B_2$$
$$I = X_3X_4X_7$$

where B_1 and B_2 are two required block variables. It is noted that the third degree of freedom associated with the four blocks is $B_1B_2 = X_1X_2X_3X_4$. The B_1B_2 interaction in this case is confounded with a four-factor interaction, which most likely will be negligible. The effect of B_1B_2 is actually given the same status as that of a main effect, since any two of the three degrees of freedom B_1, B_2, B_1B_2 can be used to generate the four blocks needed in the experiment.

The defining relation of the 2^{7-3} fractional design with four blocks under consideration is

$$\begin{aligned}
I &= X_1X_2X_5 = X_1X_3X_6 = X_3X_4X_7 = X_1X_4B_1 = X_2X_3B_2\\
&= X_2X_3X_5X_6 = X_1X_2X_3X_4X_5X_7 = X_2X_4X_5B_1\\
&= X_1X_3X_5B_2 = X_1X_4X_6X_7\\
&= X_3X_4X_6B_1 = X_1X_2X_6B_2 = X_1X_3X_7B_1 = X_2X_4X_7B_2\\
&= X_1X_2X_3X_4(B_1B_2) = X_2X_4X_5X_6X_7 = X_1X_2X_3X_4X_5X_6B_1\\
&= X_5X_6B_2 = X_2X_3X_5X_7B_1 = X_1X_4X_5X_7B_2 = X_3X_4X_5(B_1B_2)\\
&= X_6X_7B_1 = X_1X_2X_3X_4X_6X_7B_2 = X_2X_4X_6(B_1B_2) = X_1X_2X_7(B_1B_2)\\
&= X_1X_2X_5X_6X_7B_1 = X_3X_4X_5X_6X_7B_2 = X_1X_4X_5X_6(B_1B_2)\\
&= X_5X_7(B_1B_2) = X_1X_2X_3X_6X_7(B_1B_2) = X_1X_3X_5X_6X_7(B_1B_2).
\end{aligned}$$

As an illustration of the results that can be obtained from the above defining relation, we will consider the alias structure for main and block effects. Assuming that (a) block variables and their interactions do not interact with other factors, and (b) three-factor and higher-order interactions are negligible, the following alias structures are obtained:

$$\begin{aligned}
X_1 &= X_2X_5 = X_3X_6 & X_6 &= X_1X_3\\
X_2 &= X_1X_5 & X_7 &= X_3X_4\\
X_3 &= X_1X_6 = X_4X_7 & B_1 &= X_1X_4 = X_6X_7\\
X_4 &= X_3X_7 & B_2 &= X_2X_3 = X_5X_6\\
X_5 &= X_1X_2 & B_1B_2 &= X_5X_7.
\end{aligned}$$

Table 11.9

Block	X_1	X_2	X_3	X_4	$X_5=X_1X_2$	$X_6=X_1X_3$	$B_1=X_1X_4$	$B_2=X_2X_3$	X_2X_4	$X_7=X_3X_4$	$X_1X_2X_3$	$X_1X_2X_4$	$X_1X_3X_4$	$X_2X_3X_4$	B_1B_2
1	−	−	−	−	+	+	+	+	+	+	−	−	−	−	+
2	+	−	−	−	−	−	−	+	+	+	+	+	+	−	+
3	−	+	−	−	−	+	+	−	+	+	+	+	−	+	+
4	+	+	−	−	+	−	−	−	−	+	−	−	+	+	+
3	−	−	+	−	+	−	+	−	+	−	+	−	+	+	−
4	+	−	+	−	−	+	−	−	+	−	−	+	−	+	+
1	−	+	+	−	−	−	+	+	−	−	−	+	+	−	+
2	+	+	+	−	+	+	−	+	−	−	+	−	−	−	−
2	−	−	−	+	+	+	−	+	−	−	−	+	+	+	−
1	+	−	−	+	+	−	+	+	−	−	+	−	−	+	+
4	−	+	−	+	−	+	−	−	+	−	+	−	+	−	+
3	+	+	−	+	+	−	+	−	+	−	−	+	−	−	−
4	−	−	+	+	+	−	−	−	−	+	+	+	+	−	+
3	+	−	+	+	+	+	+	−	−	+	−	−	+	−	−
2	−	+	+	+	−	−	−	+	+	+	−	−	−	+	+
1	+	+	+	+	+	+	+	+	+	+	+	+	+	+	+

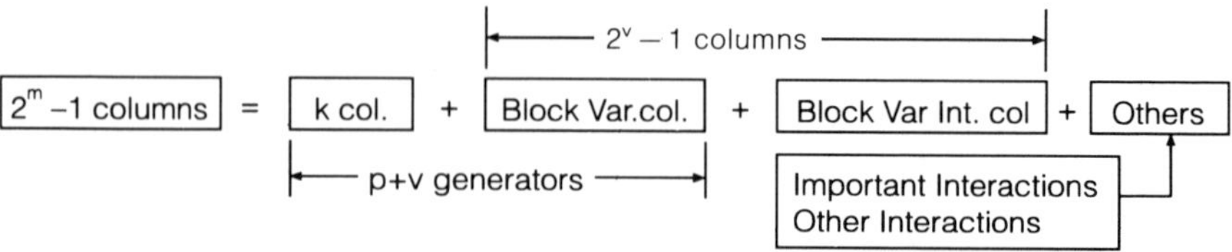

Figure 11.2 Blocking in fractional factorial design.

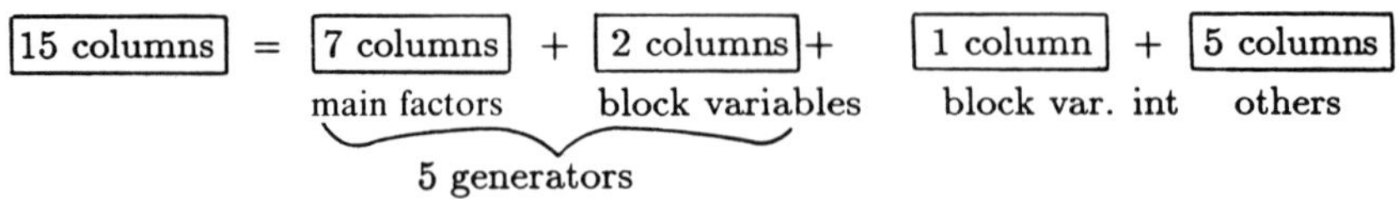

Figure 11.3 An illustration for a 2^{7-3} design.

Proceeding as we did in section 11.3, it is possible to develop a second fractional factorial design that would confound the above main effects with the same interactions but having opposite signs. In this manner it would be possible to obtain clear estimates for all main effects.

Table 11.9 shows the orthogonal contrasts belonging to the effects associated with the design being considered. This table also indicates which experimental conditions should be put together within the same block.

Figure 11.2 indicates the number of columns (orthogonal contrasts) needed for the proper study of a fractional factorial design. As can be seen in this figure, a minimum of $k + (2^v - 1)$ columns are needed to investigate main effects and block effects. As the number of effects (main and block) increases, more contrasts need to be made available, which implies that a larger design with additional experimental conditions might be needed.

As an illustration of the results shown in Figure 11.2, the experimental situation of Example 11.1 will be considered: $k = 7$, $p = 3$, $v = 2$, $b = 4$. In this example, $m = k - p = 7 - 3 = 4$, the base design is a 2^4 design, the number of runs is 16 and the number of runs per block is 4. Figure 11.3 shows that a total of 15 columns is available to accommodate 7 columns for main effects and 3 columns for block effects. In this case, the total number of generators is 5.

11.7 MAJOR AND MINOR VARIABLES IN FRACTIONAL FACTORIALS

In certain well-defined experimental situations it is possible to classify the factors being considered into two groups: major and minor variables. A major variable is a factor whose main effect and interaction effects with other major factors must be estimated. A minor variable is a factor whose main effect must be estimated, but whose interactions with other factors (major and minor) are assumed to have negligible effects.

In a tool life study for a particular operation, such as the cutting operation in a lathe, several factors may have significant effects. In Chapter 1, we identified some of these factors: feed rate, depth of cut, spindle speed, type of lathe, type of material, type of tool, operator skills, etc. Perhaps an adequate classification of these factors is that shown in Table 11.10.

Assuming that it is desired to investigate the relevant effects of the above factors using eight runs, a suitable design would be a 2^{7-4} fractional factorial design. In order to simplify our presentation, we will use the same four generators considered in section 11.1:

$$I = X_1X_2X_4 = X_1X_3X_5 = X_2X_3X_6 = X_1X_2X_3X_7.$$

In this relationship, X_1 is defined as cutting speed, X_2 as feed rate, X_3 as spindle speed, X_4 as operator skills, X_5 as type of lathe, X_6 as type of material and X_7 as type of tool.

Since interactions between a minor variable (X_4, X_5, X_6, X_7) and all other variables are assumed to be negligible, the alias structures found in section 11.2 for this 2^{7-4} fractional factorial design are reduced to

$$\begin{aligned} X_1 &= X_1 & X_5 &= X_1X_3 \\ X_2 &= X_2 & X_6 &= X_2X_3 \\ X_3 &= X_3 & X_7 &= X_7. \\ X_4 &= X_1X_2 \end{aligned}$$

Similarly, from the mirror image fraction considered in section 11.3, the corresponding alias structures would simplify to

$$\begin{aligned} X_1 &= X_1 & X_5 &= -X_1X_3 \\ X_2 &= X_2 & X_6 &= -X_2X_3 \\ X_3 &= X_3 & X_7 &= X_7. \\ X_4 &= -X_1X_2 \end{aligned}$$

Therefore, combining the results from both the principal and the mirror image design, clear estimates are obtained for all main effects and all interactions between major variables.

Table 11.10

Major variables	*Minor variables*
Cutting speed	Operator
Feed rate	Lathe
Spindle speed	Type of material
	Type of tool

EXERCISES

1. Consider a 2^{9-4} fractional factorial design.
   (a) How many variables does the design have?
   (b) How many runs are there in this design?
   (c) How many levels are used for each of the variables?
   (d) How many independent generators are there?
   (e) How many words are there in the defining relation?
2. A fume generation experiment was designed to test the effect of five variables on the rate of cadmium oxide production at a radiator brazing process. Analyze the results given in Table 11.11.
   (a) Describe completely your approach to solve this problem.
   (b) Interpret your findings.
3. An industrial engineer is assigned the responsibility to determine which of seven factors is the most important to develop a low-cost, high-reliability system for supplying parts to the production areas. Because of cost and time constraints, a 2^{7-2} design with generators $I = X_1X_2X_3X_7$ and $I = X_1X_2X_6$ will be considered.
   (a) Determine which main effects are confounded with variable X_1.
   (b) Determine which two-factor interactions are confounded with variable X_1.
   (c) Determine which three factor interactions are confounded with variable X_1.

Table 11.11

Trial No.	X_1	X_2	X_3	X_4	X_5	*Concentration* $(mg\,m^{-3})$
1	−1	−1	−1	−1	+	0.40
2	+1	−1	−1	−1	−	0.20
3	−1	+1	−1	−1	−	0.17
4	+1	+1	−1	−1	+	0.12
5	−1	−1	+1	−1	−	0.32
6	+1	−1	+1	−1	+	0.19
7	−1	+1	+1	−1	+	0.22
8	+1	+1	+1	−1	−	0.20
9	−1	−1	−1	+1	−	0.36
10	+1	−1	−1	+1	+	0.25
11	−1	+1	−1	+1	+	0.34
12	+1	+1	−1	+1	−	0.11
13	−1	−1	+1	+1	+	0.37
14	+1	−1	+1	+1	−	0.21
15	−1	+1	+1	+1	−	0.29
16	+1	+1	+1	+1	+	0.19

(d) How would you determine if other interactions are important to the study?

(e) What improved design would you suggest? Show its advantages.

4. Show how to generate a 2^{5-1} design.
5. The results for a three-factor experiment with two-level factors and two replications for each experimental condition are shown in standard order in Table 11.12. Perform an ANOVA study.
6. Set up a 'mirror design' for the experiment given in Exercise 5. Show all alias combinations and compare them to those of the original design.
7. Consider the design in Table 11.13, where -1 and 1 indicate two coded levels for each factor. Estimate the effects and indicate all possible confoundings.
8. A fractional factorial design was run to determine the relative effects of five twist drill parameters on the drill temperature as measured at a fixed point on the drill flank. The best experiment was performed in 16 runs. The parameters are shown in Table 11.14 (Wang and DeVries [2]) and the data were recorded for two replications as shown in Table 11.15. Perform an analysis on the data using Yates algorithm to determine the effects.

Table 11.12

Experimental conditions	*Replication 1*	*Replication 2*
1	33	71
2	84	103
3	74	109
4	79	47
5	43	112
6	44	88
7	58	40
8	125	16

Table 11.13

X_1	X_2	X_3	X_4	y
1	-1	-1	-1	105
-1	1	-1	-1	107
-1	-1	1	-1	102
-1	-1	-1	1	104
1	1	1	-1	114
1	1	-1	1	111
1	-1	1	1	105
-1	1	1	1	107

Table 11.14

Variable	*Parameter*	*Test levels*	
X_1	Drill point configuration	Crankshaft type grind	Thin webbed
X_2	Margin width (in)	0.03	0.060
X_3	Lip height (in)	0.00	0.004
X_4	Helix angle (degrees)	14.00	34.00
X_5	Surface condition	Coated	Noncoated

Table 11.15

Order of run	X_1	X_2	X_3	X_4	$X_5 = X_1X_2X_3X_4$	R_1	R_2
1	+	−	−	−	−	263	269
2	−	+	+	−	+	329	341
3	+	−	+	−	+	269	269
4	−	−	−	+	−	312	318
5	−	−	−	−	+	322	322
6	+	+	+	−	−	286	281
7	+	−	−	+	−	269	269
8	−	+	+	+	−	269	269
9	−	+	−	−	−	318	315
10	−	−	+	−	−	302	302
11	+	−	−	+	+	254	269
12	+	+	+	+	+	286	276
13	−	−	+	+	+	302	309
14	−	+	−	+	+	286	289
15	+	+	−	−	+	302	292
16	+	−	+	+	−	269	266

9. (a) Is it possible to develop a resolution V two-level fractional factorial for 6 variables that requires only 16 tests? Explain your answer.
   (b) What is the maximum number of variables that could be examined in a 64-run two-level fractional factorial design and maintain at least a resolution V?
   (c) If it is desired to consider two additional variables in the scheme you propose in (b), will the number of tests have to increase to maintain a resolution V design? How many tests are now required?
10. Obtain a design for examining 6 two-level variables in 4 blocks each containing 4 runs. The treatments in each block were put together according to the block generators $\boldsymbol{B}_1 = \boldsymbol{X}_1\boldsymbol{X}_2\boldsymbol{X}_4$ and $\boldsymbol{B}_2 = \boldsymbol{X}_1\boldsymbol{X}_3\boldsymbol{X}_4$. Indicate the additional (main) generators, the defining relation, and show what estimates can be found. Conduct an ANOVA test and specify your

Table 11.16

Serial No. of weighing	Objects placed on: Right pan	Objects placed on: Left pan	Standard weights to balance: Right pan	Standard weights to balance: Left pan
1	abcdefgh	–	–	4.296
2	abcd	efgh	0.269	–
3	aceg	bdfh	0.050	–
4	abef	cdgh	0.009	–
5	acfh	bdeg	0.090	–
6	abgh	cdef	0.312	–
7	adeh	bcfg	–	0.235
8	adfg	bceh	–	0.166
9	–	abcdefgh	4.521	–
10	efgh	abcd	–	0.383
11	bdfh	aceg	–	0.127
12	cdgh	abef	0.023	–
13	bdeg	acfh	–	0.198
14	cdef	abgh	–	0.209
15	bcfg	adeh	0.104	–
16	bceh	adfg	0.238	–

assumptions. Use the following data, corresponding to two replications of the design:

Block 1	12	14	16	19	Block 1	23	33	43	55
Block 2	20	24	22	30	Block 2	30	40	35	45
Block 3	56	59	48	40	Block 3	55	67	77	88
Block 4	16	15	17	28	Block 4	12	22	15	19

11. Table 11.16 gives the results in grams of 16 weighings of 8 objects a, b, c, d, e, f, g and h in a chemical balance. Estimate the weights of the objects, the standard errors of the estimates and the error variance of a single weighing.

REFERENCES

1. Finney, D.J. (1945) The fractional replication of factorial arrangements, *Annals of Eugenics*, **12**, 291–301.
2. Wang, K. and DeVries, M. (1969) Investigation of manufacturing processes by statistical experimental design techniques. *Technical Paper, University of Wisconsin Engineering Experiment Station, MR69-263*, reprint No. 1369.

FURTHER READING

Winer, B. (1962) *Statistical Principles in Experimental Design*, McGraw-Hill Book Company, New York.

National Bureau of Standards (1957) *Fractional Factorial Designs for Factors at Two Levels*, National Bureau of Standards Applied Mathematics Series, Washington D.C.

Davies, O. (ed.) (1956) *The Design and Analysis of Industrial Experiments*, Hafner Publishing Company, New York.

Box, G. and Hunter, J. (1961) The 2^{k-p} fractional factorial designs. *Technometrics*, **3**, 311–52, 449–58.

Three-level factorial designs 12

12.1 INTRODUCTION

The material to be studied in this chapter can be conceptually divided into two parts. The first part, including sections 12.2 and 12.3, will consider the full three-level factorial design. The second part includes Sections 12.4 through 12.6 and focuses on the analysis and one application of the fractional three-level factorial design.

In many cases, even if the number of factors is relatively small, the number of experimental conditions may turn out to be large. For example a three-level design with eight factors will require $3^8 = 6561$ experimental conditions. Of the associated 6560 degrees of freedom only 16 correspond to main effects. For this reason in many practical situations it is more convenient to select a fractional factorial design. In general, we may choose $(\frac{1}{3})^p$ fraction of a complete 3^k design with $p < k$; in this case the fraction contains $(\frac{1}{3})^p 3^k$ runs, or 3^{k-p} runs. For this reason the fraction is referred to as a 3^{k-p} fractional factorial design. As an illustration, a 3^{k-2} design is a one-ninth fraction of a full three-level design.

The three-level design has been extensively investigated during the last 35 years. In 1959 Connor and Zelen [1] wrote a classical paper including criteria for selecting three-level fractional factorial designs involving 4–10 factors. A good portion of this chapter is based on their article.

In order to simplify the analysis, it is assumed that certain high-order interactions (three-factor interactions or higher-order interactions) are negligible, so that estimates of the main effects and two-factor interaction effects may be obtained by running the experiment with a fraction of the complete factorial experiment. When the number of experimental conditions is reduced because of technical, economic or other reasons, a loss of information may occur. This loss of information is due to the fact that some main effects and interactions may be 'aliased' with other main effects or interactions. It is noted that two-factor interactions are usually called or referred to as first-order interactions, three-factor interactions as second-order interactions, and so on. If the use of blocks is desired once the fractional factorial design has been obtained, it should be pointed out that there is usually an additional loss of information. This loss of information results from the fact that certain interactions and their aliases become confounded with the blocks of the experiment.

12.2 THE 3^2 DESIGN

The simplest 3^k type design is the 3^2 factorial design. In this design we are interested in testing the significance of the effects of two factors, each having three levels, and their interaction effect. The experimental conditions for this design are shown in Figure 12.1, where the numbers 0, 1, and 2 represent the levels of either factor. Since there are $3^2 = 9$ treatments (level combinations), there are eight degrees of freedom among these treatments. These eight degrees of freedom can be further subdivided into two for each main effect and four for their interaction effect. Therefore, if there are r replications for each treatment, there will be $(r3^2 - 1) - 8 = 3^2(r - 1)$ degrees of freedom for the experimental error. The statistical model for this experiment is formulated as:

$$Y_{ijk} = \mu + A_i + B_j + (AB)_{ij} + \varepsilon_{ijk}$$

where the following notation is used: μ is the common effect; A_i the effect of level i of factor $A, i = 1, 2, 3$; B_j the effect of level j of factor $B, j = 1, 2, 3$; $(AB)_{ij}$ the interaction effect due to factors A and B when A is set at level i and B is set at level $j, i = 1, 2, 3; j = 1, 2, 3$ and ε_{ijk} is the experimental error, $i = 1, 2, 3$; $j = 1, 2, 3$; $k = 1, 2, \ldots, r$.

12.2.1 Main effects

The two factors investigated in the ANOVA can be either quantitative or qualitative. Quantitative factors are factors whose levels can be associated with points on a numerical scale, such as temperature, pressure or time.

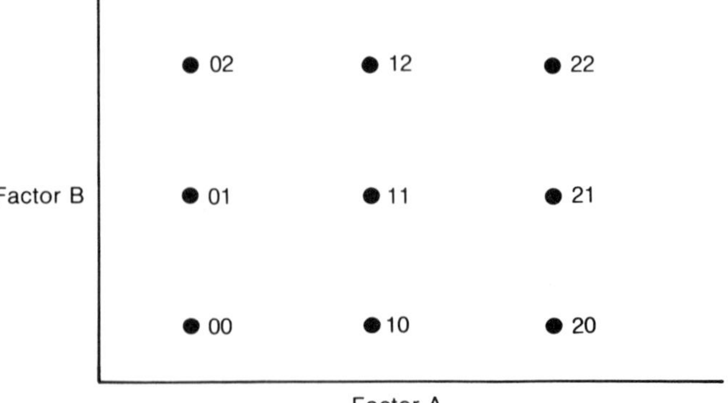

Figure 12.1 Experimental conditions for a 3^2 design.

Qualitative factors, on the other hand, are factors whose levels cannot be arranged in order of magnitude, such as operators, batches or shifts. The main effect of any quantitative factor may be partitioned into a linear and a quadratic component, each component having a single degree of freedom. In the case of qualitative factors, no quadratic component can be considered.

Now we will consider the coefficients of the orthogonal contrast of a quadratic effect for any factor with three quantitative levels specified as $X_1, X_1 + k$, and $X_1 + 2k$. Here k is the difference between the first level and the second level, and between the second and third level. According to the definition of orthogonality

$$c_2(X_1 + k) = c_1 X_1 + c_3(X_1 + 2k).$$

Consequently the orthogonal contrast coefficients of the quadratic effect are $c_1 = 1, c_2 = -2$ and $c_3 = 1$; these coefficients are shown in Table 12.1.

The design matrix is shown in Table 12.2, where $a_{(c)}$, $b_{(c)}$, $c_{(c)}$ and $d_{(c)}$ represent the coefficients in the A_L, A_Q, B_L and B_Q columns, respectively, corresponding to experimental condition c. The first column shows the level combinations corresponding to the treatments or experimental conditions. Based on the columns associated with the linear (L) and quadratic (Q) components of the main effects, it is possible to generate the remaining

Table 12.1 Orthogonal contrast coefficients

Levels	*Linear (L)*	*Quadratic (Q)*
Low (0)	−1	+1
Medium (1)	0	−2
High (2)	+1	+1

Table 12.2 Design matrix for a 3^2 experiment

Treatments	A_L	A_Q	B_L	B_Q	A_LB_L	A_LB_Q	A_QB_L	A_QB_Q
1(00)	−1	+1	−1	+1	+1	−1	−1	+1
2(01)	−1	+1	0	−2	0	+2	0	−2
3(02)	−1	+1	+1	+1	−1	−1	+1	+1
4(10)	0	−2	−1	+1	0	0	+2	−2
5(11)	0	−2	0	−2	0	0	0	+4
6(12)	0	−2	+1	+1	0	0	−2	−2
7(20)	+1	+1	−1	+1	−1	+1	−1	+1
8(21)	+1	+1	0	−2	0	−2	0	−2
9(22)	+1	+1	+1	+1	+1	+1	+1	+1
	$a_{(c)}$	$b_{(c)}$	$c_{(c)}$	$d_{(c)}$				

columns shown in Table 12.2 by multiplying, for each experimental condition, the values in the two columns indicated by the corresponding headings.

The linear and quadratic components of the main effects are estimated as follows:

$$E_{A(L)} = \frac{L_{A(L)}}{3^{n-1}} \qquad E_{A(Q)} = \frac{L_{A(Q)}}{3^{n-1}} \tag{12.1}$$

$$E_{B(L)} = \frac{L_{B(L)}}{3^{n-1}} \qquad E_{B(Q)} = \frac{L_{B(Q)}}{3^{n-1}} \tag{12.2}$$

The corresponding contrasts for factors A and B are defined as follows:

$$L_{A(L)} = \sum_c a_{(c)} \bar{Y}_{(c)} \qquad L_{A(Q)} = \sum_c b_{(c)} \bar{Y}_{(c)}$$

$$L_{B(L)} = \sum_c c_{(c)} \bar{Y}_{(c)} \qquad L_{B(Q)} = \sum_c d_{(c)} \bar{Y}_{(c)}$$

The sum of squares for any main effect can be partitioned into a sum of squares due to the linear component and a second one due to the quadratic component, each sum having one degree of freedom. The sum of squares associated with the main effect is equal to the sum of the two related sums of squares. The results are shown below for factors A and B:

$$SS_A = \frac{rL^2_{A(L)}}{\sum_c a^2_{(c)}} + \frac{rL^2_{A(Q)}}{\sum_c b^2_{(c)}} \tag{12.3}$$

$$SS_B = \frac{rL^2_{B(L)}}{\sum_c c^2_{(c)}} + \frac{rL^2_{B(Q)}}{\sum_c d^2_{(c)}} \tag{12.4}$$

12.2.2 Interaction effect

The sum of squares of the interaction effect can be partitioned into four components, each one corresponding to one degree of freedom. The orthogonal contrasts belonging to the interaction components associated with factors A and B are:

$$L_{A(L)B(L)} = \sum_c a_{(c)} c_{(c)} \bar{Y}_{(c)} \qquad L_{A(L)B(Q)} = \sum_c a_{(c)} d_{(c)} \bar{Y}_{(c)}$$

$$L_{A(Q)B(L)} = \sum_c b_{(c)} c_{(c)} \bar{Y}_{(c)} \qquad L_{A(Q)B(Q)} = \sum_c b_{(c)} d_{(c)} \bar{Y}_{(c)}$$

The sum of squares for the interaction effects of factors A and B can be calculated as the sum of the four associated sums of squares, as indicated in equation 12.5:

$$SS_{AB} = \frac{rL^2_{A(L)B(L)}}{\sum_c a^2_{(c)} c^2_{(c)}} + \frac{rL^2_{A(L)B(Q)}}{\sum_c a^2_{(c)} d^2_{(c)}} + \frac{rL^2_{A(Q)B(L)}}{\sum_c b^2_{(c)} c^2_{(c)}} + \frac{rL^2_{A(Q)B(Q)}}{\sum_c b^2_{(c)} d^2_{(c)}} \tag{12.5}$$

where each term is a sum of squares with one degree of freedom.

EXAMPLE 12.1

In this example we illustrate the use of a full three-level factorial design in the analysis of flexible pavement (road surface) performance. Performance is measured in terms of an index known as the **present serviceability index** (PSI). The PSI varies between 0 ('impassable' road) and 5 ('perfect' road) and measures the smoothness of riding conditions in terms of empirical evaluations of roughness, cracking, patching and rut depth of the pavement. This application is based on a project developed at the Texas Transportation Institute (TTI) to investigate the loss in pavement serviceability as a result of traffic loads under different types of environmental and structural conditions [2].

Specifically, we want to study the effect that two environmental factors known as the Thornthwaite index (factor A) and the average annual temperature (factor B) have on the service life of a hot mix asphaltic concrete or bituminous-base road. For any traffic mix consisting of several vehicle classes, the service life of a road is defined as the cumulative number of equivalent 18-kip single-axle load (ESAL) applications since the time the road was built (or rehabilitated). Extensive studies have provided factors for converting various axle loads to 18-kip ESAL applications. The levels to be considered for the two factors are:

Thornthwaite Index (A)	5	25	45
Temperature (B)	50 °F	60 °F	70 °F

In order to evaluate the present serviceability index for any level of cumulative traffic loads, the relationship presented in Equation 3.69 of Section 3.3.1 will be used:

$$P = P_0 - (P_0 - P_f)e^{-(\rho/W)^\beta}.$$

For this application, W represents the traffic load in millions of 18-kip ESAL applications, P_0 is the initial PSI value, P_f is the PSI value at which the performance curve has an asymptote and β and ρ are parameters which depend on environmental and structural road conditions. From a flexible pavement data base available at TTI the values $P_0 = 4$ and $P_f = 2$ were found to be typical of the road under consideration. Moreover, using procedures developed by Garcia-Diaz *et al.* [2], we can determine the parameters β and ρ of the pavement performance model for each combination of values for the Thornthwaite index and the average annual temperature. Once these values are available, the service life of a road can be computed from the performance model after setting P equal to a critical value P_c (which represents the value at which the road has finished its designed service life):

$$W = \rho\left(\frac{P_0 - P_c}{P_0 - P_f}\right)^{1/\beta}.$$

Since $P_0 = 4$ and $P_f = 2$, it is concluded that the service life of the type of road under consideration can be estimated from $W = \rho(\frac{1}{2})^{1/\beta}$.

We chose four roads of the same type for each of the nine experimental conditions defined by the level combinations of the Thornthwaite index and average annual temperature. The corresponding results in thousands of 18-kip ESAL applications are shown in Table 12.3. Using a level of significance $\alpha = 0.05$ we want to test the significance of each factor and their interaction.

The totals for the nine experimental conditions, in the same order as the level combinations are listed in Table 12.2, and are 727, 658, 650, 532, 441, 332, 382, 364 and 319. After dividing these totals by 4 and using the coefficients given in Table 12.2, the contrasts belonging to the linear and quadratic components of the main effect of factor A, for example, can be computed:

$$L_{A_{(L)}} = \sum_c a_{(c)} \bar{Y}_{(c)} = -242.5 \qquad L_{A_{(Q)}} = \sum_c b_{(c)} \bar{Y}_{(c)} = 122.5.$$

Thus, from equation (12.3), $SS_A = 4(-242.5)^2/6 + 4(122.5)^2/18 = 42\,538.89$. Proceeding in a similar fashion, we can compute $SS_B = 4\,820.23$, and $SS_{AB} = 1\,615.94$. Continuing with our analysis, the sum of squares of the totals can be calculated as follows:

$$SS_{\text{total}} = \sum_i \sum_j \sum_k Y_{ijk}^2 - \frac{T_{\cdots}^2}{N} = 53\,904.31$$

Therefore:

$$SS_{\text{error}} = SS_{\text{total}} - SS_A - SS_B - SS_{AB} = 4\,929.25$$

The results from the ANOVA tests are summarized in Table 12.4. From the ANOVA results, it is concluded that the main effects of factors A (Thornthwaite index) and B (temperature), as well as their interaction effect, are significant. The effect of the Thornthwaite index is indeed extremely significant. As an exercise, the reader is asked to plot the average road service life *versus* temperature to help in picturing the results concerning the interaction components, particularly that of $A_Q B_L$.

Table 12.3 Pavement service life results

Thornthwaite index	*Temperature*					
	50°F		*60°F*		*70°F*	
5	188	175	173	140	160	170
	184	180	180	165	172	148
25	119	138	116	104	95	84
	149	126	106	115	78	75
	94	110	81	74	73	104
45	81	97	120	89	82	60

Table 12.4 ANOVA results for example 12.1

Source	*df*	*SS*	*MS*	*F*	*Significant?*
A	2	42538.89	21269.45	116.50	Yes
A_L	1	(39204.17)	39204.17	214.75	Yes
A_Q	1	(3334.72)	3334.72	18.27	Yes
B	2	4820.23	2410.11	13.20	Yes
B_L	1	(4816.67)	4816.67	26.38	Yes
B_Q	1	(3.56)	3.56	0.02	No
AB	4	1615.94	403.98	2.21	Yes
A_LB_L	1	(12.25)	12.25	0.07	No
A_LB_Q	1	(161.33)	161.33	0.88	No
A_QB_L	1	(1408.33)	1408.33	7.71	Yes
A_QB_Q	1	(34.03)	34.03	0.19	No
Error	27	4929.25	182.56		
Total	35	53904.31			

12.3 THE 3^k DESIGN

In this section we will extend the procedures discussed in section 12.2 to the case of a 3^3 factorial experiment and then generalize the results to study a 3^k experiment.

12.3.1 Notation

The 3^3 factorial design is a factorial experiment involving three factors A, B and C, each having three levels represented by the digits 0, 1 and 2. The corresponding 27 experimental conditions or treatments of this design are shown in Figure 12.2. The 26 degrees of freedom associated with the treatments can be subdivided into six degrees for main effects, 12 degrees for two-factor interactions and eight degrees for the three-factor interaction. If each treatment is replicated r times, there will be $3^3(r-1)$ degrees of freedom associated with the experimental error.

The statistical model for this experiment is formulated as:

$$Y_{ijk} = \mu + A_i + B_j + (AB)_{ij} + C_k + (AC)_{ik} + (BC)_{jk} + (ABC)_{ijk} + \varepsilon_{ijk}$$

where the following notation is used: μ is the common effect; A_i the effect of level i of factor A, $i = 1, 2, 3$; B_j the effect of level j of factor B, $j = 1, 2, 3$; C_k the effect of level k of factor C, $k = 1, 2, 3$; $(AB)_{ij}$ the interaction effect due to factors A and B when A is set at level i and B at level j: $i, j = 1, 2, 3$; $(AC)_{ik}$ the interaction effect due to factors A and C when A is set at level i and C at level k: $i, k = 1, 2, 3$; $(BC)_{jk}$ the interaction effect due to factors B and C when B is set at level j and C at level k: $j, k = 1, 2, 3$; $(ABC)_{ijk}$ the interaction

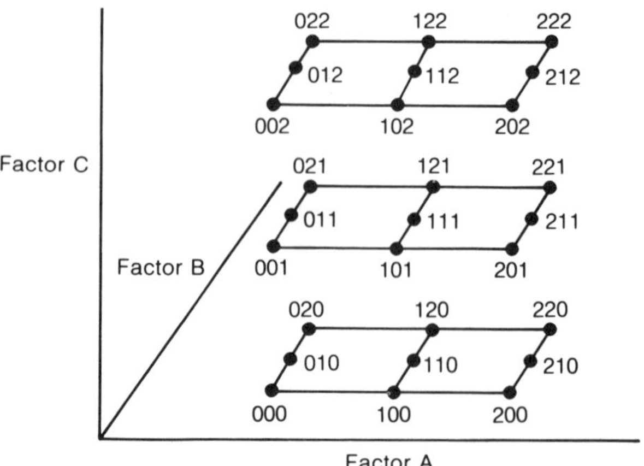

Figure 12.2 Experimental conditions for a 3^3 design.

effect due to factors A, B and C when A is set at level i, B at level j, and C at level k; $i, j, k = 1, 2, 3$; and ε_{ijk} is the experimental error when factors A, B and C are set at levels i, j and k, respectively: $i, j, k = 1, 2, 3, \ldots, r$.

12.3.2 Design matrix of a 3^3 experiment with quantitative factors

The design matrix of a 3^2 experiment can be expanded to accommodate the orthogonal contrasts associated with a 3^3 experiment, using the same procedure described in section 12.2.1 to generate Table 12.2. The corresponding results are summarized in Table 12.5. The first column lists the 27 experimental conditions of the design in standard order. The remaining portion of the table provides coefficients for 26 orthogonal contrasts defined on the averages of the treatment samples: six contrasts associated with components of main effects; twelve associated with components of two-factor interactions; and eight belonging to components of the three-factor interactions.

12.3.3 Main effects of a 3^3 experiment

The following orthogonal contrasts can be evaluated for the effects of factors A, B and C, where $a_{(c)}$, $b_{(c)}$, $c_{(c)}$, $d_{(c)}$, $e_{(c)}$ and $f_{(c)}$ are the coefficients in Table 12.5 for treatment c in the A_L, A_Q, B_L, B_Q, C_L and C_Q columns, respectively:

$$L_{A_{(L)}} = \sum_c a_{(c)} \bar{Y}_{(c)} \qquad L_{A_{(Q)}} = \sum_c b_{(c)} \bar{Y}_{(c)}$$

$$L_{B_{(L)}} = \sum_c c_{(c)} \bar{Y}_{(c)} \qquad L_{B_{(Q)}} = \sum_c d_{(c)} \bar{Y}_{(c)}$$

$$L_{C_{(L)}} = \sum_c e_{(c)} \bar{Y}_{(c)} \qquad L_{C_{(Q)}} = \sum_c f_{(c)} \bar{Y}_{(c)}$$

Using these results, the sums of squares associated with the three factors of the design are calculated as indicated in equations (12.6), (12.7) and (12.8):

$$SS_A = \frac{rL^2_{A(\mathrm{L})}}{\sum_{(c)} a^2_{(c)}} + \frac{rL^2_{A(\mathrm{Q})}}{\sum_{(c)} b^2_{(c)}} \tag{12.6}$$

$$SS_B = \frac{rL^2_{B(\mathrm{L})}}{\sum_{(c)} c^2_{(c)}} + \frac{rL^2_{B(\mathrm{Q})}}{\sum_{(c)} d^2_{(c)}} \tag{12.7}$$

$$SS_C = \frac{rL^2_{C(\mathrm{L})}}{\sum_{(c)} e^2_{(c)}} + \frac{rL^2_{C(\mathrm{Q})}}{\sum_{(c)} f^2_{(c)}} \tag{12.8}$$

12.3.4 Interaction effects of a 3^3 experiment

There are four orthogonal contrasts belonging to the four components (or degrees of freedom) of the interaction between any two factors. The four orthogonal contrasts for the interaction between factors A and B are:

$$L_{A(\mathrm{L})B(\mathrm{L})} = \sum_c a_{(c)} c_{(c)} \bar{Y}_{(c)} \qquad L_{A(\mathrm{L})B(\mathrm{Q})} = \sum_c a_{(c)} d_{(c)} \bar{Y}_{(c)}$$

$$L_{A(\mathrm{Q})B(\mathrm{L})} = \sum_c b_{(c)} c_{(c)} \bar{Y}_{(c)} \qquad L_{A(\mathrm{Q})B(\mathrm{Q})} = \sum_c b_{(c)} d_{(c)} \bar{Y}_{(c)}$$

Similarly, the four orthogonal contrasts associated with the components of the interaction effect due to factors A and C are:

$$L_{A(\mathrm{L})C(\mathrm{L})} = \sum_c a_{(c)} e_{(c)} \bar{Y}_{(c)} \qquad L_{A(\mathrm{L})C(\mathrm{Q})} = \sum_c a_{(c)} f_{(c)} \bar{Y}_{(c)}$$

$$L_{A(\mathrm{Q})C(\mathrm{L})} = \sum_c b_{(c)} e_{(c)} \bar{Y}_{(c)} \qquad L_{A(\mathrm{Q})C(\mathrm{Q})} = \sum_c b_{(c)} f_{(c)} \bar{Y}_{(c)}$$

The four contrasts associated with the interaction effect due to the factors B and C are:

$$L_{B(\mathrm{L})C(\mathrm{L})} = \sum_c c_{(c)} e_{(c)} \bar{Y}_{(c)} \qquad L_{B(\mathrm{L})C(\mathrm{Q})} = \sum_c c_{(c)} f_{(c)} \bar{Y}_{(c)}$$

$$L_{B(\mathrm{Q})C(\mathrm{L})} = \sum_c d_{(c)} e_{(c)} \bar{Y}_{(c)} \qquad L_{B(\mathrm{Q})C(\mathrm{Q})} = \sum_c d_{(c)} f_{(c)} \bar{Y}_{(c)}$$

The sum of squares due to two-factor interactions are calculated as indicated below:

$$SS_{AB} = \frac{rL^2_{A(\mathrm{L})B(\mathrm{L})}}{\sum_c (a_{(c)} c_{(c)})^2} + \frac{rL^2_{A(\mathrm{L})B(\mathrm{Q})}}{\sum_c (a_{(c)} d_{(c)})^2} + \frac{rL^2_{A(\mathrm{Q})B(\mathrm{L})}}{\sum_c (b_{(c)} c_{(c)})^2} + \frac{rL^2_{A(\mathrm{Q})B(\mathrm{Q})}}{\sum_c (b_{(c)} d^{\delta}_{(c)})^2} \tag{12.9}$$

$$SS_{AC} = \frac{rL^2_{A(\mathrm{L})C(\mathrm{L})}}{\sum_c (a_{(c)} e_{(c)})^2} + \frac{rL^2_{A(\mathrm{L})C(\mathrm{Q})}}{\sum_c (a_{(c)} f_{(c)})^2} + \frac{rL^2_{A(\mathrm{Q})C(\mathrm{L})}}{\sum_c (b_{(c)} e_{(c)})^2} + \frac{rL^2_{A(\mathrm{Q})C(\mathrm{Q})}}{\sum_c (b_{(c)} f_{(c)})^2} \tag{12.10}$$

$$SS_{BC} = \frac{rL^2_{B(\mathrm{L})C(\mathrm{L})}}{\sum_c (c_{(c)} e_{(c)})^2} + \frac{rL^2_{B(\mathrm{L})C(\mathrm{Q})}}{\sum_c (c_{(c)} f_{(c)})^2} + \frac{rL^2_{B(\mathrm{Q})C(\mathrm{L})}}{\sum_c (d_{(c)} e_{(c)})^2} + \frac{rL^2_{B(\mathrm{Q})C(\mathrm{Q})}}{\sum_c (d_{(c)} f_{(c)})^2} \tag{12.11}$$

Table 12.5 Design matrix for a 3^3 experiment

Treatments	A_L	A_Q	B_L	B_Q	A_LB_L	A_LB_Q	A_QB_L	A_QB_Q	C_L	C_Q	A_LC_L	A_LC_Q	A_QC_L	A_QC_Q	B_LC_L	B_LC_Q
1(000)	−1	1	−1	1	1	−1	−1	1	−1	1	1	−1	−1	1	1	−1
2(001)	−1	1	−1	1	1	−1	−1	1	0	−2	0	2	0	−2	0	2
3(002)	−1	1	−1	1	1	−1	−1	1	1	1	−1	−1	1	1	−1	−1
4(010)	−1	1	0	−2	0	2	0	−2	−1	1	1	−1	−1	1	0	0
5(011)	−1	1	0	−2	0	2	0	−2	0	−2	0	2	0	−2	0	0
6(012)	−1	1	0	−2	0	2	0	−2	1	1	−1	−1	1	1	0	0
7(020)	−1	1	1	1	−1	−1	1	1	−1	1	1	−1	−1	1	−1	1
8(021)	−1	1	1	1	−1	−1	1	1	0	−2	0	2	0	−2	0	−2
9(022)	−1	1	1	1	−1	−1	1	1	1	1	−1	−1	1	1	1	1
10(100)	0	−2	−1	1	0	0	2	−2	−1	1	0	0	2	−2	1	−1
11(101)	0	−2	−1	1	0	0	2	−2	0	−2	0	0	0	4	0	2
12(102)	0	−2	−1	1	0	0	2	−2	1	1	0	0	−2	−2	−1	−1
13(110)	0	−2	0	−2	0	0	0	4	−1	1	0	0	2	−2	0	0
14(111)	0	−2	0	−2	0	0	0	4	0	−2	0	0	0	4	0	0
15(112)	0	−2	0	−2	0	0	0	4	1	1	0	0	−2	−2	0	0
16(120)	0	−2	1	1	0	0	−2	−2	−1	1	0	0	2	−2	−1	1
17(121)	0	−2	1	1	0	0	−2	−2	0	−2	0	0	0	4	0	−2
18(122)	0	−2	1	1	0	0	−2	−2	1	1	0	0	−2	−2	1	1
19(200)	1	1	−1	1	−1	1	−1	1	−1	1	−1	1	−1	1	1	−1
20(201)	1	1	−1	1	−1	1	−1	1	0	−2	0	−2	0	−2	0	2
21(202)	1	1	−1	1	−1	1	−1	1	1	1	1	1	1	1	−1	−1
22(210)	1	1	0	−2	0	−2	0	−2	−1	1	−1	1	−1	1	0	0
23(211)	1	1	0	−2	0	−2	0	−2	0	−2	0	−2	0	−2	0	0
24(212)	1	1	0	−2	0	−2	0	−2	1	1	1	1	1	1	0	0
25(220)	1	1	1	1	1	1	1	1	−1	1	−1	1	−1	1	−1	1
26(221)	1	1	1	1	1	1	1	1	0	−2	0	−2	0	−2	0	−2
27(222)	1	1	1	1	1	1	1	1	1	1	1	1	1	1	1	1

Table 12.5 *(cont'd)*

Treatments	B_QC_L	B_QC_Q	$A_LB_LC_L$	$A_LB_LC_Q$	$A_LB_QC_L$	$A_LB_QC_Q$	$A_QB_LC_L$	$A_QB_LC_Q$	$A_QB_QC_L$	$A_QB_QC_Q$
1(000)	−1	1	−1	1	1	−1	1	−1	−1	1
2(001)	0	−2	0	−2	0	2	0	2	0	−2
3(002)	1	1	1	1	−1	−1	−1	−1	1	1
4(010)	2	−2	0	0	−2	2	0	0	2	−2
5(011)	0	4	0	0	0	−4	0	0	0	4
6(012)	−2	−2	0	0	2	2	0	0	−2	−2
7(020)	−1	1	1	−1	1	−1	−1	1	−1	1
8(021)	0	−2	0	2	0	2	0	−2	0	−2
9(022)	1	1	−1	−1	−1	−1	1	1	1	1
10(100)	−1	1	0	0	0	0	−2	2	2	−2
11(101)	0	−2	0	0	0	0	0	−4	0	4
12(102)	1	1	0	0	0	0	2	2	−2	−2
13(110)	2	−2	0	0	0	0	0	0	−4	4
14(111)	0	4	0	0	0	0	0	0	0	−8
15(112)	−2	−2	0	0	0	0	0	0	4	4
16(120)	−1	1	0	0	0	0	2	−2	2	−2
17(121)	0	−2	0	0	0	0	0	4	0	4
18(122)	1	1	0	0	0	0	−2	−2	−2	−2
19(200)	−1	1	1	−1	−1	1	1	−1	−1	1
20(201)	0	−2	0	2	0	−2	0	2	0	−2
21(202)	1	1	−1	−1	1	1	1	−1	1	1
22(210)	2	−2	0	0	2	−2	0	0	2	−2
23(211)	0	4	0	0	0	4	0	0	0	4
24(212)	−2	−2	0	0	−2	−2	0	0	−2	−2
25(220)	−1	1	−1	1	−1	1	−1	1	−1	1
26(221)	0	−2	0	−2	0	−2	0	−2	0	−2
27(222)	1	1	1	1	1	1	1	1	1	1

Finally, it can be verified that the following orthogonal contrasts belong to the interaction components shown:

$$\sum_c a_{(c)}c_{(c)}e_{(c)}\bar{Y}_{(c)} \qquad A_L B_L C_L$$

$$\sum_c a_{(c)}c_{(c)}f_{(c)}\bar{Y}_{(c)} \qquad A_L B_L C_Q$$

$$\sum_c a_{(c)}d_{(c)}e_{(c)}\bar{Y}_{(c)} \qquad A_L B_Q C_L$$

$$\sum_c a_{(c)}d_{(c)}f_{(c)}\bar{Y}_{(c)} \qquad A_L B_Q C_Q$$

$$\sum_c b_{(c)}c_{(c)}e_{(c)}\bar{Y}_{(c)} \qquad A_Q B_L C_L$$

$$\sum_c b_{(c)}c_{(c)}f_{(c)}\bar{Y}_{(c)} \qquad A_Q B_L C_Q$$

$$\sum_c b_{(c)}d_{(c)}e_{(c)}\bar{Y}_{(c)} \qquad A_Q B_Q C_L$$

$$\sum_c b_{(c)}d_{(c)}f_{(c)}\bar{Y}_{(c)} \qquad A_Q B_Q C_Q$$

From these results, it is possible to compute the sum of squares of the three-factor interaction as indicated in equation 12.12:

$$\begin{aligned} SS_{ABC} = {} & \frac{rL^2_{A(L)B(L)C(L)}}{\sum_c (a_{(c)}c_{(c)}e_{(c)})^2} + \frac{rL^2_{A(L)B(L)C(Q)}}{\sum_c (a_{(c)}c_{(c)}f_{(c)})^2} + \frac{rL^2_{A(L)B(Q)C(L)}}{\sum_c (a_{(c)}d_{(c)}e_{(c)})^2} \\ & + \frac{rL^2_{A(L)B(Q)C(Q)}}{\sum_c (a_{(c)}d_{(c)}f_{(c)})^2} + \frac{rL^2_{A(Q)B(L)C(L)}}{\sum_c (b_{(c)}c_{(c)}e_{(c)})^2} + \frac{rL^2_{A(Q)B(L)C(Q)}}{\sum_c (b_{(c)}c_{(c)}f_{(c)})^2} \\ & + \frac{rL^2_{A(Q)B(Q)C(L)}}{\sum_c (b_{(c)}d_{(c)}e_{(c)})^2} + \frac{rL^2_{A(Q)B(Q)C(Q)}}{\sum_c (b_{(c)}d_{(c)}f_{(c)})^2} \end{aligned} \tag{12.12}$$

EXAMPLE 12.2

The results in Table 12.6 correspond to a factorial experiment with two replications for each experimental condition. It was designed and run to determine the significance of the effects due to type of material (factor *A*), temperature of heat treatment (factor *B*) and type of laminating machine (factor *C*), as well as possible interaction effects, on the abrasive wear of a laminated sheet. Each factor has three levels as indicated:

Factor *A*: machine 1 (M_1), machine 2 (M_2), machine 3 (M_3)
Factor *B*: 250 °C (T_1), 350 °C (T_2), 450 °C (T_3)
Factor *C*: material 1 (S_1), material 2 (S_2), material 3 (S_3)

Table 12.6 Depth of bearing wear (in $\times 10^{-3}$)

	M_1			M_2			M_3		
	S_1	S_2	S_3	S_1	S_2	S_3	S_1	S_2	S_3
T_1	36	37	31	30	27	48	28	38	36
	33	32	34	28	32	41	28	36	37
T_2	31	32	28	27	33	44	28	38	34
	24	29	33	25	31	37	26	33	33
T_3	28	28	22	22	27	44	23	33	28
	27	32	32	20	24	37	25	31	28

The samples used in the analysis received a heat treatment before they were laminated by one of three machines. For each sample, the machine ran for a fixed amount of time and the corresponding depth of bearing wear was measured and recorded.

It is noted that factor A (machine) and factor C (material) are qualitative and factor B (temperature) is quantitative. For this reason, quadratic components can be considered only for the main effect of factor B. As a result, in the ANOVA, the interaction effect should not be broken into components (as we did in Example 12.1).

In order to compute the sums of squares associated with the linear and quadratic components of the main effect of factor B, we first calculate the value of the orthogonal contrasts belonging to these components. The totals for the 27 experimental conditions, in the same order as they are listed in Table 12.5, are: 69, 58, 56, 55, 52, 54, 55, 42, 48, 69, 59, 74, 61, 71, 60, 51, 64,

Table 12.7 ANOVA results for example 12.2

Source	*df*	*SS*	*MS*	*F*	*Significant?*
A	2	21.78	10.89	1.22	No
B	2	284.11	142.05	15.88	Yes
B_L	1	(283.36)	283.36	31.70	Yes
B_Q	1	(0.75)	0.75	0.08	No
AB	4	18.78	4.69	0.52	No
C	2	537.33	268.67	30.04	Yes
AC	4	609.89	152.47	17.05	Yes
BC	4	12.22	3.06	0.34	No
ABC	8	53.22	6.65	0.74	No
Error	27	241.50	8.94		
Total	53	1778.83			

65, 89, 73, 61, 81, 67, 54, 81, 56. Using the corresponding contrast coefficients shown in Table 12.5, we obtain the following results:

$$L_{B(L)} = \sum_c c_{(c)} \bar{Y}_{(c)} = -50.5 \qquad L_{B(Q)} = \sum_c d_{(c)} \bar{Y}_{(c)} = -4.5.$$

Therefore, $SS_{B(L)} = 2(50.5)^2/18 = 283.361$, and $SS_{B(Q)} = 2(4.5)^2/54 = 0.750$, and from equation 12.6, $SS_B = 284.111$.

A summary of the results is shown in Table 12.7. If we consider a level of significance $\alpha = 0.05$, the critical value for testing each main effect is 3.35; in the case of two-factor interactions the critical value is 2.73. In conclusion, the main effects of material and temperature are highly significant, although the effect due to variations in machine is not significant. However, the interaction effect due to machine and material is quite significant.

12.3.5 The general 3^k design

The concepts utilized in the 3^2 and 3^3 designs can be readily extended to the case of k factors each at three levels, that is, a 3^k factorial design. The usual digital notation is employed for the treatment combinations, so that 0120 represents a treatment or level combination in a 3^4 design with A and D at the low levels, B at the medium level and C at the high level. There are 3^k treatments or level combinations, with $3^k - 1$ degrees of freedom between them. There are k main effects, each with 2 degrees of freedom; $\binom{k}{2}$ two-factor interactions, each with 4 degrees of freedom; $\binom{k}{3}$ three-factor interactions, each with 8 degrees of freedom,... and one k-factor interaction with 2^k degrees of freedom. Sums of squares for main effects and interaction effects are computed following identical procedures to those followed to develop equations (12.3)–(12.12).

As a general rule, three-factor and higher interactions are not broken down into their one-degree-of-freedom components. However, any h-factor interaction can be broken down into 2^{h-1} orthogonal components, each having two degrees of freedom. For example, the four-factor interaction $ABCD$ has $2^{4-1} = 8$ orthogonal two-degrees-of-freedom components, denoted by $ABCD^2$, ABC^2D, AB^2CD, $ABCD$, ABC^2D^2, AB^2C^2D, AB^2CD^2 and $AB^2C^2D^2$. In case the first letter does not have an exponent equal to one, then the entire word can be squared and the exponents can be reduced modulo 3. To illustrate this, consider the word A^2BCD. In this case, $A^2BCD = (A^2BCD)^2 = A^4B^2C^2D^2 = AB^2C^2D^2$.

12.3.6 Yates algorithm for the 3^k design

Yates algorithm can be modified for use in the 3^k factorial design. The procedure will be illustrated using the data in Table 12.3, Example 12.1. The steps of the procedure are displayed in Table 12.8. As can be verified, the results in this table are identical to those shown in Table 12.4.

Table 12.8 Yates algorithm for the 3^2 design in example 12.1

A	*B*	*(0)*	*(1)*	*(2)*	*Effect*	*Divisor*	*SS*
0	0	727	1641	4405	–	–	–
1	0	532	1463	−970	A_L	24	39204.17
2	0	382	1301	490	A_Q	72	3374.72
0	1	658	−345	−340	B_Q	24	4816.67
1	1	441	−294	14	A_LB_L	16	12.25
2	1	364	−331	260	A_QB_L	48	1408.33
0	2	650	45	16	B_Q	72	3.56
1	2	332	140	−88	A_LB_Q	48	161.33
2	2	319	305	70	A_QB_Q	144	34.03

In Table 12.8 column (0) contains the totals of all observations taken for the nine treatments arranged in standard order. This column can be considered as consisting of three groups, with the first three values in group 1, the second set of three values in group 2 and the third set of three values in group 3. The entries in column (1) are computed as follows: the first group of three values in this column has the sums of values in the three groups of column (0). The three values in the second group are computed as follows: the jth value is obtained by subtracting the first total from the third value in the jth group of column (0), setting $j = 1$, then $j = 2$ and finally $j = 3$. The jth value of the third group of column (1) is obtained by taking the sum of the first and third values minus twice the second value in group j of column (0), for $j = 1, 2, 3$. For example, in column (1) the second, fifth and eight entries are $658 + 441 + 364 = 1463$, $364 - 658 = -294$ and $658 + 364 - 2(441) = 140$, respectively. Column (2) is obtained from column (1) following the same procedure outlined before. In general, k columns are generated following this procedure.

The entries in the *SS* column are finally obtained by squaring the values in column (2) and dividing each one by $r2^e3^f$, where r is the number of replications, e is the number of factors in the effect considered and f is the number of factors in the experiment minus the number of linear terms in the effect. In the current example, $r = 4$. Thus, in the case of B_Q, $e = 1$, $f = 2 - 0 = 2$, and the divisor is equal to 72.

12.4 THREE-LEVEL FRACTIONAL FACTORIAL DESIGN

The purpose of this section is to introduce basic concepts, principles, assumptions and procedures for selecting fractional three-level factorial designs. The section is divided into four subsections: basic concepts, determination of generators, determination of the fraction, and determination of aliases.

12.4.1 Basic concepts

(a) Generator

The relationship used to indicate what original factors are confounded with which interactions of the base 3^{k-p} design.

(b) Defining relation

A set of relationships that defines all possible aliases that can be obtained from the generators of the design. In order to obtain the defining relation, the procedure followed is similar to that followed in the case of a two-level fractional factorial design.

(c) Aliases

If the estimate of factor A is confounded with the estimate of the two-factor interaction effect BC the two effects are said to be aliased.

(d) Resolution of fractional factorial design

This is a concept that aids in the selection of a fractional factorial design. As defined in Chapter 11, a design with resolution equal to R is a design in which no q-factor effect, where $q < R$, is confounded with any other effect containing fewer than $R - q$ factors.

In the construction of a fractional design two basic principles should be followed: (a) no main effects should be aliased with other main effects or aliased with important two-factor interactions; (b) the number of two-factor interactions aliased with other two-factor interactions should be as low as possible. Two-factor interactions that are only aliased with higher order interactions are said to be **measurable**.

The relationship established by the generators serves two purposes. It can be used to select the appropriate subset of treatments from the full factorial design. It can also be used to determine the manner in which the various main effects and interactions are aliased with one another as a consequence of taking only a subset of measurements from the full factorial.

12.4.2 Determination of the generators

In general, a $(\frac{1}{3})^p$ fraction of a 3^k design, where $p < k$, contains 3^{k-p} experimental conditions or runs. Such a design is called a 3^{k-p} fractional factorial design. As an illustration, a 3^{k-2} design is a (1/9)th fraction, a 3^{k-4} design is a (1/81)th fraction, and so on.

The number of required generators is equal to p. For example in a $\frac{1}{3}$

fractional replicate of a 3^k factorial design we need $p = 1$ generator. If the required fraction is equal to 1/9 we need $p = 2$ generators, and so on. Each generator can be represented by the relation:

$$I_i = A^{a_i} B^{b_i} C^{c_i} \cdots \tag{12.13}$$

where $i = 1, \ldots, p$; $a_i, b_i, c_i, \ldots$ can take on values 0, 1, 2; and the factors are represented by capital letters. It is important to note that the exponent associated with the first letter is always equal to one.

12.4.3 Determination of the fraction

After the defining relationship has been established, an appropriate fraction can be determined. The defining relationship has p generators that are, as stated before, products of factors, with each factor having an exponent equal to 0, 1 or 2. From these p generators we obtain p simultaneous equations with unknowns defined as the levels of the factors in the experimental conditions chosen for the fraction to be used.

The general form of the system of simultaneous equations is:

$$a_i X_1 + b_i X_2 + c_i X_3 + \cdots = 0 \ (\text{mod } 3), \quad i = 1, 2, \ldots, p \tag{12.14}$$

where X_1 is the level of factor A, X_2 is the level of factor B, and so on. Similar equations in which 1 or 2 is used in place of the right-hand side 0 are equally valid [7]. It is noted that $a_i, b_i, c_i, \ldots$ are the exponents of the generators defined in equation 12.13.

The way in which the p simultaneous equations are obtained will be illustrated using as an example a (1/27) fractional replicate of a 3^7 factorial design. For this design $k = 7$ and $p = 3$. The three chosen generators are:

$$I_1 = ACDEF^2G \tag{12.15}$$

$$I_2 = BC^2EF^2G \tag{12.16}$$

$$I_3 = ABCEG^2 \tag{12.17}$$

From these generators the following three simultaneous equations are obtained:

$$X_1 + X_3 + X_4 + X_5 + 2X_6 + X_7 = 0 \quad (\text{mod } 3) \tag{12.18}$$

$$X_2 + 2X_3 + X_5 + 2X_6 + X_7 = 0 \quad (\text{mod } 3) \tag{12.19}$$

$$X_1 + X_2 + X_3 + X_5 + 2X_7 = 0 \quad (\text{mod } 3) \tag{12.20}$$

Equation 12.18 corresponds to the generator defined in equation 12.15; equation 12.19 corresponds to equation 12.16; and equation 12.20 to equation 12.17. These simultaneous equations can be alternatively set equal to 1 or 2 instead of 0 and the results are equally valid [7]. The solution to the equations indicate the levels of the factors in each experimental condition included in the fraction. There are 81 solutions each one corresponding to an experimental condition (seven factor levels in each one). Three of these 81 solutions are, for example, the levels of the following experimental conditions $(X_1 X_2 X_3 X_4 X_5 X_6 X_7)$: 0221011, 0221102, 0221220. In their excellent article, Connor and Zelen [7] list the experimental conditions for 41 experimental plans considering from four to ten factors and fractional replications from 1/3 to 1/243.

12.4.4 Determination of the aliases

Based on the assumption that three-factor and higher order interactions are negligible, we can determine the aliases following the two steps outlined below:

Step 1. Determine the defining relation of the design

The defining relation consists of $\frac{1}{2}(3^p - 1) - 1$ words connected by equal signs. If, for example, $p = 2$ the relation consists of 4 words (not including I). These four words are given by $I_1, I_2, I_1 I_2$, and $I_1 I_2^2$. Thus, the defining relation can be expressed as $I = I_1 = I_2 = I_1 I_2 = I_1 I_2^2$. The exponents associated with the letters are reduced modulo 3 such that the first letter in each word has exponent equal to one.

As an illustration, let us consider a 1/9 replicate of a three-level experiment with eight factors A, B, C, D, E, F, G and H. Let $I_1 = ABCDEH^2$ and $I_2 = CD^2EF^2G^2$. It can be easily verified that in this case $I_1 I_2 = ABC^2E^2F^2G^2H^2$ and $I_1 I_2^2 = ABD^2FGH^2$. Therefore, the defining relation for this design is given by:

$$I = ABCDEH^2 = CD^2EF^2G^2 = ABC^2E^2F^2G^2H^2 = ABD^2FGH^2$$

In some cases the exponent of the first letter of a word is not equal to one as a result of the (mod 3) reduction. In such cases the word can be squared before taking (mod 3).

Step 2. Determine the aliases of main effects or parts of a two-factor interaction

To determine the aliases of any main effect or interaction component, all the words and the square of the words in the defining relation are multiplied by that particular main effect or interaction. After doing so it is found that each main effect or interaction is aliased with other $3^p - 1$ interactions.

The procedure is illustrated below for a $\frac{1}{3}$ fractional replicate of a 3^4 factorial design. Here we have only one generator, since $p = 1$. Then the defining relation has also only one word. Using $I = ABCD$, the aliases are computed as follows:

(a) Aliases for main effects

As previously indicated, each word of the defining relation (in this example it has one word) and the square of each word are multiplied by the main effect under consideration. The leading exponent of every word is equal to one and all exponents are reduced modulo 3. Let us now consider factor A:

$$\begin{aligned} I &= ABCD = (ABCD)^2 \\ A &= A(ABCD) = A(ABCD)^2 \\ &= A^2BCD = B^2C^2D^2 \\ &= AB^2C^2D^2 = BCD \end{aligned}$$

In a similar way we will obtain:

$$\begin{aligned} B &= AB^2CD = ACD \\ C &= ABC^2D = ABD \\ D &= ABCD^2 = ABC \end{aligned}$$

(b) Aliases for two-factor interactions

Proceeding as in the case of main effects, we obtain the following results. It is noted that *AB*, *CD*, *AC*, *BD*, *AD* and *BC* are not measurable, since they are confounded with other two-factor interaction components:

$$\begin{aligned} AB &= ABC^2D^2 = CD \\ AC &= AB^2CD^2 = BD \\ AD &= AB^2C^2D = BC \\ AB^2 &= AC^2D^2 = BC^2D^2 \\ AC^2 &= AB^2D^2 = BC^2D \\ AD^2 &= AB^2C^2 = BCD^2 \\ BC^2 &= AB^2D = AC^2D \\ BD^2 &= AB^2C = ACD^2 \\ CD^2 &= ABC^2 = ABD^2 \end{aligned}$$

12.5 ANOVA FOR FRACTIONAL DESIGNS

The mathematical model for the response value corresponding to an experimental condition $X_1X_2X_3\ldots$ of a 3^k factorial experiment can be formulated as follows:

$$Y(X_1X_2X_3\ldots) = \eta(X_1X_2X_3\ldots) + \beta_i + \varepsilon(X_1X_2X_3\ldots) \qquad (12.21)$$

In equation 12.18 the following notation is used:

$\eta(X_1X_2X_3\ldots)$: true value of the observation

β_i: effect associated with the block in which the measurement is made ($i = 1, \ldots, b$)

$\varepsilon(X_1X_2X_3\ldots)$: uncorrelated random error.

If the design to be analyzed is a full factorial the true value will include all main effects and other interactions. For a fractional design only the measurable interactions in addition to the main effects will be included in the true value. Consequently, the true value of a treatment or level combination can be written as the algebraic sum of components associated with the common effect, main effects and measurable interactions.

In the case of a 3^{k-p} fractional factorial design the following estimates are needed to conduct the ANOVA tests of both linear and quadratic components of main effects and interactions between these components:

$$\hat{\mu} = \frac{1}{3^{k-p}} \sum Y(X_1X_2X_3\ldots)$$

$$(\hat{A})_i = \frac{1}{3^{k-p-1}} \sum_{X_1=i} Y(X_1X_2X_3\ldots) - \hat{\mu}$$

$$(\hat{A})_i = \frac{1}{3^{k-p-1}} \sum_{X_1=i} Y(X_1X_2X_3\ldots) - \hat{\mu}$$

$$\vdots$$

$$(\hat{B})_i = \frac{1}{3^{k-p-1}} \sum_{X_2=i} Y(X_1X_2X_3\ldots) - \hat{\mu}$$

$$(\widehat{AB})_i = \frac{1}{3^{k-p-1}} \sum_{X_1+X_2=i\,(\text{mod } 3)} Y(X_1X_2X_3\ldots) - \hat{\mu}$$

$$(\widehat{AB^2})_i = \frac{1}{3^{k-p-1}} \sum_{X_1+2X_2=i\,(\text{mod } 3)} Y(X_1X_2X_3\ldots) - \hat{\mu}$$

$$\vdots$$

where $i = 0, 1, 2$ for each estimate, and all summations refer only to those 3^{k-p} measurements selected for the fraction.

The above quantities estimate the parameters $\mu, (A)_i, (B)_i, (C)_i, \ldots, (AB)_i, (AB^2)_i, (AC)_i, (AC^2)_i, \ldots$. All these parameters are defined by the above relationships after substituting parameters for estimates. The quantities $(A)_i$, $(B)_i$, $(C)_i$ are parameters associated with the main effects of factors $A, B, C, \ldots$, respectively. The quantities $(AB)_i$ and $(AB^2)_i$ are parameters associated with the two-factor interaction components AB and AB^2, respectively. Similarly,

Table 12.9 Degrees of freedom for completely randomized fractional fractorial design

Sources of variation	*Degrees of freedom*
Main effects	$2k$
Measurable interactions	$2q$
Non-measurable interactions (error)	$3^{k-p} - 1 - 2k - 2q$
Total	$3^{k-p} - 1$

the parameters $(AC)_i$, $(AC^2)_i$ are associated with AC and AC^2. Every estimate will be biased in a way that depends upon its aliases [1].

12.5.1 Degrees of freedom

After computing the estimates for all main effects and measurable interactions, we can proceed with the ANOVA in a relatively simple way. For the sum of squares (SS) of, for example, factor A, we use equation 12.22:

$$SS_A = \sum_{i=0}^{2} (\hat{A})_i^2 \tag{12.22}$$

The SS associated, for example, with the interaction of components AB and AB^2 can be computed as indicated in equations 12.23 and 12.24:

$$SS_{AB} = \sum_{i=0}^{2} (\widehat{AB})_i^2 \tag{12.23}$$

$$SS_{AB^2} = \sum_{i=0}^{2} (\widehat{AB^2})_i^2 \tag{12.24}$$

Once the required SS values are computed, F-tests on the main effects and measurable two-factor interactions can be carried out. Although higher-order interactions are treated as negligible, depending on the experiment, the relatively large number of such higher-order interactions has the potential of biasing a particular treatment mean square.

For a completely randomized fractional factorial experiment, the degrees of freedom are computed as shown in Table 12.9. In this table, k is the number of factors and q is the number of measurable two-factor interaction components. The estimate of the residual error is supplied by the mean square associated with the higher order interaction components.

12.5.2 Blocking in fractional factorial designs

Occasionally a fractional factorial design requires so many runs that all the observations cannot be obtained under homogeneous conditions. In addition, changes in the environment of the experiment can introduce extraneous influences on the measured response that can, in turn, affect the estimate of the effects of the variables under consideration. Examples of such influences can be present in the form of raw material differences, operator, machine differences or environmental differences. In situations where these differences affect the homogeneity of the data to be analysed, the runs of the design should be grouped into blocks to increase the sensitivity of the significance tests concerning main effects and lower-order interaction effects.

When blocking is considered, the degrees of freedom are as shown in Table 12.10. In this table, b is the number of blocks, k is the number of factors and q is the number of measurable two-factor interaction components.

To arrange the 3^{k-p} experimental conditions into $b = 3^s$ blocks, s additional generators must be formulated. These generators are used to determine $\frac{1}{3}(3^s - 1)$ additional interaction components, which become confounded with the blocks.

Let us now assume that the block generators are $I_1, I_2, \ldots, I_s$. Moreover, let $I_i = A^{a_i}B^{b_i}C^{c_i}\ldots$, for $i = 1, 2, \ldots, s$. Therefore, the 3^{k-p} treatments of the fractional factorial design are grouped into 3^s blocks in such a way that the level combinations in the jth block will satisfy the following s simultaneous equations:

$$a_iX_1 + b_iX_2 + c_iX_3 + \cdots = \alpha_{ij} \text{ (mod 3)}, \quad i = 1, 2, \ldots, s \qquad (12.25)$$

where $\alpha_{ij} = 0, 1$ or 2. The 3^s blocks are obtained by considering all possible combinations of the α_{ij} over the numbers 0, 1, 2. For example, in the case of a 3^{6-2} fractional factorial design generated by $I_1 = ACDE$ and $I_2 = BC^2DE^2F$, if the interaction components AC, BC and BF are used to generate blocks, the 81 treatments are placed into 27 blocks according to the following three simultaneous equations (for each value $j = 1, 2, 3$):

$$X_1 + X_3 = \alpha_{1j} \quad \text{(mod 3)}$$
$$X_2 + X_3 = \alpha_{2j} \quad \text{(mod 3)}$$
$$X_2 + X_6 = \alpha_{3j} \quad \text{(mod 3)}$$

In the above equations, $(\alpha_{1j}, \alpha_{2j}, \alpha_{3j})$ is equal to (0, 0, 0) for block 1, (0, 0, 1) for block 2, (0, 0, 2) for block 3, (0, 1, 0) for block 4 and finally (2, 2, 2) for block 27.

Table 12.10 Degrees of freedom of fractional factorial design using blocks

Source of variation	*Degrees of freedom*
Blocks	$b - 1$
Main effects	$2k$
Measurable interactions	$2q$
Non-measurable interactions (error)	$3^{k-p} - b - 2k - 2q$
Total	$3^{k-p} - 1$

12.6 APPLICATION

The example considered in this section has been chosen to illustrate the use in the textile industry of a $(\frac{1}{3})^p$ fractional replicate of a 3^k factorial design with blocks. The section consists of a statement of the problem, a fractional factorial design, a blocking arrangement for this design and results with brief recommendations.

Statement of the problem

The EDE Company, a textile factory involved in the production of threads, has just signed a contract with a finishing textile factory. EDE agrees to provide this finishing factory with 1000 kg each quarter of any given year. One of the specifications in the contract is that the strength of the thread must be good enough to resist the sewing of hard materials. The strength of the thread depends mainly on the quality of the fibers used. The kind of thread to be produced is a combination of four raw materials, acrylic, polyester, wool and alpaca, used in the following proportions: acrylic (5%), polyester (10%), wool (35%), alpaca (50%).

Since this is a newly designed product, EDE would like to assess the significance of the effect that each of these components has on the breaking strength of the final product. The synthetic fibers (acrylic and polyester) have a standard fiber length, and the supplier of these raw materials offers three types of fibers classified as extra, high and regular. The wool and alpaca supplier classifies his fibers according to the length of the fiber, as shown below:

Classification	Length of fiber (cm)
Wool AA	5.20
Wool AAA	6.10
Wool AAAA	7.00
Alpaca old	6.90
Alpaca regular	7.30
Alpaca baby	7.96

Time is a relevant constraint for the trial, since the preparation of the mixture and the processing of the thread and final packing need to be accomplished in no more than 15 days. Money is also an important consideration. Preparation and use of the machines for each run are expensive. Due to the budget available it is necessary to run the lowest possible number of experiments. The production manager has given the quality control department the task of designing and running an appropriate trial.

Fractional factorial design

The quality control department has decided to run a three-level fractional factorial design. In this design, the four types of fiber will be considered as the factors of the experiment and the three types of each fiber will be considered as the three levels of the factors. Owing to budgetary and time limitations, not more that 27 runs can be included in the trial, with no more than nine runs conducted in a given day. Table 12.11 shows the different factors and levels chosen.

The observed response chosen for the trial is the strength of the thread measured in $g\,m^{-1}$. It is desired to examine the effects of all factors and as many two-factor interactions as possible. Since the entire fractional design could be run in three days, with nine runs per day, it is decided to use a 3^{4-1} fractional factorial design with three blocks of nine treatments each.

Fractional factorial design with blocks

Plan 3.4.9 (page 11 of [1]) will be used. For this plan we know the following information:

(a) The generator of the design is given by $I = ABCD$.
(b) According to equation 12.14, the 27 experimental runs can be determined from the equation $X_1 + X_2 + X_3 + X_4 = 0 \pmod 3$. These 27 experimental conditions are shown in Table 12.12.
(c) The alias structure for this problem is given in section 12.4.4, step 2.
(d) The measurable two-factor interaction components are AB^2, AC^2, AD^2, BC^2, BD^2 and CD^2.
(e) Blocks are confounded with AB. According to equation 12.22, treatments are placed into the jth block ($j = 1, 2, 3$) according to the solution of the equation $X_1 + X_2 = \alpha_{1j} \pmod 3$, where $\alpha_{1j} = 0$ for block 1, $j = 1$ for block 2, and $j = 2$ for block 3. The experimental conditions in each block are also shown in Table 12.12.

The SSs of the main effects and the measurable interactions can now be calculated. The SS due to factor A and the SS due to the interaction

Table 12.11 Factors and their levels

Factors	*Symbol*	*Levels*
Alpaca	A	old, regular, baby
Wool	B	*AA, AAA, AAAA*
Acrylic	C	extra, high, regular
Polyester	D	extra, high, regular

Table 12.12 Experimental conditions of a 3^{4-1} design with blocks

Block 1				*Block 2*				*Block 3*			
A	*B*	*C*	*D*	*A*	*B*	*C*	*D*	*A*	*B*	*C*	*D*
0	0	0	0	0	1	0	2	0	2	0	1
0	0	2	1	0	1	2	0	0	2	2	2
0	0	1	2	0	1	1	1	0	2	1	0
1	2	1	2	1	0	1	1	1	1	1	0
1	2	0	0	1	0	0	2	1	1	0	1
1	2	2	1	1	0	2	0	1	1	2	2
2	1	2	1	2	2	2	0	2	0	2	2
2	1	1	2	2	2	1	1	2	0	1	0
2	1	0	0	2	2	0	2	2	0	0	1

component AB^2, for example, are equal to:

$$SS(A) = \frac{A_0^2 + A_1^2 + A_2^2}{n} - \frac{T^2}{N}$$

$$SS(AB^2) = \frac{(AB^2)_0^2 + (AB^2)_1^2 + (AB^2)_2^2}{n} - \frac{T^2}{N}$$

where A_i is the sum of all observations for experimental conditions having factor A at level i and $i = 0, 1, 2$, n is the number of observations at each level, T is the sum of all the observations, N is the total number of observations and $(AB^2)_i$ is the sum of all observations for experimental conditions satisfying the equation $X_1 + 2X_2 = i \pmod 3$, $i = 0, 1, 2$.

Using the relationships previously given for $SS(A)$ and $SS(AB^2)$ and the experimental results summarized in Appendix 12.A, we obtain:

$$SS(A) = \frac{(1167.7)^2 + (1048.3)^2 + (1264.7)^2}{9} - 448\,713.796 = 2610.901$$

and

$$SS(AB^2) = \frac{(1146.2)^2 + (1209.8)^2 + (1124.7)^2}{9} - 448\,713.796 = 435.156$$

The SS for all other main effects and measurable interactions are computed in a similar manner using the results given in Appendix 12.A. The ANOVA for the fractional factorial design under consideration is summarized in Table 12.13. The ANOVA study was conducted using a level of significance $\alpha = 0.05$. For this level of significance and two and four degrees of freedom for the numerator and denominator, respectively, the critical value of the F-statistic

Table 12.13 ANOVA results for the fractional design with blocks

Source	*df*	*SS*	*MS*	*F*	*Significant?*
A	2	2610.901	1305.45	3.13	No
B	2	12941.340	6470.67	15.50	Yes
C	2	312.747	156.37	0.37	No
D	2	1146.267	573.13	1.37	No
AB^2	2	435.156	217.58	0.52	Yes
AC^2	2	1269.556	634.78	1.52	No
AD^2	2	1140.383	570.19	1.37	No
BC^2	2	172.187	86.09	0.21	No
BD^2	2	93.361	46.68	0.11	No
CD^2	2	2191.352	1095.68	2.62	No
Blocks	2	1262.054	631.03	1.51	No
Error	4	1670.153	417.54		
Total	26	25245.457			

is 6.94. It is, therefore, concluded that factor B has a very significant effect on the breaking strength of the kind of thread studied in this example. All other factors seem to have a rather limited effect, with the possible exception of factor A.

If all measurable interaction components and the blocks are considered as components of the error, taking into account the previous ANOVA tests, we obtain the results shown in Table 12.14. It is noted that in Table 12.14 the effect of factor A would be significant if a level of significance equal to $\alpha = 0.10$ were used, since in this case the critical value of the F-statistic is 2.62. Based on these results, the quality control department recommended that any of the three synthetic fibers be used (assuming that the fibers cost approximately the same) with long wool (AAAA-type) and alpaca (baby-type)

Table 12.14 ANOVA table for main effects

Source	*df*	*SS*	*MS*	*F*
A	2	2610.901	1305.45	2.85
B	2	12941.340	6470.67	14.14
C	2	312.747	156.37	0.34
D	2	1146.267	573.13	1.25
Error	18	8234.202	457.46	
Total	26	25245.457		

fibers. Since the effect of factor B is quite significant, it was recommended that the current product design be examined with a view to increasing the preparation of alpaca.

APPENDIX 12.A: EXPERIMENTAL RESULTS FOR THE APPLICATION

Block 1					Block 2					Block 3				
A	B	C	D	Data	A	B	C	D	Data	A	B	C	D	Data
0	0	0	0	90.6	0	1	0	2	99.6	0	2	0	1	189.6
0	0	2	1	108.6	0	1	2	0	123.8	0	2	2	2	144.6
0	0	1	2	112.4	0	1	1	1	111.6	0	2	1	0	186.9
1	2	1	2	123.4	1	0	1	1	89.7	1	1	1	0	97.6
1	2	0	0	156.8	1	0	0	2	86.9	1	1	0	1	167.6
1	2	2	1	141.7	1	0	2	0	97.7	1	1	2	2	86.9
2	1	2	1	134.8	2	2	2	0	161.1	2	0	2	2	117.8
2	1	1	2	167.8	2	2	1	1	166.6	2	0	1	0	123.5
2	1	0	0	111.8	2	2	0	2	154.8	2	0	0	1	126.5
Total				1147.9	Total				1091.8	Total				1241.0

From the above results we obtain $SS_{\text{total}} = 25\,245.45$ (with 26 degrees of freedom) and $SS_{\text{block}} = 1262.054$ (with two degrees of freedom). From these results we can also compute the totals of the observations corresponding to the levels of main effects and the levels of combinations for interaction components. For main effects the three totals for each level are:

Factor A: 1167.7, 1048.3, 1264.7
Factor B: 953.7, 1101.5, 1425.5
Factor C: 1184.2, 1179.5, 1117.0
Factor D: 1149.8, 1236.7, 1094.2

In the case of each of the measurable interactions AB^2, AC^2, AD^2, BC^2, BD^2 and CD^2, we must also find three totals. The first total corresponds to experimental conditions having levels X_i and X_j such that $X_i + 2X_j = 0$ (mod 3), the second total corresponds to experimental conditions having levels X_i and X_j such that $X_i + 2X_j = 1$ (mod 3) and the third total corresponds to experimental conditions having levels X_i and X_j such that $X_i + 2X_j = 2$ (mod 3). In these equations, $i = 1$ and $j = 2$ for AB^2; $i = 1$ and $j = 3$ for AC^2 and so on. In summary, the three totals correspond to the following groups of experimental conditions:

Group	X_i	X_j	$X_i + 2X_j$ (mod 3)
	0	0	0
1	1	1	0
	2	2	0
	0	2	1
2	1	0	1
	2	1	1
	0	1	2
3	1	2	2
	2	0	2

Proceeding as indicated, the following totals can be found:

For AB^2: 1146.2, 1209.8, 1124.7
For AC^2: 1104.2, 1246.2, 1130.3
For AD^2: 1240.7, 1136.6, 1103.4
For BC^2: 1128.4, 1180.0, 1172.3
For BD^2: 1148.6, 1148.2, 1183.9
For CD^2: 1076.4, 1134.4, 1269.9

EXERCISES

1. Determine which two-factor interaction components are measurable in the case of a 3^{4-1} fractional factorial design, using the generator $I = ABCD$.
2. Repeat Exercise 1, assuming that the 27 experimental conditions are grouped into nine blocks, using as block generators AB and AC^2. Indicate which treatments are included in each block.
3. Conduct an ANOVA for the application presented in Section 12.6, ignoring the need for blocks.
4. Consider a 3^{6-2} fractional factorial design with generators $I = ACDE$ and $I = BC^2DE^2F$. It is desired to have 27 blocks of three treatments each. (a) Write the simultaneous equations needed to determine the experimental conditions of the design. (b) Determine the aliases of the main effect A. (c) Write the simultaneous equations needed to group the treatments into blocks.
5. Analyze the results given for Exercise 8 in Chapter 7, considering only treatments A, B and C.
6. An experiment was designed to study the effect that temperature and level of humidity have on a gluing operation. For this purpose a completely

randomized experiment was used in which the temperature was set at three levels (cold, ambient, hot) and the humidity controlled at three levels corresponding to 55, 75 and 95%. The force in pounds needed to separate two glued items was considered as the experimental response. Do the ANOVA using the following results for two replications per treatment:

		Temperature		
		Cold	Ambient	Hot
	55%	1.2	3.4	3.8
		2.2	4.1	2.9
Humidity	75%	4.5	3.4	1.7
		3.8	4.0	2.3
	95%	3.2	3.9	4.3
		1.9	2.7	5.6

7. Consider the experimental situation analyzed in example 12.1. Interpret and explain the ANOVA results by constructing a graph of average service-life *versus* temperature.

REFERENCES

1. Connor, W. S. and Zelen, M. (1959) Fractional factorial experiment designs for factors at three levels, National Bureau of Standards, Washington, DC, *Applied Math. Series*, **54**.
2. Garcia-Diaz, A., Riggins, M. R. and Liu, S. J. (1984) *Development of Performance Equations and Servivor Curves for Flexible Pavements, Report 284–5*, Texas Transportation Institute, Texas A&M University, College Station, Texas.

FURTHER READING

Box, G. E. P., Hunter, W. G. and Hunter, J. S. (1978) *Statistics for Experimenters*, John Wiley & Sons, New York.

Davies, O. L. (1971) *Design and Analysis of Industrial Experiments*, Oliver and Boyd, Edinburgh.

Hicks, C. R. (1982) *Fundamental Concepts in the Design of Experiments* (3rd edn), Holts, Rinehart and Winston, New York.

Kempthorne, O. (1952) *The Design and Analysis of Experiments*, John Wiley & Sons, New York.

13 Response surface methodology

The material discussed in this book can be conceptually divided into three major parts:

(a) basic concepts of statistical design and analysis;
(b) factor-screening procedures to determine the significance of each factor;
(c) model formulation procedures to investigate how important factors affect the response of an experiment.

Chapters 1 through 6 contain the ANOVA concepts that served as the foundation for further developments. Chapters 7 through 12 included various procedures to estimate the effects of factors and interactions, as well as to test the significance of these effects. Chapter 13 will consider the development of mathematical relationships that link the factors to the response, and the use of these relationships to optimize an experimental response.

13.1 INTRODUCTION

The response surface methodology (RSM) discussed in this chapter has been extensively and successfully applied in a large number of experimental situations. Tool life testing [1–3], inertial welding processes [4], plant experimentation [5], human performance research [6] and chemical reaction studies [7] are typical areas where RSM has been successfully used. The technique was first developed and described in reference [8]. An excellent review of the methodology is given in reference [9].

In most RSM problems the mathematical relationship describing the experimental response in terms of the factors being considered is not known. Thus, the first step is to find a suitable approximation for this relationship. The nature of the approximating relationship is not behavioral, since it does not include any causal capabilities. Instead it only predicts a likely value of the response for a specified experimental condition.

Once a fitted relationship is obtained, it can be used to determine the most efficient operating conditions of the system being studied. This is achieved by using the fitted relationship to predict values of the response within a climbing optimization procedure. The complete procedure to identify the optimal operating conditions will be illustrated in section 13.4. Section 13.2 describes the basic procedure followed in response surface analysis in the case of a two-factor experimental response. Section 13.3 summarizes the procedure into a sequence of general steps.

13.2 BASIC RSM PROCEDURE

RSM is a sequential procedure. Essentially, it starts with a factorial experiment in a localized region of the response surface to determine a series of experimental conditions yielding increasing (more attractive) values of the response. When the yield of the experimental conditions shows reductions instead of additional increases, a second factorial experiment is designed, and a series of experimental conditions is generated and tested. This procedure is repeated until the experimental response can be improved very slowly and only by modest quantities. At this point the region of the optimum is reached and a more elaborate factorial design is used to identify the optimum.

As an illustration of the basic procedure, let us consider a system whose experimental response is affected by two-level factors. Assuming that the first factorial experiment is conducted in a localized region that is far from the optimum, a plausible fitting relationship would be a first-order polynomial such as

$$Y = b_0 + b_1 X_1 + b_2 X_2 \tag{13.1}$$

where the coded values $X_i = -1$ and $X_i = +1$ define the experimental region under consideration. In Chapter 11 it was shown that if E_i represents the effect of factor X_i, then the coefficient of this factor in the above equation is equal to $\frac{1}{2}E_i$. The coefficient b_0 can be estimated by the average of the four experimental values of the response associated with the 2^2 design. Figure 13.1 shows the four experimental conditions of the initial 2^2 design. The nearly parallel lines of this figure are contours representing all experimental conditions yielding the same value of the response. The direction along which the response is increased at a maximum rate is perpendicular to the contours, as indicated in the figure. This direction is known as the **path of steepest ascent**.

In order to illustrate graphically the procedure to reach the optimal region, a top view of the entire response surface is shown in Figure 13.2. This figure indicates how the nearly parallel lines of Figure 13.1 are actually segments of elongated contours. In Figure 13.2 both X_1 and X_2 are measured in their original units. For the first design, $X_1 = A_1$ and $X_1 = A_2$ correspond to the coded levels -1 and $+1$ respectively. Similarly, the values B_1 and B_2 correspond to the coded levels of X_2. As can clearly be seen in this figure,

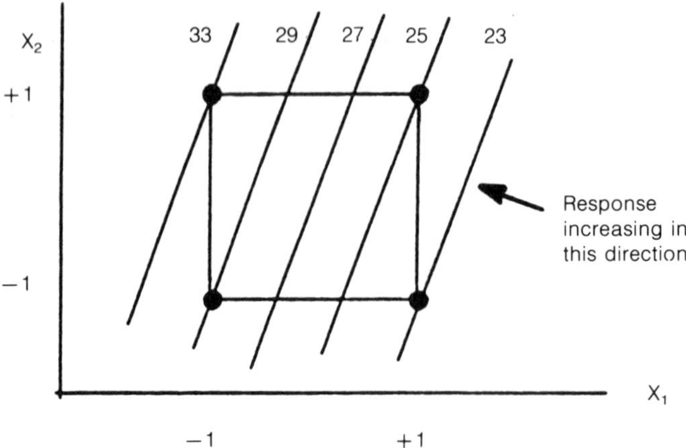

Figure 13.1 Path of steepest ascent.

the path of steepest ascent obtained from the initial factorial design yields increasing values of the response until the contour corresponding to a value approximately equal to 45 is reached. For this reason, a second path of steepest ascent must be generated.

To generate a second path, new values A_3, A_4, B_3 and B_4 are chosen for the two factors and another factorial design is run to compute new effect estimates and, in turn, new coefficients for the model formulated in equation (13.1). These coefficients are finally used, as will be indicated in section 13.3, to define the path of steepest ascent.

When the region containing the optimum is approached, the steepest ascent method will begin to yield only minor increases in the response, as a result of the curvature of the surface. At this moment a more elaborate analysis is needed to model this curvature. Typically, a second-degree polynomial is chosen to fit the response surface in the region of the optimum. A three-level design can be used to estimate the coefficients of the model containing second degree terms. In the case of two factors, the fitted polynomial is formulated as follows:

$$Y = b_0 + b_1X_1 + b_2X_2 + b_{11}X_1^2 + b_{22}X_2^2 + b_{12}X_1X_2. \tag{13.2}$$

The coefficients of the model formulated in equation (13.2) can be estimated using nine experimental response values associated with a 3^2 factorial design. In this design, it is possible to use coded values -1, 0, $+1$ for each factor. The corresponding table of coefficients for each term of equation (13.2) is shown in Table 13.1.

It is noted that in Table 13.1 the sum of the terms in each column is equal to 9, 0, 0, 6, 6 and 0 for X_0, X_1, X_1^2, X_2^2 and X_1X_2, respectively. Therefore,

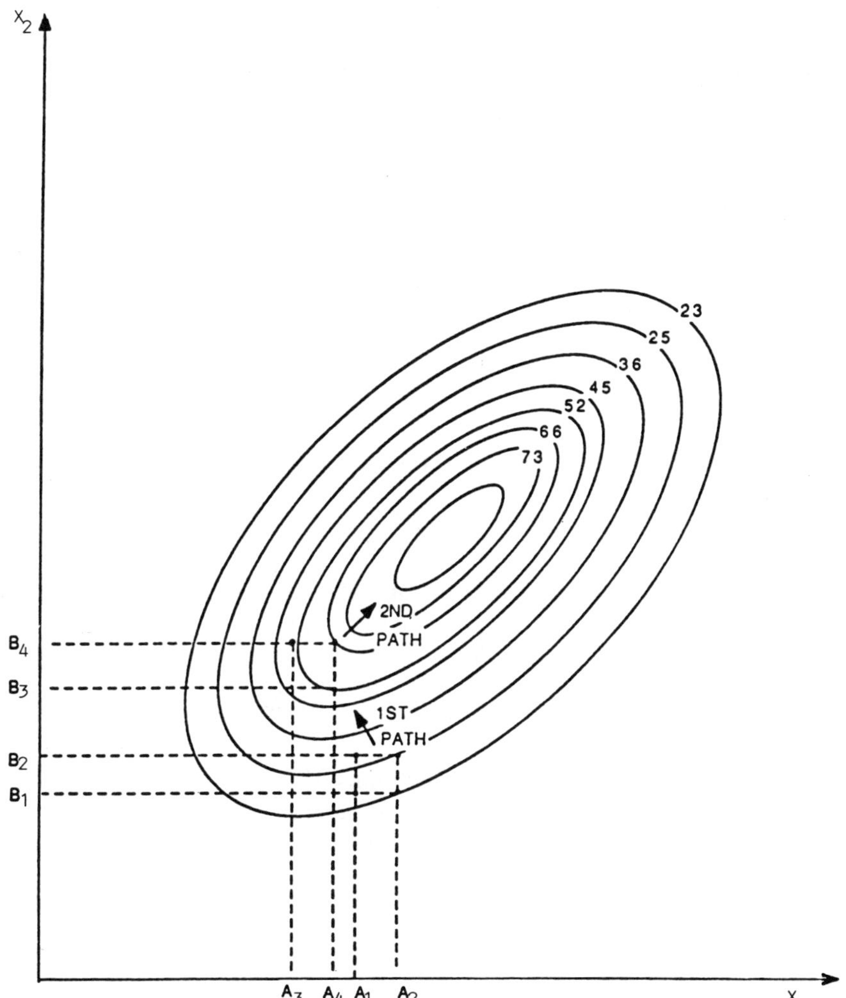

Figure 13.2 Basic RSM procedure.

using the model given in equation (13.2), we conclude that

$$\sum_{c=1}^{9} Y_c = 9b_0 + 6b_{11} + 6b_{22}. \tag{13.3}$$

The result shown in equation (13.3) can be generalized for a 3^n design. The corresponding result is shown as follows:

$$\sum_{c=1}^{N} Y_c = Nb_0 + \left(\frac{2N}{3}\right)b_{11} + \left(\frac{2N}{3}\right)b_{22} \tag{13.4}$$

Table 13.1 Three-level design coefficients

Run	X_0	X_1	X_2	X_1^2	X_2^2	X_1X_2	*Response*
1	1	−1	−1	1	1	1	Y_1
2	1	0	−1	0	1	0	Y_2
3	1	1	−1	1	1	−1	Y_3
4	1	−1	0	1	0	0	Y_4
5	1	0	0	0	0	0	Y_5
6	1	1	0	1	0	0	Y_6
7	1	−1	1	1	1	−1	Y_7
8	1	0	1	0	1	0	Y_8
9	1	1	1	1	1	1	Y_9

where $N = 3^n$. From equation (13.4) it is concluded that

$$b_0 = \bar{Y} - \tfrac{2}{3}b_{11} - \tfrac{2}{3}b_{22}. \tag{13.5}$$

Substituting the result from equation (13.5) for b_0 in equation (13.2), a modified model is obtained for Y, as follows:

$$Y = \bar{Y} + b_1X_1 + b_2X_2 + b_{11}(X_1^2 - \tfrac{2}{3}) + b_{22}(X_2^2 - \tfrac{2}{3}) + b_{12}X_1X_2. \tag{13.6}$$

The table of coefficients corresponding to the terms of the model given in equation (13.6) can be obtained from Table 13.1. The results are shown in Table 13.2.

The reader can now verify that the columns of Table 13.2 have coefficients that can be used to generate a set of orthogonal contrasts. Based on this fundamental result, the coefficients of the model given in equation (13.6) can be computed as the sum of the products of the response values with the

Table 13.2 Modified three-level design coefficients

Run	X_0	X_1	X_2	$(X_1^2 - \frac{2}{3})$	$(X_2^2 - \frac{2}{3})$	X_1X_2	*Response*
1	1	−1	−1	$\frac{1}{3}$	$\frac{1}{3}$	1	Y_1
2	1	0	−1	$-\frac{2}{3}$	$\frac{1}{3}$	0	Y_2
3	1	1	−1	$\frac{1}{3}$	$\frac{1}{3}$	−1	Y_3
4	1	−1	0	$\frac{1}{3}$	$-\frac{2}{3}$	0	Y_4
5	1	0	0	$-\frac{2}{3}$	$-\frac{2}{3}$	0	Y_5
6	1	1	0	$\frac{1}{3}$	$-\frac{2}{3}$	0	Y_6
7	1	−1	1	$\frac{1}{3}$	$\frac{1}{3}$	−1	Y_7
8	1	0	1	$-\frac{2}{3}$	$\frac{1}{3}$	0	Y_8
9	1	1	1	$\frac{1}{3}$	$\frac{1}{3}$	1	Y_9

coefficients of the appropriate column of Table 13.2 divided by the sum of squares of the elements in that column. That is,

$$b_i = \frac{L_i}{\sum_c a_{ic}^2} \tag{13.7}$$

$$b_{ii} = \frac{L_{ii}}{\sum_c a_{iic}^2} \tag{13.8}$$

$$b_{ij} = \frac{L_{ij}}{\sum_c a_{ijc}^2} \tag{13.9}$$

where L_i, L_{ii} and L_{ij} are the orthogonal contrasts belonging to the effects of X_i, X_i^2 and X_iX_j, respectively, and a_{ic}, a_{iic} and a_{ijc} are the coefficients of Y_c in contrasts L_i, L_{ii} and L_{ij} respectively.

Finally, if σ^2 is the error variance, the variance of the coefficients is given by σ^2 divided by the sum of squares of the elements associated with the corresponding contrasts. In symbols,

$$V[b_i] = \frac{\sigma^2}{\sum_c a_{ic}^2} \tag{13.10}$$

$$V[b_{ii}] = \frac{\sigma^2}{\sum_c a_{iic}^2} \tag{13.11}$$

$$V[b_{ij}] = \frac{\sigma^2}{\sum_c a_{ijc}^2}. \tag{13.12}$$

In the above relationships, an estimate s^2 of σ^2 can be used instead of the error variance, if this variance is unknown. The procedure described in this section will be illustrated on a numerical example (section 13.4).

13.3 RSM STEPS

The purpose of this section is to summarize the RSM procedure into a sequence of well-defined steps. The overall methodology consists of two phases.

Phase I performs a first-order analysis of the response surface and uses the gradient method to identify paths of steepest ascent to reach the region of the optimum. Phase II performs a second-order analysis of the response by taking into consideration the curvature in the neighborhood of the optimum. It uses a three-level factorial design to generate the coefficients of the second-degree polynomial. Once the polynomial is obtaned, elementary optimization concepts are invoked to identify the optimum.

13.3.1 Phase I: first-order analysis

(a) Select the factors to be investigated.
(b) Design and run a two-level factorial experiment in a localized region of the response surface.
(c) Compute the estimates of the effects and thereby calculate the coefficients of the linear model $Y = b_0 + b_1X_1 + b_2X_2 + \cdots + b_nX_n$.
(d) Select a reference factor to be used as a guide in determining the appropriate steps along the direction of each factor in order to continue moving along the path of steepest ascent.
(e) Select a few experimental conditions along the path of steepest ascent and run trials to determine if the response continues to increase. If the response ceases to increase, a new path should be generated.
(f) If a new path is needed, design and run a new two-level factorial experiment. All previous steps are repeated until no substantial improvement in the response is obtained. At this point phase II is conducted.

13.3.2 Phase II: second-order analysis

(a) Design and run a three-level factorial experiment in the region where the path of the steepest ascent yields no substantial improvement in the response.
(b) Compute the coefficient of the model

$$Y = b_0 + b_1X_1 + b_2X_2 + \cdots + b_{11}X_1^2 + b_{22}X_2^2 + \cdots$$
$$+ b_{12}X_1X_2 + \cdots + b_{n-1,n}X_{n-1}X_n.$$

(c) Using the above model, determine the nature of the stationary point of the response surface. The stationary point is one where the gradient vanishes.

13.4 AN APPLICATION OF RSM TO PAVEMENT DESIGN

This section will consider the problem of evaluating an existing pavement (road surface) structure to determine the most effective overlay design. The most economic way of evaluating a pavement section is by nondestructive testing. One of the most popular nondestructive techniques used by pavement engineers is known as the falling weight deflectometer (FWD) test system. This technique allows the pavement engineers to determine quickly and accurately the structural condition in terms of the elastic modulus of each layer of the pavement.

With the Dynatest 8000 FWD test system the structural condition of a pavement may be determine through nondestructive FWD tests, and the required overlay may be calculated from analytical structural design methods.

The most visible unit of the testing equipment is the FWD, which is a pavement loading device used to produce a load impulse that approximately has the same effect as a moving wheel load. Figure 13.3 shows a typical FWD unit.

The FWD unit has seven transducers that evaluate the deflection basin resulting from the deflection of the pavement surface. When the thickness of each pavement layer is known, the stiffness of each layer can be estimated by matching the deflection basin computed by theoretic layer elastic models to the deflection profile determined by the FWD device. The evaluation of the stiffness of each layer is an essential step in the overall overlay design.

Figure 13.3 Falling weight deflectometer.

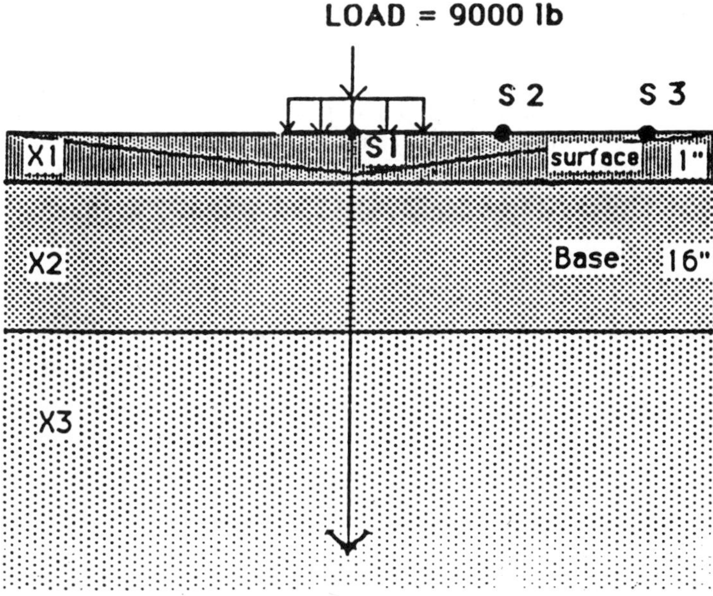

Figure 13.4 Pavement (road surface) structure.

Since the deflection basin for a given pavement section can be obtained from the FWD unit, a suitable experiment to estimate the stiffness of each layer consists of inputting a set of layer stiffness values to a layer elastic computer model and comparing the theoretic deflection pattern against that identified by the FWD unit. Using the RSM procedure, it is possible to adjust sequentially the stiffness values until the observed and the theoretical deflections are matched.

In this application a nondestructive test was performed on a three-layer pavement structure with an asphalt–concrete surface layer of 1 in (25 mm) thickness, a 16 in (400 mm) granular layer, and an infinite clay subgrade. This pavement structure is shown in Figure 13.4. Also the measured deflections at the seven sensor locations are given below in microns (10^{-6} m):

S_1	S_2	S_3	S_4	S_5	S_6	S_7
14.5	9.66	7.06	3.99	2.88	2.10	1.63

Following section 13.3, the results from this application can be summarized into two phases. The steps involved in the execution of each phase are shown below.

13.4.1 Phase I: first-order analysis

(a) Select the factors

In order to find the first path of the steepest ascent a 2^3 factorial experiment was designed considering the stiffness values of the layers as the three factors. The localized region for this experiment is designed in terms of the quantities associated with the coded levels -1 and $+1$. Table 13.3 indicates the stiffness values in pounds per square inch corresponding to a change of one coded unit for each factor. This is represented by u_i in the table.

(b) Design and run the experiment

Table 13.4 summarizes the results obtained for the initial factorial experiment. The response for the experiment was defined as the degree of closeness

Table 13.3 Coded levels for initial design

Factor	*Description*	-1	$+1$	u_i
X_1	Surface stiffness value	500 000	550 000	25 000
X_2	Base stiffness value	50 000	55 000	2 500
X_3	Subgrade stiffness value	20 000	21 000	500

Table 13.4 First factorial design

X_1	X_2	X_3	*Response*
−1	−1	−1	59.99
1	−1	−1	59.91
−1	1	−1	71.72
1	1	−1	71.69
−1	−1	1	61.00
1	−1	1	61.99
−1	1	1	71.99
1	1	1	71.99

between the theoretical basin deflection and the observed basin deflection. As previously mentioned, the theoretical deflection profile is determined by a computerized elastic layer deflection procedure. The particular program used in this application is known as the BISAR [10] program.

From previous experience it is known that the sum of squares associated with the difference between the two deflection basins does not exceed the value 100; for this reason, the response was defined as

$$Y = 100 - \sum_{i=1}^{7} e_i^2$$

where e_i is the deviation observed between the two deflection basins at the ith sensor location. The goal therefore is to maximize the response, since this would minimize the difference between the computed and the measured deflections.

(c) Compute effect estimates and coefficients

Using the results given in Table 13.4, the following estimates of the effects were obtained: $E_1 = 0.220$, $E_2 = 11.124$, $E_3 = 0.916$; also, the common effect was estimated as the average $\bar{Y} = 66.285$. Therefore, the coefficients of the fitted first-degree polynomial for the localized region defined in Table 13.3 are given by $b_0 = 66.285$, $b_1 = 0.110$, $b_2 = 5.562$ and $b_3 = 0.458$. The corresponding linear model is

$$Y = 66.285 + 0.110X_1 + 5.562X_2 + 0.458X_3. \tag{13.13}$$

(d) Select a reference factor

Factor X_1 was considered as the reference factor to monitor the response along the path of steepest ascent. This factor was used to determine appropriate changes in the other factors when X_1 was changed by a specified quantity. In this application this quantity was chosen equal to 5000 psi (350 kg cm^{-2}).

A change of d psi in the direction of the reference factor must be made simultaneously with a change of $(u_2b_2(u_1b_1)^{-1})d$ psi in the direction of X_2, and a change of $(u_3b_3(u_1b_1)^{-1})d$ in the direction of X_3, so that the experimental condition associated with the changed values will be on the path of steepest ascent.

(e) Run trials on selected experimental conditions

Considering $d = 5000$ psi, the values of the factor X_2 and X_3 must be changed by the following quantities:

$$d\left(\frac{u_2b_2}{u_1b_1}\right) = 25\,282 \text{ psi } (1770 \text{ kg cm}^{-2})$$

$$d\left(\frac{u_3b_3}{u_1b_1}\right) = 416 \text{ psi } (29 \text{ kg cm}^{-2}).$$

Using the above results it is possible to generate a sequence of points (X_1, X_2, X_3) along the path of steepest ascent. Starting with the values $X_1 = 525\,000$, $X_2 = 52\,500$, $X_3 = 20\,500$, the sequence of points on the path of steepest ascent can be generated (Table 13.5).

Additional trials were run for the above experimental conditions. The corresponding values for the response of the experiment were equal to 67.02, 86.79 and 81.00. As can be seen, the first trial resulted in an increase over the average value of the response in the localized region of the initial experiment ($b_0 = 66.285$). The second trial resulted in an increase over the response value obtained for the first trial. However, the third trial indicated that the response decreased, and, therefore, a new path of steepest ascent needs to be found.

(f) Generate a new path of steepest ascent

Since the effect of X_1 is reltively low according to equation (13.13), it was decided to move the base level (midpoint between the two levels) away from the path of steepest ascent, keeping the value of u_1 at approximately its previous magnitude. The base levels of X_2 and X_3 were chosen near the

Table 13.5

X_1	X_2	X_3
525 000	52 500	20 500
530 000	77 782	20 916
535 000	103 064	21 332

values 77 782 and 20 916, respectively, since these values are associated with the response value 86.79, which is the largest value up to this point. Table 13.6 summarizes the coded levels for the second factorial design. This table also shows the new u_1, u_2 and u_3 values. Table 13.7 shows the results obtained for the response for the experimental conditions of the factorial experiment designed in Table 13.6. Using these results, and following the same procedure used in finding the first path of steepest ascent, a new path was found:

$$Y = 86.37 + 0.01X_1 + 0.192X_2 - 1.0X_3. \tag{13.14}$$

From equation (13.14) it can be concluded that in the region of the second factorial experiment the average value of the response is substantially larger than that obtained from the first design. In the region of the second experiment, the coefficients of all factors are rather low. This indicates that perhaps the steepest ascent method is generating experimental conditions near the optimal region. In order to shed more light on the nature of the response surface, two additional factorial experiments were run. The results from these experiments are summarized in Table 13.8.

Table 13.8 shows the design matrix for the third and fourth factorial designs, as well as the response values and the coefficients of the fitted first-degree polynomials. The coefficients in this table correspond to the variables with a coded level equal to 1 in the design matrix. The first value in the column of coefficients is the average of eight runs. An inspection of the last

Table 13.6 Coded levels for second design

Factor	*Description*	*−1*	*+1*	u_i
X_1	Surface stiffness value	430 000	450 000	10 000
X_2	Base stiffness value	74 000	76 000	1 000
X_3	Subgrade stiffness value	20 500	21 500	500

Table 13.7 Second factorial design

X_1	X_2	X_3	*Response*
−1	−1	−1	87.04
1	−1	−1	87.28
−1	1	−1	87.61
1	1	−1	87.55
−1	−1	1	85.22
1	−1	1	85.17
−1	1	1	85.57
1	1	1	85.52

Table 13.8 Additional factorial designs

X_1	X_2	X_3	*Response for third design*	*Coefficients for third design*	*Response for fourth design*	*Coefficients for fourth design*
−1	−1	−1	89.21	88.444	90.48	90.11
1	−1	−1	89.17	0.010	90.30	−0.06
−1	1	−1	89.54	0.150	90.79	0.12
1	1	−1	89.49	–	90.56	–
−1	−1	1	87.27	−0.910	89.54	−0.43
1	−1	1	87.51	–	89.61	–
−1	1	1	87.70	–	89.84	–
1	1	1	87.66	–	89.73	–

Table 13.9 Coded levels for additional designs

	Third design		*Fourth design*	
i	$X_i = -1$	$X_i = +1$	$X_i = -1$	$X_i = +1$
1	460 000	480 000	450 000	650 000
2	76 000	78 000	78 000	80 000
3	19 500	20 500	19 000	19 500

two models,

$$Y = 88.44 + 0.01X_1 + 0.15X_2 - 0.91X_3$$

and

$$Y = 90.11 - 0.06X_1 + 0.12X_2 - 0.43X_3$$

indicates that the improvement in the average response is very small, and that first-order effects are not very significant. This was interpreted as being in the region of the optimum. The location of this region is perhaps around the localized regions considered for the third and fourth designs. These regions are defined in Table 13.9.

13.4.2 Phase II: second-order analysis

(a) Three-level design

After some experimentation it was decided that the response was most sensitive to changes in the value of the subgrade modulus, but almost insensitive when the modulus of the thin surface asphalt–concrete layer was varied. Hence it was decided that the three-level factorial should include only factors X_2 and X_3, considering X_1 set equal to 500 000 psi. The design matrix for the 3^2 factorial experiment including X_2 and X_3 as factors is given in Table 13.10. In this table the coded levels $-1, 0$ and 1 indicate three values chosen for each factor. The three stiffness values for E_2 were chosen as 85 000, 90 000 and 95 000; for E_3 the three values were set equal to 15 500, 16 000 and 16 500 psi. It should be noted that since the subgrade modulus has a larger effect on the response value than the base modulus, the difference between the three levels of subgrade modulus (=500 psi) was kept smaller than that of the base modulus (=5000 psi) such that the result would be more accurate.

(b) Second-order polynomial

From the results given in Table 13.10, equations (13.7)–(13.9) can be used to compute the coefficients of the second-order polynomial. The results are as follows (rounding off coefficients to four decimal places):

$$b_2 = 0.9133$$
$$b_3 = 0.0217$$
$$b_{22} = -0.2500$$
$$b_{33} = -0.1450$$
$$b_{23} = -0.2425.$$

Using the above results and the value of the average $\bar{Y} = 90.52$ for the nine experimental conditions of the three-level design, equation (13.5) is used to obtain $b_0 = 94.73$.

(c) Nature of the stationary point

The postulated model is formulated as

$$Y = 94.73 + 0.9133X_2 + 0.0217X_3 - 0.250X_2^2 - 0.145X_3^2 - 0.2425X_2X_3. \quad (13.15)$$

Table 13.10 Three-level factorial design

X_2	X_3	*Response*
-1	-1	93.16
0	-1	94.56
1	-1	95.47
-1	0	93.56
0	0	94.74
1	0	95.39
-1	1	93.69
0	1	94.60
1	1	95.03

The above model can be transformed into a new model free of first-order terms by translating the center of the original coordinate system to the center of the canonical form. A subsequent rotation of the axes would further transform the model into another model which is free of cross products, using the canonical analysis procedure outlined in Appendix 13.A.

The center of the canonical form is obtained by solving the following system of simultaneous equations:

$$\frac{\partial Y}{\partial X_2} = 0.9133 - 0.500X_2 - 0.2425X_3 = 0$$

$$\frac{\partial Y}{\partial X_3} = 0.0217 - 0.290X_3 - 0.2425X_2 = 0.$$

Solving the above equations, we get $X_2(\mathrm{s}) = 3.012$, $X_3(\mathrm{s}) = -2.443$. Substituting the values of $X_2(\mathrm{s})$ and $X_3(\mathrm{s})$ into equation (13.15), we have $Y(\mathrm{s}) = 95.95$. The new model without first-order terms is, therefore, given by

$$Y = 95.95 - 0.25X_2^2 - 0.145X_3^2 - 0.2425X_2X_3.$$

We can see that after the translation the first-order terms have been eliminated. Next we shall eliminate the cross-product term by performing a rotation of the axes. The latent roots $\lambda_1 = -0.3297$, $\lambda_2 = -0.065$ (coefficients of the canonical form) are obtained by solving

$$\begin{vmatrix} -0.25 - \lambda & -0.1213 \\ -0.1213 & -0.145 - \lambda \end{vmatrix} = 0.$$

Therefore, the canonical form of the experimental response being considered is given by

$$Y - 95.95 = -0.3297Y_2^2 - 0.065Y_3{}^2. \tag{13.16}$$

As indicated in Appendix 13.A, the latent vectors $\boldsymbol{U}_1$ and $\boldsymbol{U}_2$ are obtained from the solution of

$$\begin{pmatrix} -0.25 - \lambda_1 & -0.1213 \\ -0.1213 & -0.145 - \lambda_1 \end{pmatrix}\begin{pmatrix} U_{11} \\ U_{21} \end{pmatrix} = \begin{pmatrix} 0 \\ 0 \end{pmatrix}$$

$$\begin{pmatrix} -0.25 - \lambda_2 & -0.1213 \\ -0.1213 & -0.145 - \lambda_2 \end{pmatrix}\begin{pmatrix} U_{12} \\ U_{22} \end{pmatrix} = \begin{pmatrix} 0 \\ 0 \end{pmatrix}.$$

It can be easily verified that the system of equations yields the solution $U_{11} = 1.522U_{21}$ and $U_{12} = -0.657U_{22}$. In order to normalize the latent vectors, each one is divided by its length. As an illustration, the normalized latent vector $\boldsymbol{q}_1$ in the direction of $\boldsymbol{U}_1$ has components q_{11} and q_{21} given by

$$q_{11} = \frac{U_{11}}{(U_{11}^2 + U_{21}^2)^{1/2}} \qquad q_{21} = \frac{U_{21}}{(U_{11}^2 + U_{21}^2)^{1/2}}.$$

Since $U_{11}U_{21}^{-1} = 1.522$, it can be verified that $q_{11} = 0.8357$ and $q_{21} = 0.5491$. Similarly, for the unit latent vector $\boldsymbol{q}_2$ in the direction of $\boldsymbol{U}_2$, we would obtain $q_{12} = -0.5491$ and $q_{22} = 0.8357$. Therefore, the matrix with its columns being unit latent vectors is found to be

$$\mathbf{Q} = \begin{pmatrix} q_{11} & q_{12} \\ q_{21} & q_{22} \end{pmatrix} = \begin{pmatrix} 0.8357 & 0.5491 \\ -0.5491 & 0.8357 \end{pmatrix}.$$

The inverse of this matrix is

$$\boldsymbol{Q}^{-1} = \begin{pmatrix} \cos\theta & \sin\theta \\ -\sin\theta & \cos\theta \end{pmatrix} = \begin{pmatrix} 0.8357 & 0.5491 \\ -0.5491 & 0.8357 \end{pmatrix}.$$

Therefore, $\cos\theta = 0.8357$, $\sin\theta = -0.5491$, hence the angle of rotation is $\theta = -33.3°$.

Since both λ_1 and $\lambda_2 < 0$, and $|\lambda_1| > |\lambda_2|$, the fitted response surface is an elliptical hill with its center being at the maximum. The maximum response is equal to 95.95 which occurs when $X_2 = 3.012$, $X_3 = -2.443$. In terms of

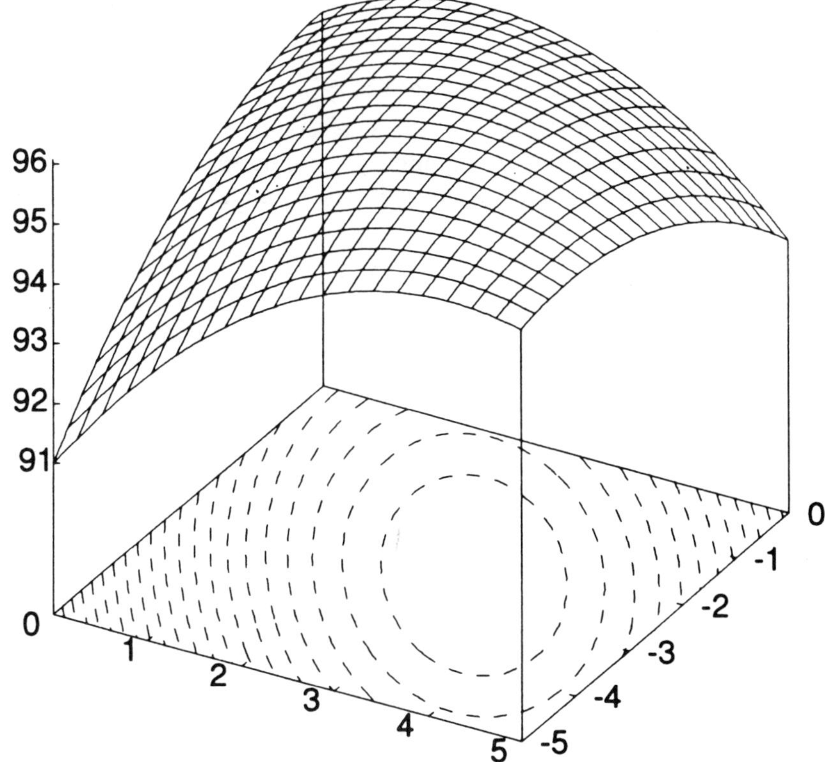

Figure 13.5 Contours of the fitted response surface.

elastic modulus, the maximum, i.e. the best fit, occurs when

$$E_2 = 90\,000 + 3.016(5000) = 105\,060\ \text{psi}$$

and

$$E_3 = 16\,000 - 2.443(500) = 14\,778\ \text{psi}.$$

The contours of the response surface of equation (13.16) are plotted in Figure 13.5. The center of the system represents a maximum point.

APPENDIX 13.A CANONICAL ANALYSIS

A polynomial of basic importance is the quadratic form. A quadratic form in the variables $X_1, X_2, \ldots, X_n$ is a polynomial

$$F = \sum_{i,j=1}^{n} a_{ij} X_i X_j$$

that is, it is homogeneous second degree polynomial in the n variables. A quadratic form may likewise be written in matrix form. If $\boldsymbol{X} = [X_1, X_2, \ldots, X_n]'$, and $\boldsymbol{A} = [a_{ij}]_n$, we have

$$F = \boldsymbol{X}'\boldsymbol{A}\boldsymbol{X}$$

The matrix $\boldsymbol{A}$ is called the **matrix of the quadratic form**.

By the replacement of each of a_{ij} and a_{ji}, $i \neq j$, by their mean $\frac{1}{2}(a_{ij} + a_{ji})$, the matrix of the quadratic form may always be made symmetric. This is fair since

$$a_{ij}X_iX_j + a_{ij}X_jX_i = [\tfrac{1}{2}(a_{ij} + a_{ji})]X_iX_j + [\tfrac{1}{2}(a_{ij} + a_{ji})]X_jX_i.$$

For any $n \times n$ symmetric matrix $\boldsymbol{A}$ with real elements there exist a scalar λ and a nonzero column vector $\boldsymbol{u}$ such that $\boldsymbol{Au} = \lambda\boldsymbol{u}$.

The scalar λ is called the **latent root** of $\boldsymbol{A}$, and $\boldsymbol{u}$ is called the corresponding **latent vector**. From the fact that $\boldsymbol{Au} = \lambda\boldsymbol{u}$ we can write that $\boldsymbol{Au} = \lambda\boldsymbol{Iu}$, where $\boldsymbol{I}$ is the identity matrix and $\boldsymbol{u}$ is a column vector. Therefore

$$\boldsymbol{Au} - \lambda\boldsymbol{Iu} = \boldsymbol{0}$$

or

$$(\boldsymbol{A} - \boldsymbol{I}\lambda) = \boldsymbol{0}.$$

If the inverse $(\boldsymbol{A} - \boldsymbol{I}\lambda)^{-1}$ exists, then we can write that

$$(\boldsymbol{A} - \boldsymbol{I}\lambda)^{-1}(\boldsymbol{A} - \boldsymbol{I}\lambda)\boldsymbol{u} = (\boldsymbol{A} - \boldsymbol{I}\lambda)^{-1}\boldsymbol{0}.$$

Therefore, $\boldsymbol{Iu} = \boldsymbol{0}$. But $\boldsymbol{Iu} = \boldsymbol{u}$, so $\boldsymbol{u} = \boldsymbol{0}$. On the other hand, by definition $\boldsymbol{u} \neq \boldsymbol{0}$. Then the inverse $(\boldsymbol{A} - \boldsymbol{I}\lambda)^{-1}$ must not exist.

Then $(\boldsymbol{A} - \boldsymbol{I}\lambda)$ must be singular, that is, its determinant must be equal to zero. In symbols,

$$|A - I\lambda| = 0.$$

The above equation is called the characteristic equation of A and it is the necessary and sufficient condition for finding the latent roots λ.

EXAMPLE 13.1

Consider the equation $F = 31X_1^2 + 10\sqrt{3}X_1X_2 + 21X_2^2 = 144$. Note $F = 144$ is the equation of an ellipse. In matrix form,

$$F = X'AX$$

$$F = (X_1 \quad X_2)\begin{pmatrix} 31 & 5(3)^{1/2} \\ 5(3)^{1/2} & 21 \end{pmatrix}\begin{pmatrix} X_1 \\ X_2 \end{pmatrix}.$$

Now

$$A - I\lambda = \begin{pmatrix} 31 & 5(3)^{1/2} \\ 5(3)^{1/2} & 21 \end{pmatrix} - \begin{pmatrix} \lambda & 0 \\ 0 & \lambda \end{pmatrix} = \begin{pmatrix} 31 - \lambda & 5(3)^{1/2} \\ 5(3)^{1/2} & 21 - \lambda \end{pmatrix}.$$

Also

$$|A - I\lambda| = (31 - \lambda)(21 - \lambda) - 75 = 0.$$

Solving for λ we obtain $\lambda_1 = 16$ and $\lambda_2 = 36$. These are the latent roots of A.

Now, let us find the latent vectors. Recall that $(A - I\lambda)u = 0$, where λ is a latent root of A. Then for $\lambda_1 = 16$,

$$\begin{pmatrix} 31 - 16 & 5(3)^{1/2} \\ 5(3)^{1/2} & 21 - 16 \end{pmatrix}\begin{pmatrix} U_1 \\ U_2 \end{pmatrix} = \begin{pmatrix} 0 \\ 0 \end{pmatrix}$$

$$\begin{pmatrix} 3 & 3^{1/2} \\ 3^{1/2} & 1 \end{pmatrix}\begin{pmatrix} U_1 \\ U_2 \end{pmatrix} = \begin{pmatrix} 0 \\ 0 \end{pmatrix}.$$

Then we must solve the following system of simultaneous equations:

$$3u_1 + 3^{1/2}u_2 = 0 \text{ (i)}$$

$$3^{1/2}u_1 + u_2 = 0 \text{ (ii)}.$$

The solution is $u_1 = -[\frac{1}{3}(3)^{1/2}]u_2 = -[3^{-1/2}]u_2$ which arises from either (i) or (ii). So, any vector u which satisfies the relationship $u_1 = -3^{-1/2}u_2$ is a latent vector. In this case, we have an infinite number of latent vectors corresponding to the latent root $\lambda_1 = 16$.

Let W be the length of the latent vector u. Then we can always define a unit vector Q_u in the same direction of the vector u. This unit vector is referred to as a normalized latent vector. In symbols, $u = [u_i]$, $Q_u = [q_i]$, where $q_i = u_iW^{-1}$.

The normalized latent vector for $\lambda_1 = 16$ is

$$\boldsymbol{Q}_1 = \begin{pmatrix} \frac{1}{2} \\ -\frac{1}{2}(3)^{1/2} \end{pmatrix}.$$

In a similar manner, the normalized latent vector for $\lambda_2 = 36$ can be found to be

$$\boldsymbol{Q}_2 = \begin{pmatrix} \frac{1}{2}(3)^{1/2} \\ \frac{1}{2} \end{pmatrix}.$$

Then, the matrix of normalized latent vectors is given by

$$\boldsymbol{Q} = \begin{pmatrix} \frac{1}{2} & \frac{1}{2}(3)^{1/2} \\ -\frac{1}{2}(3)^{1/2} & \frac{1}{2} \end{pmatrix}.$$

Now let us recall that for any latent vector **u** and associated latent root λ_u

$$\boldsymbol{Au} = \lambda_u \mathbf{u}.$$

Let us divide by the scalar W,

$$\boldsymbol{A}\left[\frac{\mathbf{u}}{W}\right] = \lambda_u \left[\frac{\boldsymbol{u}}{W}\right]$$

$$\boldsymbol{AQ}_u = \lambda_u \boldsymbol{Q}_u.$$

Therefore

$$\boldsymbol{Q}'_u \boldsymbol{AQ}_u = \boldsymbol{Q}'_u \boldsymbol{\lambda} \boldsymbol{Q}_u = \boldsymbol{Q}'_u \boldsymbol{Q}_u \lambda_u = \lambda_u.$$

Now, define $\boldsymbol{Q}$ as the matrix formed with the unit vectors $\boldsymbol{Q}_u$ as columns, and $\boldsymbol{\lambda}$ as the matrix with the scalars λ_u on the main diagonal and all other elements equal to zero (that is, a diagonal matrix):

$$\boldsymbol{Q} = [\boldsymbol{Q}_u]; \quad \text{and} \quad \boldsymbol{\lambda} = [\lambda_u].$$

Then we can write the following expression:

$$\boldsymbol{Q}'\boldsymbol{AQ} = \boldsymbol{\lambda}.$$

This ability to reduce a symmetric matrix $\boldsymbol{A}$ to a diagonal matrix makes it possible to reduce a quadratic form to a functional form without cross-product terms. Such a form is called the **canonical form**.

Recall that the original quadratic form $\boldsymbol{F}$ was written as $\boldsymbol{F} = \boldsymbol{X}'\boldsymbol{AX}$. Let us now consider the nonsingular transformation $\boldsymbol{X} = \boldsymbol{QY}$ in order to express $\boldsymbol{F}$ in terms of $y_1, y_2, \ldots, y_n$ instead of $x_1, x_2, \ldots, x_n$. We obtain:

$$\begin{aligned} \boldsymbol{F} &= (\boldsymbol{QY})'\boldsymbol{A}(\boldsymbol{QY}) \\ &= \boldsymbol{Y}'(\boldsymbol{Q}'\boldsymbol{AQ})\boldsymbol{Y} \\ &= \boldsymbol{Y}'\boldsymbol{\lambda Y}. \end{aligned}$$

EXAMPLE 13.2

For the quadratic form given in Example 13.1 we have $\lambda_1 = 16$ and $\lambda_2 = 36$. Therefore

$$\boldsymbol{\lambda} = \begin{pmatrix} 16 & 0 \\ 0 & 36 \end{pmatrix}.$$

Using $\boldsymbol{Y}$ instead of $\boldsymbol{X}$,

$$\boldsymbol{F} = (y_1 \quad y_2) \begin{pmatrix} 16 & 0 \\ 0 & 36 \end{pmatrix} \begin{pmatrix} y_1 \\ y_2 \end{pmatrix} = 16y_1^2 + 36y_2^2.$$

Note that this is also an ellipse, as in Example 13.1, but the coordinate system is different. In this case all terms with cross products are eliminated.

Angle of rotation

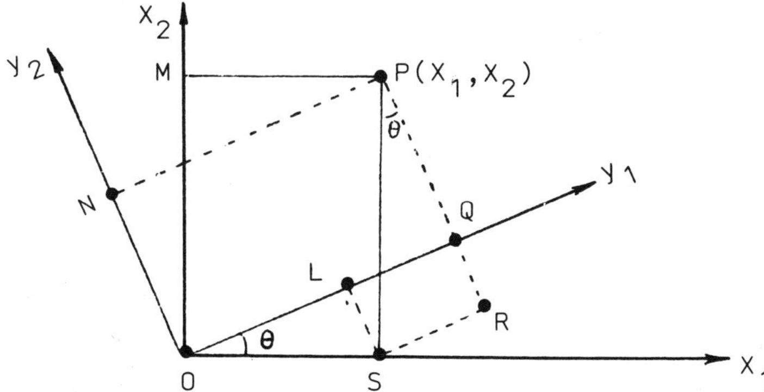

(a) $OQ = OL + LQ = OL + SR$
$y_1 = X_1 \cos\theta + X_2 \sin\theta.$

(b) $ON = PR - QR = PR - LS$
$y_2 = X_2 \cos\theta - X_1 \sin\theta.$

(c) From (a) and (b) we can write

$$\begin{pmatrix} y_1 \\ y_2 \end{pmatrix} = \begin{pmatrix} \cos\theta & \sin\theta \\ -\sin\theta & \cos\theta \end{pmatrix} \begin{pmatrix} X_1 \\ X_2 \end{pmatrix}.$$

But

$$\boldsymbol{X} = \boldsymbol{QY}.$$

Thus

$$\boldsymbol{Y} = \boldsymbol{Q}^{-1}\boldsymbol{X}.$$

Then

$$\boldsymbol{Q}^{-1} = \begin{pmatrix} \cos\theta & \sin\theta \\ -\sin\theta & \cos\theta \end{pmatrix}$$

Table 13.11

Case	*Coefficient relations*	*Signs* $b_{11}b_{22}$	*Types of contours*	*Geometric interpretation*	*Center*
1	$b_{11}=b_{22}$	$- -$	Circles	Circular hill	Maximum
2	$b_{11}=b_{22}$	$+ +$	Circles	Circular hollow	Minimum
3	$b_{11}\neq b_{22}$	$- -$	Ellipses	Elliptical hill	Maximum
4	$b_{11}=b_{22}$	$+ +$	Ellipses	Elliptical hollow	Minimum
5	$b_{11}=b_{22}$	$+ -$	Hyperbolas	Symmetrical saddle	Saddle point
6	$b_{11}=b_{22}$	$- +$	Hyperbolas	Symmetrical saddle	Saddle point
7	$b_{11}\neq b_{22}$	$+ -$	Hyperbolas	Elongated saddle	Saddle point
8	$b_{22}=0$	$-$	Straight lines	Stationary ridge	None
9	$b_{22}=0$	$+$	Straight lines	Stationary valley	None
10	$b_{22}=0$	$-$	Parabolas	Rising ridge	At ∞
11	$b_{22}=0$	$+$	Parabolas	Falling valley	At ∞

which means that we can obtain the rotation angle θ from the inverse of matrix $\boldsymbol{Q}$. As an exercise, find θ for the example being considered. In this case it can be verified that the inverse of $\boldsymbol{Q}$ is given by

$$\boldsymbol{Q}^{-1}=\begin{pmatrix} \frac{1}{2} & \frac{1}{2}(3)^{1/2} \\ -\frac{1}{2}(3)^{1/2} & \frac{1}{2} \end{pmatrix}.$$

Interpretation of the canonical function

The canonical function

$$F(Y_1, Y_2)=b_{11}Y_1^2+b_{22}Y_2^2$$

can be interpreted according to the results shown in Table 13.11.

EXERCISES

1. A fitted model was developed from a statistically designed experiment. The result

$$Y=89.30+16.48X_1+3.38X_2-16.5X_2^2-17-20X_2^2-6.99X_1X_2$$

was found to represent adequately the response function in the neighborhood of interest. The variables X_1 and X_2 are defined as follows:

$$X_1=\frac{\text{(weight of reactants in grams)} - 110\text{ g}}{10\text{ g}}$$

$$X_2=\frac{\text{(pressure in psia)} - 20\text{ psia}}{\frac{20}{3}\text{ psia}}$$

Y = observed yield.

Table 13.12

Distance X_1	*Frequency* X_2	*Errors* Y
−1	−1	4
+1	−1	4
−1	+1	2
+1	+1	3
−2	0	1
+2	0	2
0	−2	5
0	+2	1
0	0	3
0	0	5

The process conditions were limited to a maximum weight of 110 g and a maximum pressure of 20 psi for safety reasons. On the basis of these data, how would you run the process to optimize the output?

2. A human factors engineer had been studying an operator's control panel of a large drag line. Two factors were observed to affect the operator's performance:

   X_1 = distance of hand travel between controls

   X_2 = frequency of use of a control.

   A model is desired to fit the experimental response (performance), measured as a number of mistakes made by the operator, as a function of the factors. The data are given in Table 13.12.

   (a) Fit a first-order model to these data.
   (b) Draw contour lines of the fitted response surface.
   (c) Draw the direction of steepest ascent.
   (d) Criticize your analysis.

Table 13.13

Trial	X_1	X_2	%
1	−1	−1	1.7
2	0	−1	9.2
3	1	−1	10.1
4	−1	0	5.2
5	0	0	11.5
6	1	0	9.1
7	−1	1	6.3
8	0	1	10.2
9	1	1	5.8

3. An industrial engineer had been studying an assembly process to determine the workers' nonproductive time percentage. Two factors, X_1 (lighting) and X_2 (room temperature), were varied over three levels to observe the effects. After a series of tests a near-stationary region was reached. The engineer wanted to determine its nature by applying a second-degree polynomial to test its fit. A 3^2 factorial experiment was conducted and the results are shown in Table 13.13.
   (a) Estimate the variance of the estimates b_0, b_1, b_2, b_{11}, b_{22}, b_{12}; assume the variance $\sigma^2 = 0.6$.
   (b) Conduct an analysis of variance.
4. Consider the data given for problem 3:
   (a) Plot the given data in a contour diagram.
   (b) Find the canonical form of the given response.
   (c) Find the angle of rotation,
5. Find the characteristic roots and characteristic vectors for the matrix:
   (a)
$$A = \begin{pmatrix} 1 & 2 \\ 2 & 1 \end{pmatrix}$$
   (b)
$$A = \begin{pmatrix} 1 & -2 \\ 0 & 1 \end{pmatrix}.$$
   Discuss the difference between cases (a) and (b).
6. Consider the quadratic form
$$F = 2X_1^2 + X_1X_2 - 3X_1X_3 + 2X_2X_3 - X_3^2.$$
   Find the characteristic roots and characteristic vectors. Find the new expression for F without cross products.
7. Do the same as in Exercise 6, but with $F = 2X_1X_2 + 2X_2X_3 + X_3^2$.
8. When is a quadratic form said to be:
   (a) Positive definite?
   (b) Positive semi-definite?

Table 13.14

x_1	x_2	y
−1	−1	11
1	−1	4
−1	1	28
1	1	6
0	0	13

(c) Negative definite?
(d) Negative semi-definite?
(e) Indefinite?

9. Show that $F = 6X_1 + 3X_2 - 4X_1X_2 - 2X_1^2 - 3X_2^2$ is negative definite.
10. In Exercises 5, 6 and 7 find the transformation $\boldsymbol{Y} = \boldsymbol{Q}^{-1}\boldsymbol{X}$.
11. Two coded variables $x_1 = \frac{1}{10}(X_1 - 99)$ and $x_2 = \frac{1}{20}(X_2 - 17)$ are examined, on the basis of the experimental data shown in Table 13.14. Is the point $(X_1, X_2) = (57, 68)$ on the path of steepest ascent? Explain your answer.

REFERENCES

1. Wu, S. M. (1964) Tool life testing by response surface methodology, Part 1. *Journal of engineering for Industry, Transactions of the ASME*, May, 105–10.
2. Wu, S. M. (1964) Tool life testing by response surface methodology, Part 2. *Journal of engineering for Industry, Transactions of the ASME*, May, 111–16.
3. Wu, S. M. and Meyer, R. M. (1964) Cutting tool temperature predicting equation by response surface methodology. *Journal of Engineering for Industry, Transactions of the ASME*, May, 150–5.
4. Wang, K. K. and Rasmussen, G. (1972) Optimization of inertia welding process by response surface methodology. *Journal of Engineering for Industry, Transactions of the ASME*, November, 995–1006.
5. Hill, W. J. and Wiles, R. A. (1975) Plant experimentation (PLEX). *Journal of Quality Technology*, **7**(3), July, 115–22.
6. Clark, C. and Williges, R. C. (1973) Response surface methodology central composite design modifications for human performance research. *Human Factors*, **15**(4), 295–310.
7. Davies, O. L. (ed.) (1971) *The Design and Analysis of Industrial Experiments*, Hafner, New York.
8. Box, G. E. P. and Wilson, K. B. (1951) On the experimental attainment of optimum conditions. *J. Roy. Stat. Soc.*, **B13**, 1.
9. Hill, W. J. and Hunter, W. G. (1966) Response surface methodology: a review. *Tech. Rept. No. 62*, Department of Statistics, University of Wisconsin.
10. Koninklijke/Shell-Laboratorium (1972) *BISAR* (*BItumen Structures Analysis in Roads*), Computer Program: Users' Guide Manual, Amsterdam, Netherlands, July.

14 Taguchi methods

Traditionally the role of quality control has been that of eliminating defective products on the basis of statistical sampling and inspection rules. As a result of an increased emphasis on quality products and cost-effectiveness in modern complex manufacturing systems, the scope of the quality control process has been significantly revised during the 1980s. One of the most outstanding contributions in this area can be attributed to the Japanese engineer Genichi Taguchi, an international consultant in the field of quality management. Taguchi has set forth an innovative approach to total quality control and assurance in the development cycle of a product by integrating on-line design quality and off-line production quality procedures. He formulated both a philosophy and a methodology for the process of quality improvement that depends on statistical concepts, especially statistically designed experiments.

The primary goals of the Taguchi methodology can be described as:

(a) a reduction in the variation of a product or process design to improve quality and lower the loss imparted to society;
(b) a proper product or process implementation strategy which can further reduce the level of variation.

There are three off-line quality control phases involved in the design of a product or process: system design, parameter design and tolerance design. **System design** is the phase to generate a basic prototype design that performs the functions of the product with minimum deviation from target performance values. In this phase new concepts, ideas and methods are developed using current technology, processes, materials and engineering methods to provide new or improved products to consumers [1]. **Parameter design** is the phase where methods of experimental design are used to identify settings of product and process parameters in such a way that the sensitivity of the desired product characteristics to changes in the uncontrollable environmental variables is minimized. **Tolerance design** is the phase to study each parameter or factor by trading off quality loss and cost. Of all these phases the parameter design phase, which is actually an off-line quality control method, is the most important stage for achieving high quality without any substantial increase in the cost.

This chapter will be divided into four sections. A brief description of each section is given as follows. The fundamental concepts and definitions are presented in section 14.1. The experimental design aspects and the corresponding procedures are described in section 14.2. A detailed example is presented in section 14.3. Finally, a short critical analysis of the Taguchi methods is presented in section 14.4.

14.1 BASIC CONCEPTS AND DEFINITIONS

Traditional procedures recognize the importance of controlling product quality in manufacturing processes. However, only during the 1980s has it been widely acknowledged that product quality should be engineered into product design as well. This is in essence Taguchi's view of quality engineering. He contends that quality cannot be achieved economically through inspection and product screening only, and for this reason it has to be engineered into the product at the design stage. Before we discuss the aspects of quality engineering, the following basic concepts of Taguchi's strategy will be presented:

(a) Quality can be expressed in terms of the loss imparted to society.
(b) The loss to society has a functional relationship with the variation in each important product characteristic and can be quantified and expressed as a **loss function**.
(c) A quality characteristic is expressed in terms of an ideal **target value**.
(d) The quality improvement of a product is achieved by reducing the variation around the target value of the product performance characteristic in every level of product design.
(e) The sensitivity of an effect to changes in specific factors can be measured by a performance statistic known as a **signal-to-noise ratio**.
(f) The quality of a final product results from engineering design and manufacturing processes.

Taguchi defines quality as the loss imparted to society from the time a product is shipped to the market. There is a loss every time a product fails to perform adequately the function for which it was designed. Societal losses are not always tangible and measurable, as in the case of repair costs of bad products, added sales cost passed on to the consumer of defective products, and perhaps, although to a lesser degree, the market share erosion caused by a negative perception by the customers of the product quality of a company. Other types of losses like those related to air pollution, radiation or radiation leaks from a nuclear plant are more difficult to bring into the quality engineering process of a product.

14.1.1 The loss function

The loss function is a relationship that links the deviation of a performance characteristic of a product from a specified target value to a cost imparted to the users of the product as a result of such a deviation. If L is the loss in dollars, K a known cost coefficient that depends on each particular application, Y the value achieved for a quality characteristic and T is the target value for the same quality characteristic, a common type of loss function is the quadratic function defined in equation (14.1):

$$L = K(Y - T)^2. \qquad (14.1)$$

This loss function is graphically represented in Figure 14.1. In this figure LSL and USL represent a lower loss limit and an upper loss limit, respectively. This loss function is applicable to those situations in which the product characteristic is best at a specified target value and equal deviations in either direction from the target value are equally penalized. It is noted that some loss is also incurred even if the performance characteristic is within specified upper and lower limits. The thrust of the methods developed by Taguchi is to reduce this loss by reducing the variation in the critical performance characteristic. The performance characteristic may be the size of a shoe, the geometric dimensions of a part, the breaking strength of a cable or the gas mileage of a car.

A loss function can ideally be developed for each product or process based on detailed estimates of actual losses incurred due to the variation experienced by the performance characteristic. To illustrate the impact of this variation, let us consider four factories producing electric circuits that have a target

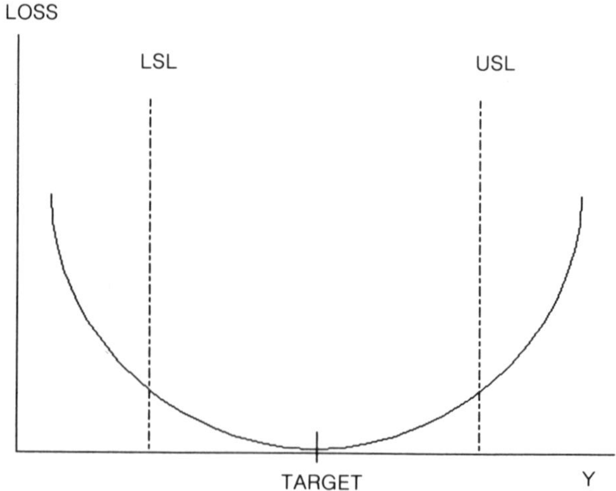

Figure 14.1 A quadratic loss function.

value of 115 V. The distributions of output voltage for the four factories are shown in Figure 14.2. Factory 1 has the largest output voltage variability and factory 4 has the lowest. This figure also indicates specified lower and upper values on the voltage. As can be observed in the figure, in all cases the target value coincides with the midpoint of the specified range for the output voltage. However, a circuit manufactured by factory 3 may be unacceptable since the mean of output voltage is far from the target value. On the other hand, when the target value is well within the specifications, a circuit manufactured in factory 4 would be more attractive than those produced in factories 1 and 2, if the price is assumed to be the same for circuits made in these factories. In this case, a consumer will choose a product from factory 4 because the deviation of the output voltage from the target value is the least as compared to the circuits manufactured by the other factories. From this example it is clear that a loss might occur not only when the product characteristic falls outside the specification limits but also when it is within the specification limits.

14.1.2 The target value

In order to determine the degree to which a product or process satisfies a quality characteristic, a target value should be specified. This target value

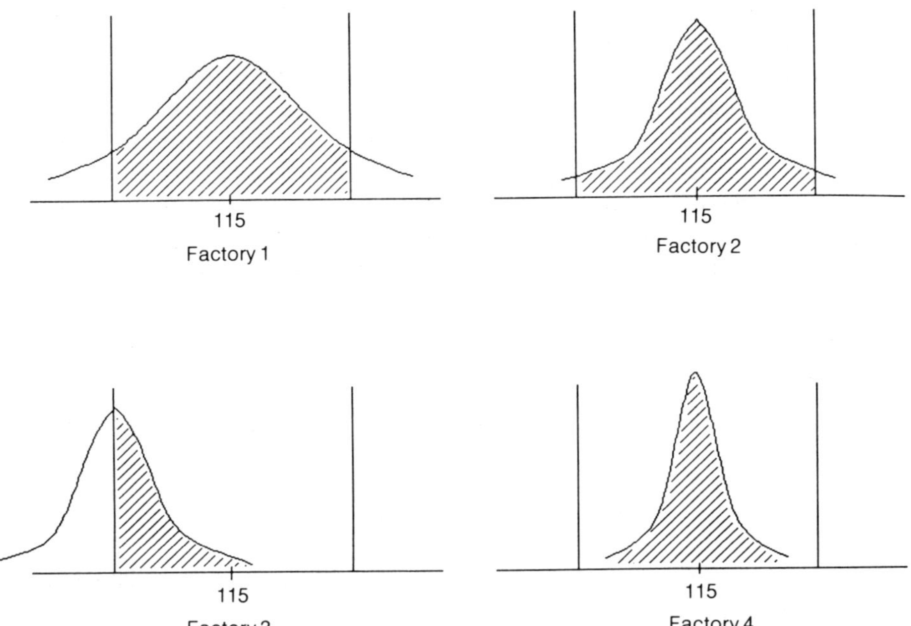

Figure 14.2 Output distributions for four factories.

represents an ideal state of the quality characteristic from the customer's point of view. There are three types of fixed targets which can be stated for any quality characteristic:

(a) The target is defined as a nominal value, and the loss increases as the characteristic deviates from the target value.
(b) The target is a value as large as possible, and the loss increases as the characteristic approaches the value of zero.
(c) The target is avalue as low as possible, and the loss increases as the characteristic increases.

For example, if an automobile door frame is supposed to be 34 in (863 mm) wide, then the closer to this nominal value the better the product. However, if the quality characteristic is miles per gallon, the larger the better. Finally, if the quality characteristic is operating cost, then the smaller the better. One should note the departure in philosophy from the traditional thinking that just being within the specification limits is acceptable.

14.1.3 Signal-to-noise ratio

Traditionally a designed experiment can be used to estimate or test the significance of certain factors on the basis of a measurable response over a set of experimental conditions. Taguchi emphasizes that in addition to this the variation of the experimental data needs to be studied. In order to facilitate this study he uses the concept of a signal-to-noise ratio. The simplest form of signal-to-noise ratio (S/N ratio) is the ratio of the mean (signal) to the standard deviation (noise), which is the inverse of the coefficient of variation. Among 60–70 available choices for measuring S/N rations, Taguchi recommends the generic forms given in equations (14.2)–(14.4) for three different types of criteria, where y_i is the value of the response in the ith experimental condition (observation) tested, with $i = 1, 2, \ldots, n$. Additionally, in equation (14.4) $\bar{Y}$ and S^2 are the sample mean and variance:

$$\text{Type 1:} \quad S/N_S = -10\log_{10}\left(\frac{\sum y_i^2}{n}\right) \tag{14.2}$$

$$\text{Type 2:} \quad S/N_B = -10\log_{10}\left(\frac{1}{n}\sum\frac{1}{y_i^2}\right) \tag{14.3}$$

$$\text{Type 3:} \quad S/N_T = 10\log_{10}\left(\frac{\bar{Y}^2}{S^2}\right). \tag{14.4}$$

The S/N ratio consolidates several repetitions (at least two data points are required) into one value which reflects the amount of variation present. The loss function attains its minimum at the target value and increases as the quality characteristic deviates from the target value. The first type of ratio, known as the smaller-is-better S/N_s ratio, is used when there is a nonnegative characteristic with an ideal value equal to zero ($T = 0$). Typical examples of its applicability are those situations where the quality characteristic is related to wear, shrinkage or deterioration. Alternatively, the second type of ratio, known as the larger-is-better S/N_B ratio, is used when there is no predetermined value for the target ($T = \infty$), and the larger the value of the characteristic, the better the product. Typical examples of its applicability are those cases where the quality characteristic is related to strength of materials, length of service life, or fuel efficiency. Finally, the third type of ratio, know as the nominal-value-is-best S/N_T ratio, is used when a nominal size, or characteristic, is preferred ($T = T_0$). Typical examples of its applicability are situations where the quality characteristic is related to a dimension, clearance, weight or viscosity, and deviations in both directions from the specified target are undesirable.

14.1.4 Factor classification

The performance characteristic of a product or process and its variation depend on both design and noise factors. Design factors are those that can be rather easily controlled in a design or manufacturing process. Examples of these factors are choices of material, motor RPM and batch size. On the other hand, noise factors are those that are difficult, if not impossible, to control. Examples of noise factors are the temperature and humidity of a working area.

All relevant design factors in a particular situation can be divided into control factors and signal factors. Control factors are those that have a significant effect on the variability of a performance characteristic, while signal factors are those that have a significant effect on the product or process characteristic but not on its variability [2].

Similarly, noise factors can be classified as inner noise, outer noise and between-product noise. Inner noise is created by variations in product or process parameters (such as tool wear and tolerances on the process parameters). Outer noise is due to the effect of environmental factors (such as ambient temperature and raw material quality variation). Between-product noise is usually due to manufacturing variations between individual products or batches [3]. Although noise factors cause the product characteristic to deviate from its target value, the strategy recommended by Taguchi is not to identify the most significant noise factors and then control them. Instead, he proposes to select the levels of all controllable factors in such a way that the final product characteristic is the least sensitive to the variation induced by noise factors. This leads to the concept of 'parameter design' to be further considered in section 14.2.

14.1.5 Orthogonal arrays

Another important feature of the Taguchi methodology is the extensive use of **orthogonal arrays** or matrices. The rows of these arrays correspond to experimental conditions and the orthogonal columns to contrasts. These arrays have been widely used in the USA in traditional experimentation for more than 60 years, but in the application of Taguchi methods only for about 15 years. In the analysis of full and fractional factorial designs in Chapters 8 through 11 we have already used orthogonal arrays with columns consisting of coefficients equal to either -1 or $+1$, representing the two coded values specified for each factor. For example, the array in Table 14.1 corresponds to a 2^{7-4} fractional factorial design with generators defined as $X_4 = -X_2X_3$, $X_5 = -X_1X_3$, $X_6 = -X_1X_2$ and $X_7 = X_1X_2X_3$:

Following the procedures presented in Chapter 11, it is possible to verify that this design has resolution III and that the main effects are confounded with the two-factor interactions shown below, after assuming that all three-factor or higher-order interactions are negligible:

$$X_1 = -X_2X_6 = -X_3X_5 = -X_4X_7$$
$$X_2 = -X_1X_6 = -X_3X_4 = -X_5X_7$$
$$X_3 = -X_1X_5 = -X_2X_4 = -X_6X_7$$
$$X_4 = -X_1X_7 = -X_2X_3 = -X_5X_6$$
$$X_5 = -X_1X_3 = -X_2X_7 = -X_4X_6$$
$$X_6 = -X_1X_2 = -X_3X_7 = -X_4X_5$$
$$X_7 = -X_1X_4 = -X_2X_5 = -X_3X_6$$

As we know, if different generators were chosen, we would obtain different alias structures for the design, as well as possibly a different value for the design resolution.

Table 14.1 2^{7-4}_{III} Fractional factorial design

Run	X_1	X_2	X_3	X_4	X_5	X_6	X_7
1	−	−	−	−	−	−	−
2	+	−	−	−	+	+	+
3	−	+	−	+	−	+	+
4	+	+	−	+	+	−	−
5	−	−	+	+	+	−	+
6	+	−	+	+	−	+	−
7	−	+	+	−	+	+	−
8	+	+	+	−	−	−	+

Table 14.2 $L_8(2^7)$ orthogonal array

Run	*A*	*B*	*C*	*D*	*E*	*F*	*G*
1	1	1	1	1	1	1	1
2	1	1	1	2	2	2	2
3	1	2	2	1	1	2	2
4	1	2	2	2	2	1	1
5	2	1	2	1	2	1	2
6	2	1	2	2	1	2	1
7	2	2	1	1	2	2	1
8	2	2	1	2	1	1	2

Taguchi refers to the fractional factorial design being considered as an $L_8(2^7)$ array. The subscript of this notation indicates the number of runs, and the quantity in parentheses the number of experimental conditions associated with a full design, which in this case has 7 two-level factors. If we relabel the factors $X_1, X_2, X_3, X_4, X_5, X_6$ and X_7 as D, B, A, C, E, F and G, respectively, and use the entries 1 and 2 instead of -1 and $+1$ to indicate the two levels of each factor, the $L_8(2^7)$ array under consideration can be rewritten as shown in Table 14.2.

For brevity, an array is usually referred to only by the number of rows. When there are two arrays with the same number of rows, the second array is identified with a prime (′) note. For example, two arrays with 36 rows are referred to as L_{36} and L'_{36}. In order to specify which columns of the orthogonal array are associated with which factors or important two-factor interactions of the experiment, Taguchi provides a set of **linear graphs** consisting of nodes (circles) and arcs (lines). For any graph associated with an orthogonal array L_n, the nodes and arcs are labeled with the numbers $1, 2, \ldots, m$, where $m = n$. The total number of both factors and interactions that can be included in the analysis must be at most n. The rules for using this graph are:

(a) *Rule 1.* Each factor associated with a node is assigned to the column of the array that corresponds to the label of the node.
(b) *Rule 2.* Each line joining two nodes corresponds to the interaction of the factors represented by the nodes. The corresponding interaction is assigned to the column that corresponds to the label of the arc.
(c) *Rule 3.* Additional factors (those to be confounded with two-factor interactions) are assigned to arcs associated with interactions assumed to be negligible.

The linear graphs proposed by Taguchi to study the $L_8(2^7)$ orthogonal array given in Table 14.2 are shown in Figures 14.3(a) and (b). These graphs can be used for assigning two-factor interactions to the columns of an

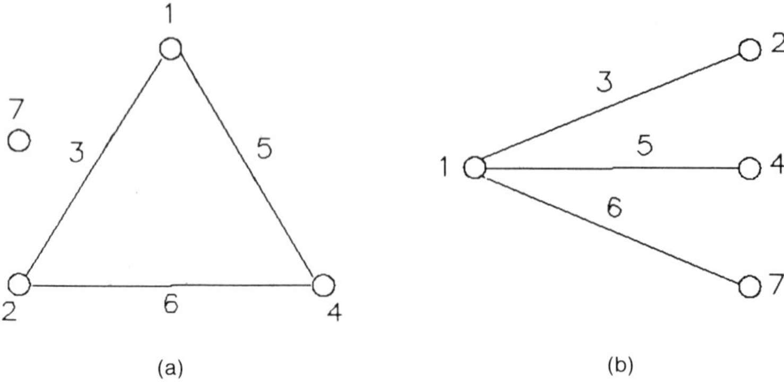

Figure 14.3 Linear graphs for $L_8(2^7)$ orthogonal array.

orthogonal array in such a way that the aliasing problem is alleviated. Figure 14.3(a) indicates that variables C, E and F are confounded with the two-factor interactions AB, AD and BD, respectively. Additionally, this figure indicates that variable G is not confounded with any two-factor interaction. Alternatively, Figure 14.3(b) shows that variables C, E and F can be associated with the interactions AB, AD and AG, respectively.

Taguchi has tabulated 18 basic orthogonal arrays that are known as the **standard orthogonal arrays**. Each orthogonal array uses a notation that indicates its number of rows and columns, as well as the number of levels in each column. For example, the $L_4(2^3)$ orthogonal array has 4 rows and 3 two-level columns. Moreover, the $L_{18}(2^1 3^7)$ orthogonal array has 18 rows, 1 two-level column and 7 three-level columns.

Table 14.3 illustrates the 18 standard orthogonal arrays [1]. As an example of how to use this table we can consider an experiment with 7 two-level factors. If only the main effects are going to be examined, the number of degrees of freedom required by the experiment is equal to 8, since there is 1 degree of freedom for the overall mean and 7 degrees for the 7 two-level factors. Therefore, the array to be used must have at least 8 rows. From Table 14.3, it is noted that the L_8 array can be selected since it has 7 two-level columns which can be assigned to the 7 factors.

As an additional example, we now consider an experiment with 1 two-level factor and 6 three-level factors. In this case, the number of degrees of freedom required by the experiment is equal to 14, since there is 1 degree of freedom for the overall mean, 1 degree for the two-level factor and 12 degrees for the 6 three-level factors. From Table 14.3 again it is noted that the smallest array with at least 14 rows is the L_{16} array. This array, however, cannot accommodate three-level factors, and thus we need to consider the next larger array, which is the L_{18} array. This array has 1 two-level column and 7 three-level columns. The two-level factor can be assigned to the two-level

Table 14.3 Standard orthogonal arrays

Orthogonal array	*Number of rows*	*Maximum number of factors*	*Maximum number of columns at these levels*			
			2	*3*	*4*	*5*
L_4	4	3	3	–	–	–
L_8	8	7	7	–	–	–
L_9	9	4	–	4	–	–
L_{12}	12	11	11	–	–	–
L_{16}	16	15	15	–	–	–
L'_{16}	16	5	–	–	5	–
L_{18}	18	8	1	7	–	–
L_{25}	25	6	–	–	–	6
L_{27}	27	13	–	13	–	–
L_{32}	32	31	31	–	–	–
L'_{32}	32	10	1	–	9	–
L_{36}	36	23	11	12	–	–
L'_{36}	36	16	3	13	–	–
L_{50}	50	12	1	–	–	11
L_{54}	54	26	1	25	–	–
L_{64}	64	63	63	–	–	–
L'_{64}	64	21	–	–	21	–
L_{81}	81	40	–	40	–	–

column and the 6 three-level factors to 6 of the 7 three-level columns, with 1 three-level column remaining unassigned.

14.2 THE PARAMETER DESIGN PROCEDURE

Once the system design phase is concluded, the next step is the parameter design procedure. This procedure consists of developing a strategy for intentionally designing quality into a product at a low cost by selecting the operating levels of a manufacturing process in such a way that the variation in the product parameters is minimized. These optimal levels of the parameters involved in the study are determined using experimental design methods. The steps followed in the parameter design procedure are summarized below.

Step 1. Identify design and noise factors

Design and noise factors that significantly affect the performance of the product or process are identified along with their entire feasible ranges. The levels of the noise factors should include the range in which the product performance is desired to be insensitive. The number of levels used for a particular factor depends on its relationship with the performance characteristic. If it is known that the performance characteristic varies as a linear function of the factor then two levels are sufficient, otherwise three or more levels are recommended to obtain some measure of curvature in the response surface [8].

Step 2. Construct the design and noise matrices and perform the experiment

The parameter design experiment essentially consists of the determination of two arrays known as the **design matrix** with rows representing product design or process settings, and the **noise matrix** with rows representing different combinations of the levels of the noise factors. The design factors and their relevant interactions are assigned to the columns of the design matrix in accordance with the linear graph of the corresponding orthogonal array. Similarly, the noise factors and their interaction are assigned to the columns of the noise matrix. Figure 14.4 shows typical design and noise matrices [4]. In this figure it is noted that for each level combination of the product factors an experiment is conducted with noise factors set at levels prescribed by each row of the noise matrix. In other words, each factor combination of the design matrix is replicated for each factor combination of the noise matrix.

Step 3. Evaluate the performance statistic

As indicated in Figure 14.4, from the designed experiment it is possible to obtain values of the performance characteristic or response and then combine them into values of a performance statistic. In general, the m values of the performance characteristic associated with each set of n replications corresponding to the m rows of the design matrix are used to calculate a performance statistic defined as an S/N ratio. Since the S/N ratio is proportional to the inverse of the coefficient of variation, maximizing the S/N ratio should lead to lower variation. The logarithm transformation used in the typical formulas of the S/N ratio often reduces curvatures and interactions in the response surface [4].

Step 4. Use the values of the performance statistic to predict new settings of the design parameters

Using the S/N ratios calculated in step 3, an analysis of variance (ANOVA) can be conducted to determine which of the factors are significant. The S/N ratios should be examined to verify the normality, independence and equal-variance assumptions. Since the ANOVA methodology is quite robust with respect to deviation from normality, more emphasis should be placed on the other two assumptions.

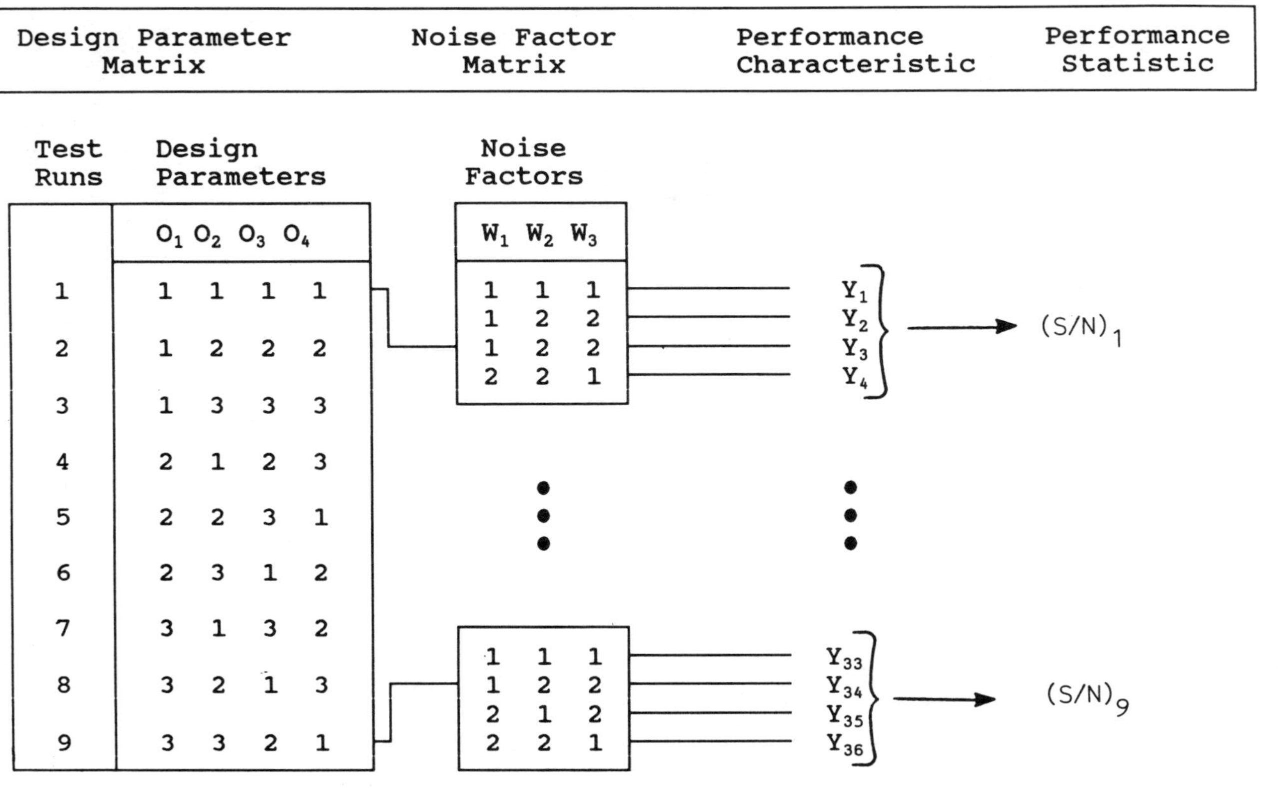

Figure 14.4 Example of parameter design experiment.

Those factors that are found to be significant with respect to S/N ratios are considered to be control factors. Not all the other factors are necessarily signal factors. For this reason, one must do an ANOVA on the means, and those noncontrol factors that turn out to be significant in this analysis are considered to be signal factors. The control factors should be set at levels at which the S/N ratios are maximized, thereby providing the maximum reduction in variation. However, the trade-off of this process is that the performance characteristic may be pushed off the target value. Then the signal factors, which do not affect the variation of the characteristic value, can be adjusted to return to the desired target value without having an adverse effect on the variation of the performance characteristic. The optimum levels of each signal and control factor can be obtained from a plot of factor level vs S/N ratio and the performance characteristic, as will be illustrated later on in Figure 14.8 for an application of the Taguchi methodology.

Step 5. Confirm that the new settings improve the performance measure

The final step is to run an experiment with the control factors set at the levels determined in step 4 and confirm that there is an improvement in the performance statistic over the initial settings. If the experiment yields satisfactory results, this outcome can be interpreted as a confirmation of the assumptions underlying the design. If the increase in performance statistic obtained is not significant, another iteration of the parameter design experiment including more factors and interactions might be needed. This can be accomplished by employing a larger orthogonal array or a more intensive experimental design.

EXAMPLE 14.1

This section describes an example taken from an article by T. B. Barker [5]. The problem concerns the manufacturing of a plastic butterfly valve used in the carburetor of a lawn mower engine. It is required to withstand the corrosive effects of the fuel, as well as the pressure of the return spring in the automatic choke. The current design is failure prone and has resulted in a high number of complaints. Figure 14.5 shows a plot of loss incurred vs break strength for the butterfly valve. A loss of \$50 is incurred for a break strength of 100 psi (7 kg cm^{-2}) and no loss is incurred for a break strength of 160 psi (11.3 kg cm^{-2}).

The reason for the failures has been attributed to the fact that the extrusion process that produces the part does not yield the required break strength. Figure 14.6 shows the schematic diagram of the extrusion process used to produce the part. The plastic raw material is first fed from a hopper into the mold cavity by a screw conveyor. Before it enters the mold cavity the plastic

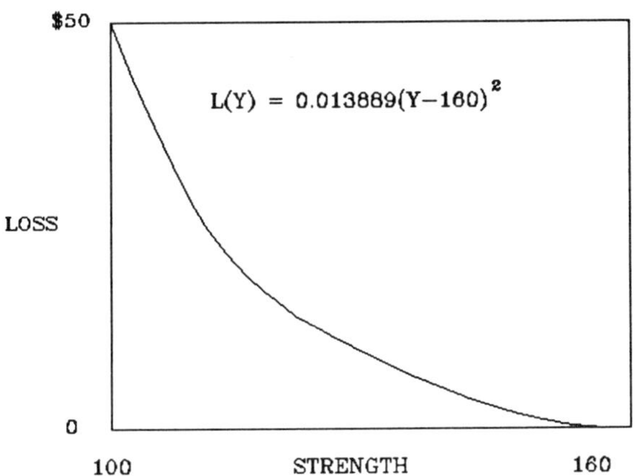

Figure 14.5 Loss function for the butterfly valve.

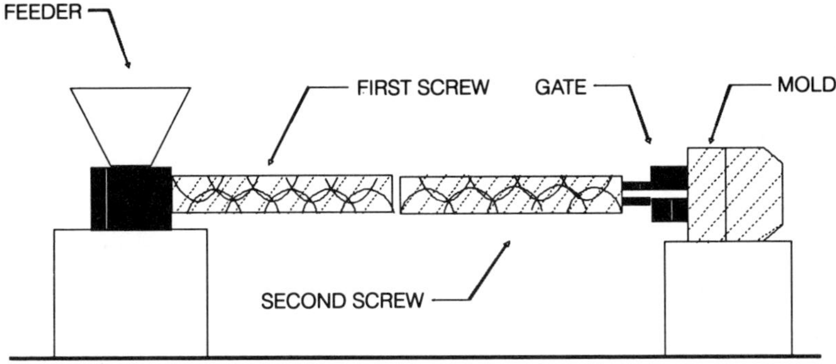

Figure 14.6 Manufacturing process for butterfly valve.

passes through two high-temperature zones where it is melted. The temperature in the first zone is lower than the temperature in the second. The amount of melted plastic entering the mold cavity is controlled by a gate. All the process parameters are within the control of the operator. The objective of the experiment was to determine the parameter settings that will produce parts with the required break strength and with minimal variation.

Step 1. Identify design and noise factors

The design factors, their range of interest and low-cost tolerances are shown in Table 14.4. Three levels are considered for each factor. In this experiment, the only type of noise factors considered are inner noise factors. Thus noise in this case is the variation of the process parameters around their set points. The range of performance is desired to be insensitive in the whole of the

Table 14.4 Design and noise factors

Parameter	*Range of interest*	*Low-cost tolerance (%)*
Feed rate GMS/MIN	1000–1400	20
First RPM	400–480	10
Second RPM	850–950	10
Gate size thousandths from nominal	– 30 to + 30	10
First temperature	280–360°	15
Second temperature	320–400°	15

low-cost tolerance range. This is because the process parameters can be maintained within that range at a relatively low cost.

Step 2. Construct design and noise matrices

Design matrix

The number of degrees of freedom needed for the ANOVA study is 13, since there is 1 degree of freedom for the overall mean and 2 for each of 6 controllable factors. In case that the interaction effect is found valuable later on, an L_{27} orthogonal array should be chosen for the design matrix. According to a standard linear graph for the array, columns 1, 2, 5, 9, 10 and 12 of L_{27} are assigned to the design factors in the same order as shown in the first column of Table 14.4. In this analysis the interaction effects have been considered to be negligible.

Noise matrix

The number of degrees of freedom for the noise matrix is equal to the same number computed for the design matrix. Since the interaction effects among the noise factors are considered to be insignificant, an L_{18} array was chosen for the noise matrix. The design matrix and the noise matrix for the first experimental condition of the former matrix are shown in Figure 14.7 [5].

For each row of the design matrix a set of experiments is conducted to cover the entire range of level combinations of the noise factors. These level combinations are specified by the rows of the noise matrix. Thus for each row of the design matrix 18 experiments are conducted and the corresponding values of the break strength are determined. The total number of experimental runs that need to be tested is equal to 486 (27 × 18).

Feed Rate	First RPM	Second RPM	Gate	First Temp	Second Temp
1000	400	850	−30	280	320
1000	400	900	0	320	360
1000	400	950	+30	360	400
1000	400	850	0	320	400
1000	400	900	+30	360	320
1000	400	950	−30	280	360

Inner Array

						Response
800	360	765	−27	238	272	27.75
800	400	850	−30	280	320	72.12
800	440	935	−33	322	368	102.18
1000	360	765	−30	280	368	96.68
1000	400	850	−33	322	272	61.00
1000	440	935	−27	238	320	118.55
1200	360	850	−27	322	320	106.24
1200	400	935	−30	238	368	143.23
1200	440	765	−33	280	272	59.01
800	360	935	−33	280	320	84.85
800	400	765	−27	322	368	65.48
800	440	850	−30	238	272	29.22
1000	360	850	−33	238	368	120.35
1000	400	935	−27	280	272	83.78
1000	440	765	−30	322	320	94.60
1200	360	935	−30	322	272	81.50
1200	400	765	−33	238	320	89.90
1200	440	850	−27	280	368	128.07

Figure 14.7 Results for first row of design matrix.

Step 3. Evaluate the performance statistic

The performance statistic or S/N ratio appropriate for this situation is the one corresponding to the 'larger-is-better' type. The S/N ratio for each test setting or experimental condition of the design matrix is calculated from the break strength values for each row of the noise matrix by using equation (14.3):

$$S/N_B = -10 \log_{10} \frac{1}{n} \sum_{i=1}^{n} \left(\frac{1}{y_i^2} \right).$$

As an illustration, the mean break strength, sample variance and the S/N

Table 14.5 Consolidated results from the parameter design experiment [5]

Feed rate	*First RPM*	*Second RPM*	*Gate size*	*First temperature*	*Second temperature*	$\bar{X}$	S	$\frac{S}{N}$
1000	400	850	−30	280	320	87.4	31.5	35.9
1000	400	900	0	320	360	15.6	24.6	40.5
1000	400	950	+30	360	400	106.2	36.3	36.8
1000	440	850	0	320	400	101.5	34.6	37.3
1000	440	900	+30	360	320	117.6	35.4	39.1
1000	440	950	−30	280	360	115.2	24.3	40.5
1000	480	850	+30	360	360	131.1	30.7	41.4
1000	480	900	−30	280	400	93.9	35.4	36.8
1000	480	950	0	320	320	121.3	35.5	40.8
1200	400	850	0	360	360	111.6	21.9	40.5
1200	400	900	+30	280	400	108.6	30.5	39.2
1200	400	950	−30	320	320	111.9	29.3	40.0
1200	440	850	+30	280	320	105.7	28.0	39.3
1200	440	900	−30	320	360	118.3	20.1	41.1
1200	440	950	0	360	400	133.1	34.0	41.5
1200	480	850	−30	320	400	104.1	33.2	38.8
1200	480	900	0	360	320	144.5	35.4	42.3
1200	480	950	+30	280	360	133.5	21.4	42.2
1400	400	850	+30	320	400	82.5	41.6	33.2
1400	400	900	−30	360	320	85.8	42.0	29.5
1400	400	950	0	280	360	120.4	33.5	40.4
1400	440	850	−30	360	360	99.3	36.7	38.0
1400	440	900	0	280	400	99.1	41.5	36.2
1400	440	950	+30	320	320	115.3	41.1	38.8
1400	480	850	0	280	320	96.2	40.9	34.1
1400	480	900	+30	320	360	121.6	36.7	40.4
1400	480	950	−30	360	400	120.8	46.4	39.2

ratio for each of the 27 test settings of the design matrix are shown in Table 14.5.

Step 4. Use the values of the performance statistic to predict the new settings of the design parameters

Two methods are involved in this step. The first one is the ANOVA methodology and the second is a graphical method of analysis. The ANOVA technique will determine which sources of variation contribute significantly to the variation in strength and which of the parameters put the strength on target. The calculations of the sums of squares will be performed here for the S/N response values only. From the 27 experimental results given in Table 14.5 it can be verified that the sum of squared observations is 40 581.04

and that the grand total is equal to 1043.8. Therefore, the total sum of squares is

$$SS_{\text{total}} = 40\,581.04 - \frac{(1043.8)^2}{27} = 228.5052.$$

The sum of squares of the main effect of the feed rate will be considered now. There are three regions associated with the three levels of the factor. A summary of the steps taken is shown below (procedure A, Chapter 7, assuming that $i = 1$ for this factor):

(a) $R_{11} = \{1, 2, 3, 4, 5, 6, 7, 8, 9\}$
$R_{12} = \{10, 11, 12, 13, 14, 15, 16, 17, 18\}$
$R_{13} = \{19, 20, 21, 22, 23, 24, 25, 26, 27\}$.

(b) $n_{11} = n_{12} = n_{13} = 9$.

(c) $T_{11} = 35.9 + 40.5 + \cdots + 40.8 = 349.1$
$T_{12} = 40.5 + 39.2 + \cdots + 42.2 = 364.9$
$T_{13} = 33.2 + 29.5 + \cdots + 39.2 = 329.8$.

(d) $SS_{\text{feed rate}} = \left(\frac{1}{9}\right)[(349.1)^2 + (364.9)^2 + (329.8)^2] - \frac{(1043.8)^2}{27} = 68.67185$

$$df = 2$$

$$MS_{\text{feed rate}} = \frac{68.67185}{2} = 34.34.$$

The contribution of the sum of squares of the feed rate to the total sum of squares is

$$100\left(\frac{68.67185}{228.5052}\right) = 30\%.$$

Using the rule that any sum of squares contributing 10% or less is not significant [2], we conclude that the feed rate is a significant source of variation.

The sums of squares associated with other factors can be calculated in the same manner. Their contributions to the total sum of squares are found to be 10.82%, 12.03, 5.07, 0.9 and 21.26% for the first and second RPM, gate, the first and second temperature, respectively. Due to their small contributions, the sums of squares of the main effects of gate and the first temperature are then pooled to form an artificial sum of squares of the error.

The results of the ANOVA methodology for the S/N ratio are shown in Table 14.6, where the pooled sum of squares for the error is shown on the last row. A similar ANOVA methodology can be conducted for the analysis of the mean strength. Table 14.7 shows the statistically significant contributors to S/N ratio and to mean strength. This table shows that all the factors have significant effects on the break strength but only four of the factors have a significant effect on the S/N ratio.

Table 14.6 ANOVA results for S/N ratios

Source	*df*	*SS*	*MS*	*F*
Feed rate	2	68.67	34.36	10.47*
First RPM	2	24.71	12.36	3.77*
Second RPM	2	27.50	13.75	4.19*
Gate size	2	11.59	5.80	1.77
First temperature	2	2.23	1.11	0.34
Second temperature	2	48.58	24.29	7.41*
Error	14	45.22	3.23	
Total	26	228.51		
(Error)	(18)	(59.04)	(3.28)	

*Significant

Table 14.7 Summary of analysis of variance

	For $\frac{S}{N}$		For $\bar{X}$	
Source	*MS*	*F*	*MS*	*F*
Feed rate	34.36	10.47*	471	10.7*
First RPM	12.36	3.77*	625	14.3*
Second RPM	13.75	4.19*	815	18.6*
Gate size	5.80	1.77	421	9.6*
First temperature	1.11	0.34	225	5.1*
Second temperature	24.29	7.41*	381	8.7*

*Significant

As a result of undergoing analysis the factors can be divided into two groups, control and signal factors, as indicated below:

(a) Control factors: feed rate, first RPM, second RPM and second temperature.
(b) Signal factors: gate size and first temperature.

The graphical method involves plotting the effects and visually identifying the factors which appear to be significant. Figure 14.8 shows (a) the plot of factor levels vs S/N ratios and (b) factor levels vs break strength for all the factors. Using these plots it is possible to set the control factors at levels that yield the highest S/N ratios, and the signal factors at levels that yield the highest break strength. Proceeding in this fashion, the factor settings obtained from the current analysis are shown in Table 14.8.

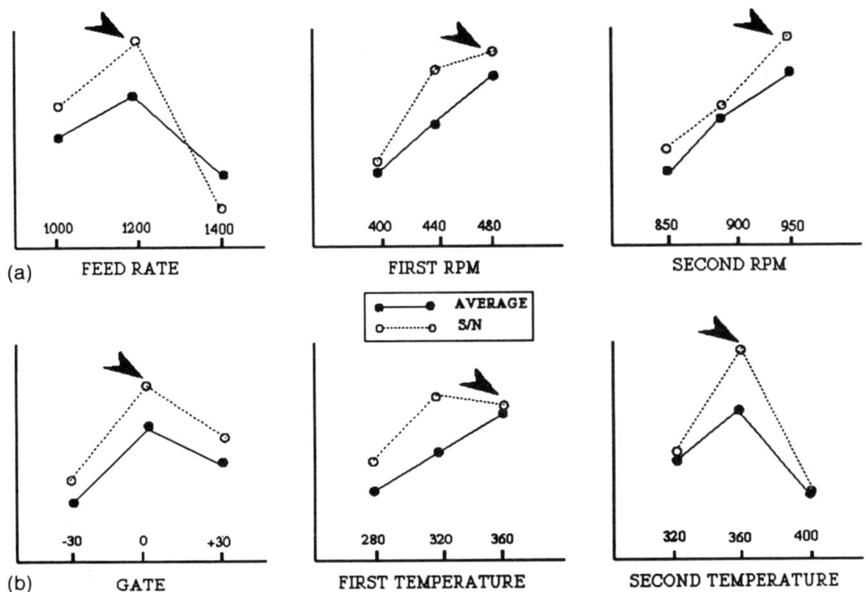

Figure 14.8 S/N ratios and average break strength plots [5].

Table 14.8 Optimal settings of design factors

Factor	*Optimal setting*
Feed rate	1200
First RPM	480
Second RPM	950
Gate size	0
First temperature	360
Second temperature	360

It is noted that the set of optimal conditions determined from the previous step did not occur in any of the 27 treatments of the first orthogonal array. This usually happens when highly fractionated designs are used. By using the same low-cost tolerances for the noise factors, a new noise matrix can be generated for the best settings, and an additional set of 18 experiments can be conducted. The results for this subsequent phase of the overall procedure are presented in Table 14.9 [5]. The average break strength obtained is 182.152 with a standard deviation of 28.0234. Even though the break strength obtained is satisfactory the variation is found to be unacceptably high. The next step is to perform a tolerance design to set rational tolerances on the parameters. The standard deviation was reduced to 9.2968. (Refer to T. B. Barker [5] for further details of this application.)

Table 14.9 Results of the confirmation experiment

Feed rate	*First RPM*	*Second RPM*	*Gate size*	*First temperature*	*Second temperature*	*Break strength*
960	432	855	0	306	306	108.77
960	480	850	0	360	360	168.68
960	528	1045	0	414	414	214.87
1200	432	855	0	360	414	129.07
1200	480	950	0	414	306	182.12
1200	528	1045	0	306	360	192.84
1440	432	950	0	414	360	154.93
1440	480	1045	0	306	414	151.94
1440	528	855	0	360	306	150.74
960	432	1045	0	360	360	168.20
960	480	855	0	414	414	148.56
960	528	950	0	306	306	140.86
1200	432	950	0	306	414	142.33
1200	480	1045	0	360	306	183.81
1200	528	855	0	414	360	212.60
1440	432	1045	0	414	306	154.59
1440	480	855	0	306	360	134.21
1440	532	950	0	360	414	168.82

14.3 CRITICAL ANALYSIS OF TAGUCHI METHODS

Taguchi's methodology is one of the most relevant contributions to modern quality control during the 1980s. It extended the scope of the field of quality engineering by promoting the philosophy that the quality of a product should be the result of a comprehensive process from its very design stage. By defining quality as 'loss to society', he set a new standard of quality to be achieved by both the designers and manufacturers. He popularized the concept of robust product design by showing how the transmission of variability due to component variation, manufacturing variation and environmental variation can be minimized [6]. Furthermore, Taguchi simplified tolerance analysis through designed experiments where approaches to statistical tolerances, including Taylor series expansions, numerical integration and Monte Carlo simulation have failed to gain widespread acceptance [6].

By stressing the importance of statistically designed experiments in parameter design, he identified a sound application of statistical methods to improve the quality of the engineering design of products and processes in industrial situations where the traditional inspection had failed. His ideas have been criticized with regards to the statistical sophistication of his methods. In spite of its statistical limitations, the Taguchi method provides a structured and useful technique in improving quality.

14.3.1 Single-to-noise ratios

Taguchi recommends the use of a 'signal-to-noise ratio' (*S/N*) to analyze experiments (in which both the average and the standard deviation are under study) as a measure of variation rather than the standard deviations. Box, Bisgaard and Fung [3] have shown that the use of the *S/N* ratio is equivalent to an analysis of the standard deviation of the logarithm of the data. The focus here is to eliminate unnecessary coupling of location as well as dispersion effects. The *S/N* analysis implies that use of a logarithmic transformation of the data can always achieve uncoupling. Box *et al.* comment that situations frequently occur where either no transformation or some transformation other than logarithmic is required to be able to produce the uncoupling [3]. As has been discussed in Pignatiello and Ramberg [7], when this transformation does not perform the task as expected, the analysis can be severely affected.

14.3.2 Loss functions

Another aspect of Taguchi's technique that has attracted some criticism is the use of the loss function. The use of the loss function is a criterion where minimizing loss is equivalent to minimizing the mean square deviation from a target value. Box *et al.* argue that this idea constitutes a 'revolutionary departure' in the context of certain tolerance specifications. Their alternate suggestion is to emphasize minimization of the mean square error.

14.3.3 Accumulation analysis

A common procedure used for analysing data where no exact measurements are available (like saltiness, sweetness, etc.) is assigning an appropriate score to such categories and then proceeding with the use of standard statistical techniques as if the numbers were actual measurements. A typical gauge in the food industry, for example, for taste, sweetness, saltiness, etc. is on a scale from 1 to 7. Taguchi uses a method called accumulation analysis for analyzing such data. (The reader is referred to Phadke [6] for further details about this.) It has been shown by a few researchers [8–10] that this method is inefficient and cumbersome.

EXERCISES

1. Case study of robust design methodology in reliability improvement. Some important considerations in engineering design include increasing the longevity of a product, the time between maintenance of a manufacturing process, etc. Life tests are performed to identify settings of design factors for increasing product life. The following case describes a study of improving router bits using life studies.

Table 14.10

Factors	*Levels*			
	1	*2*	*3*	*4*
Suction (in of Hg)	1	2		
X–Y feed (in min^{-1})	60	80		
In-feed (in min^{-1})	10	50		
Type of bit	1	2	3	4
Spindle position	1	2	3	4
Speed (rev min^{-1})	30 000	40 000		

Printed wiring boards are usually cut from panels by stamping or by the routing process. The routing process gives good dimensional control and smooth edges. The router bit gets dull after some use. Changing the router bit is expensive. The objective here is to increase the life of the router bits.

The routing machine has four spindles, synchronized in their rotational speed, X–Y feed and vertical (in-feed) feed. The cutting process involves lowering the spindle to an edge of the board, cutting the board using X–Y spindle feed and lifting the spindle. This procedure is repeated for each spindle. Consider the following noise factors: out-of-center spindle rotation, variation of one router bit from another, material variation within a panel and between panels and variation in the speed of the drive motor. Additionally, assume the control factors and levels given in Table 14.10.

Develop a design matrix using Taguchi principles. All the main effects as well as the key two-factor interactions must be considered.

2. Factories A and B are manufacturers of electrical wire sections of a specified length. The thickness of a plastic coating is an important factor affecting the wear strength of the wire. When a section is found to be defective in the factory a treatment costing $1.20 per section can be applied to correct the deficiency. Factory A has a normal distribution with mean equal to the target value and standard deviation equal to $\frac{1}{10}$ of the tolerance with the latter equal to 0.02 mm. Factory B has a uniform distribution within the tolerance limits. Compare the losses per wire section caused by deviations from the target.
3. A particular product has a quality characteristic with target value equal to 100. The product becomes defective if the characteristic exceeds the tolerance limits of ± 25. Before shipping the product, a defective unit can be corrected by an operation that costs $2.00. Find the manufacturer's specifications.
4. Explain the procedure for selecting orthogonal arrays. Provide examples.
5. How are the data analyzed according to Taguchi's method in terms of the inner array/outer array structure for a designed experiment?

REFERENCES

1. Farnum, Nicholas, R. (1994) *Modern Statistical Quality Control and Improvement*, Duxubry Press, Belmont, California.
2. Box, G. and Bisgaard, S. (1988) Statistical tools for improving designs. *Mechanical Engineering*, January, 32–40.
3. Box, G. E. P., Bisgaard, S. and Fung, C. (1988) An explanation and critique of Taguchi's contributions to quality engineering. *Research Report*, Center for Quality and Productivity Improvement, University of Wisconsin-Madison, Madison, Wisconsin.
4. Gunter, B. (1987) A perspective on the Taguchi methods. *Quality Assurance*, **13**(3), September, 81–7.
5. Barker, T. B. (1987) Quality engineering by design: Taguchi's philosophy. *Quality Assurance*, **13**(3), September, 72–9.
6. Phadke, M. S. (1989) *Quality Engineering Using Robust Design*, Prentice-Hall, New Jersey.
7. Pignatiello, J. J. and Ramberg, J. S. (1991) Top ten triumphs and tragedies of Genichi Taguchi. *Quality Engineering*, **4**(2), June, 211–25.
8. Kackar, R. N. (1986) Taguchi's quality philosophy: analysis and commentary. *Quality Progress*, **19**(12), December, 21–8.
9. Leon, R. V., Shoemaker, A. C. and Kackar, R. N. (1987) Performance measures independent of adjustment. *Technometrics*, **29**(3), August, 253–65.
10. Montgomery, D. C. (1988) Experiment design and product and process development. *Manufacturing Engineering*, September, 57–63.

FURTHER READING

Box, G. E. P. and Jones, S. (1986) An investigation of the method of accumulation analysis. *Report No. 19*, Center for Quality and Productivity Improvement, University of Wisconsin-Madison.

Burgam, P. M. (1985) Design of experiments – the Taguchi way. *Manufacturing Engineering*, May, 44–7.

Byrne, D. M. and Taguchi, S. (1986) The Taguchi approach to parameter design. *ASQC 40th Anniversary Quality Congress Transactions*, Anaheim, California, May 19–21, pp. 168–77.

Hamada, M. and Wu, C. F. J. (1986) Discussion of testing in industrial experiments with ordered categorical data by V. N. Nair. *Technometrics*, **28**, 295–301.

Kackar, R. N. (1985) Off-line quality control, parameter design, and the Taguchi method. *Journal of Quality Technology*, **17**(4), October, 176–88.

Nair, V. N. (1986) Testing in industrial experiments with ordered categorical data. *Technometrics*, **28**, 283–91.

Ross, P. J. (1988) *Taguchi Techniques for Quality Engineering*, McGraw-Hill, New York.

Ross, N. E. (1987) Tapping into Taguchi. *Manufacturing Engineering*, May, 43–6.

Taguchi, G. and Wu, I. (1980) *Off-line Quality Control Systems*, Central Japan Quality Control Association, Nagoya, Japan.

15 Nested factor and split plot experimental designs

In this chapter, the design structure and analysis of two special types of experiments will be explored. The first experiment type is a nested (or hierarchical) design. This type of experimental design involves situations where all the levels of one factor cannot be tested within each of the levels of a second factor. The second type of experimental design covered is the split plot design. The split plot design involves situations where multiple factors must be treated randomly and independently. Typically, all the levels of one factor are contained in each level of another factor.

15.1 NESTED DESIGNS

When designing an experiment, nesting occurs if the levels of one factor change within each level of a second factor. In this case, the levels of the first factor are said to be nested within the levels of the second factor [1]. The models in these experiments may be fixed effect, random effect or a combination of the two. Among others, Anderson and McLean [2] have summarized general results on expected mean squares (EMS) to conduct ANOVA tests in nested designs.

EXAMPLE 15.1

An example of a three-factor nested experiment is illustrated. In this example, the data collected may represent the number of defective items in a sample of final product. From each of the three manufacturing cells available, two functionally similar machines are sampled at random and four samples of product are collected at random from each machine. In this experiment, cell is the major factor, machine is a factor nested within cell and product sample is a nested factor within machine. An example of the design is shown in Table 15.1.

15.1.1 Definitions and assumptions

The following definitions and assumptions will be used in the discussion of nested factor experiments:

Table 15.1 Three-factor nested design example

Cell 1		*Cell 2*		*Cell 3*	
Machine 1	*Machine 2*	*Machine 1*	*Machine 2*	*Machine 1*	*Machine 2*
Sample 1	Sample 1	Sample 1	Sample 1	Sample 1	Sample 1
Sample 2	Sample 2	Sample 2	Sample 2	Sample 2	Sample 2
Sample 3	Sample 3	Sample 3	Sample 3	Sample 3	Sample 3
Sample 4	Sample 4	Sample 4	Sample 4	Sample 4	Sample 4

(a) *Major factor:* the main factor of interest. It is a factor which is not nested within another factor, but may contain factors nested within it. In Example 15.1, cell is the major factor.
(b) *Nested factor:* when each level of factor A contains different levels of factor B, it can be said that factor B is nested within factor A [3]. In Example 15.1, machine is a factor nested within state, and sample is a factor nested within machine.

Letting cells be represented by factor A, machines by factor B and product samples by factor C, the model for this experiment follows:

$$Y_{ijk} = \mu + \tau_j + \beta_{i(j)} + \varepsilon_{ijk} \tag{15.1}$$

for $j = 1, 2, \ldots, a,\quad i = 1, 2, \ldots, b,\quad k = 1, 2, \ldots, c$ where a, b and c are the number of levels of factors A, B and C respectively, and μ is the overall mean, τ_j the effect due to the jth level of factor A, $\beta_{i(j)}$ the effect due to the ith level of factor B within the jth level of factor A, and ε_{ijk} the residual component of the ijkth observation.

Assuming cell to be a fixed effect, the following assumptions accompany the model:

(a) $\sum_j \tau_j = 0$.
(b) $\beta_{i(j)}$ are distributed as NID$(0, \sigma_\beta^2)$ for each j.
(c) ε_{ijk} are distributed as NID$(0, \sigma_\varepsilon^2)$ for each i and j.
(d) $\beta_{i(j)}$ and ε_{ijk} are distributed independently of each other.

Some items of interest with regard to nested experiments are:

(a) Level of nested factors do not appear in all combinations with the factor it is nested in.
(b) Levels of most factors change for every level of the other factors, therefore there can be no interactions [3].
(c) Effects of the lowest-level factor cannot be determined without replication. (In Example 15.1, factor C would be considered to be the lowest-level factor.)

15.1.2 Analysis of variance calculations

In Example 15.1, the data in Table 15.2 were collected. To conduct an ANOVA, we will define the correction term

$$G = \frac{(\sum_i \sum_j \sum_k Y_{ijk})^2}{abc} \tag{15.2}$$

where $abc = N =$ total number of observation and $\sum_i \sum_j \sum_k Y_{ijk} = T_{...}$, therefore

$$G = \frac{T_{...}^{\,2}}{N}. \tag{15.3}$$

For the example,

$$G = \frac{(297)^2}{24} = 3675.375.$$

The total sum of squares is given by the equation

$$SS_{\text{total}} = \sum_i \sum_j \sum_k Y_{ijk}^2 - G \tag{15.4}$$

$$= (9)^2 + (12)^2 + (15)^2 + \cdots + (16)^2 - 3675.375$$
$$= 3817 - 3675.375$$
$$= 141.625$$

and the sum of squares of factor A can be found by

$$SS_A = bc \sum_j \bar{Y}_{.j.}^2 - N\bar{Y}_{...}^2 \tag{15.5}$$

$$= \frac{\sum_j (T_{.j.})^2}{bc} - G \tag{15.6}$$

$$= \frac{(92)^2 + (94)^2 + (111)^2}{(2)(4)} - 3675.375$$
$$= 27.25.$$

Table 15.2 Cellular manufacturing example

Product sample	*Cell 1*		*Cell 2*		*Cell 3*	
	Machine 1	*Machine 2*	*Machine 1*	*Machine 2*	*Machine 1*	*Machine 2*
1	9	8	7	10	15	13
2	12	14	15	12	16	15
3	15	9	11	12	10	14
4	13	12	13	14	15	13
Machine totals $T_{ij.}=$	49	43	46	48	56	55
Cell totals $T_{.j.}=$	92		94		111	
Grand total $T_{...}=$			297			

The sum of squares of factor B within factor A can be found by

$$SS_{B(A)} = c\sum_i \sum_j \bar{Y}_{ij.}{}^2 - bc\sum_j \bar{Y}_{.j.}{}^2 \quad (15.7)$$

$$= \frac{\sum_i \sum_j (T_{ij.})^2}{c} - \frac{\sum_j (T_{.j.})^2}{bc} \quad (15.8)$$

$$= \frac{(49)^2 + (43)^2 + (46)^2 + (48)^2 + (56)^2 + (55)^2}{4}$$

$$- \frac{(92)^2 + (94)^2 + (111)^2}{(2)(4)}$$

$$= 5.125.$$

Therefore, the sum of squares for the error is

$$SS_{\text{error}} = SS_{\text{total}} - SS_A - SS_{B(A)} \quad (15.9)$$

$$= 141.625 - 27.25 - 5.125$$

$$= 109.25.$$

The ANOVA table is shown in Table 15.3, and the results from the ANOVA test are summarized in Table 15.4.

Table 15.3 ANOVA for a three-factor nested design

Source	*df*	*SS*	*MS*	*EMS*
Factor A	$a-1$	SS_A	MS_A	$\sigma_\varepsilon^2 + c\sigma_\beta^2 + \frac{bc_r}{a-1}\Sigma\tau_j^2$
Factor $B(A)$	$a(b-1)$	$SS_{B(A)}$	$MS_{B(A)}$	$\sigma_\varepsilon^2 + c\sigma_\beta^2$
Error	$ab(c-1)$	SS_{error}	MS_{error}	σ_ε^2
Total	$abc-1$	SS_{total}		

Table 15.4 ANOVA for the cellular manufacturing example

Source	*df*	*SS*	*MS*	*F*	*Significant? ($\alpha = 0.10$)*
A = Cells	2	27.25	13.625	7.977	Yes
$B(A)$ = Machines within cells	3	5.125	1.708	0.281	No
Error	18	109.25	6.069		
Total	23	141.625			

15.1.3 Hypothesis testing

The procedures for testing the fixed-effect factor (factor A) are the same as those described in Chapter 3. The null hypothesis is $H_0: \tau_j = 0$ for all j and the alternate hypothesis is $H_1: \tau_j \neq 0$ for at least one j. Note the expected mean square for factor A, in Table 15.3, contains the term σ_β^2, therefore the resultant F-test would be based on $F_0 = MS_A(MS_{B(A)})^{-1}$. For the random-effect factor (factor B), the null hypothesis is $H_0: \sigma_\beta^2 = 0$ and the alternate is $H_1: \sigma_\beta^2 > 0$. The F-test would be based on $F_0 = MS_{B(A)}(MS_{\text{error}})^{-1}$.

15.2 SPLIT PLOT DESIGNS

Contrary to the nested designs described previously, a split plot design would have every level of one factor within each level of another factor. The assignment of these factors would be random and independent of each other.

EXAMPLE 15.2

Suppose a farmer wants to test the effects of two factors (A – irrigation methods and B – fertilizers) independently but only has enough fields (or time) to perform one experiment. The farmer would get valuable data if in each plot of land assigned for an irrigation method, he developed subplots for each fertilizer. Provided the farmer has two replicates of the irrigation method, he can determine the effects of the irrigation methods and the effects of fertilizers. This can be shown graphically by Table 15.5.

In the above example, the irrigation methods would be described as the **whole plots** factor and the fertilizers would be the **split plots** factor. Total residual (error) is now split into two components for the whole 'and split plots. Each of these error components will be used to determine the significance of their respective factors.

Table 15.5 Split plot design example

Replication 1			*Replication 2*		
Irrigation method 1	*Irrigation method 2*	*Irrigation method 3*	*Irrigation method 3*	*Irrigation method 2*	*Irrigation method 1*
Fertilizer 1	Fertilizer 3	Fertilizer 2	Fertilizer 4	Fertilizer 3	Fertilizer 1
Fertilizer 4	Fertilizer 2	Fertilizer 1	Fertilizer 1	Fertilizer 2	Fertilizer 4
Fertilizer 2	Fertilizer 4	Fertilizer 3	Fertilizer 3	Fertilizer 4	Fertilizer 3
Fertilizer 3	Fertilizer 1	Fertilizer 4	Fertilizer 2	Fertilizer 1	Fertilizer 2

15.2.1 Definitions and assumptions

The model of a simple split plot design is

$$Y_{ijk} = \mu + \rho_i + \tau_j + \delta_{ij} + \beta_k + (\tau\beta)_{jk} + \varepsilon_{ijk} \quad (15.10)$$

for $i = 1, 2, \ldots, r$, $j = 1, 2, \ldots, a$, $k = 1, 2, \ldots, b$, where μ is the overall mean, ρ_i the effect due to the ith replicate, τ_j the effect of the jth level of A, δ_{ij} the error component for the whole plot, β_k the effect of the kth level of B, $(\tau\beta)_{jk}$ the interaction effect of A and B and ε_{ijk} the error component for the split plot.

The basic assumptions needed for a valid analysis of a split plot design are the following:

(a) δ_{ij} are distributed as NID$(0, \sigma_\delta^2)$ for all i and j.
(b) ε_{ijk} are distributed with $N(0, \sigma_\varepsilon^2)$ for all i, j and k.
(c) δ_{ij} and ε_{ijk} are distributed independently of each other.

15.2.2 Analysis of variance calculations

For the purposes of explanation, an example will be used to illustrate the calculations.

The superintendent of a semiconductor manufacturing firm decided to investigate the effect of three pressure levels, (a) high (b) medium and (c) low, on the deposition thickness of a special kind of wafer. However, he was also concerned about the effect of the amount of time the wafer was kept inside the machine. Once a pressure setting was selected at random, two wafers from a homogeneous set of 18 wafers were chosen and placed in the machine. After this the chamber pressure was set at the selected level and one of the wafers was mechanically removed when a short settting of time was reached. The second wafer was kept inside until a longer setting of time was up. This experiment was repeated 9 times until all pressure/time combinations were exhausted. The three groups of six observations corresponding to the replications of the experiment are shown in Table 15.6.

Table 15.6 Wafer manufacturing example

		High pressure	*Medium pressure*	*Low pressure*
Replication 1	Short time	9.0	8.2	7.9
	Long time	9.5	8.7	8.5
Replication 2	Short time	9.2	8.5	8.2
	Long time	9.4	8.8	8.7
Replication 3	Short time	8.9	8.4	8.0
	Long time	9.7	9.0	8.8

The totals used in the ANOVA calculations are shown below:

$$\text{Replication } T_{i..} = 51.8 \quad 52.8 \quad 52.8$$
$$\text{Pressure level } T_{.j.} = 55.7 \quad 51.6 \quad 50.1$$
$$\text{Replication} \times \text{pressure } T_{ij.} = 18.5, 16.9, 16.4, 18.6, 17.3, 16.9, 18.6, 17.4, 16.8$$
$$\text{Deposition time } T_{..k} = 76.3 \quad 81.1$$
$$\text{Pressure} \times \text{time } T_{.jk} = 27.10, 25.10, 24.10, 28.60, 26.50, 26.00$$
$$\text{Grand total } T_{...} = 157.4.$$

To conduct an ANOVA, we define the correction factor

$$G = \frac{(\sum_i \sum_j \sum_k Y_{ijk})^2}{rab} \tag{15.11}$$

where $rab = N =$ total number of observations and $\sum_i \sum_j \sum_k Y_{ijk} = T_{...}$; therefore

$$G = \frac{T_{...}^2}{N}. \tag{15.12}$$

For the example,

$$G = \frac{(157.4)^2}{18} = 1376.38.$$

The total sum of squares is given by equation (15.4):

$$\begin{aligned} SS_{\text{total}} &= \sum_i \sum_j \sum_k Y_{ijk}^2 - G \\ &= (9.0)^2 + (9.5)^2 + (8.2)^2 + \cdots + (8.8)^2 - 1376.38 \\ &= 1380.76 - 1376.38 \\ &= 4.38 \end{aligned}$$

and the sum of squares for replications can be found by

$$SS_R = ab \sum_i \bar{Y}_{i..}^2 - N\bar{Y}_{...}^2 \tag{15.13}$$

$$= \frac{\sum_i (T_{i..})^2}{ab} - G \tag{15.14}$$

$$\begin{aligned} &= \frac{(51.8)^2 + (52.8)^2 + (52.8)^2}{(2)(3)} - 1376.38 \\ &= 1376.49 - 1376.38 \\ &= 0.11. \end{aligned}$$

The sums of squares of factor A (pressure) can be found by

$$SS_A = rb \sum_j \bar{Y}_{.j.}^2 - N\bar{Y}_{...}^2 \tag{15.15}$$

$$= \frac{\sum_j (T_{.j.})^2}{rb} - G \tag{15.16}$$

$$= \frac{(55.7)^2 + (51.6)^2 + (50.1)^2}{(2)(3)} - 1376.38$$

$$= 2.80.$$

The sum of squares for the subclassing of replication and factor A can be found by

$$SS_{RA} = b \sum_i \sum_j \bar{Y}_{ij.}^2 - N \bar{Y}_{...}^2 \tag{15.17}$$

$$= \frac{\sum_i \sum_j (T_{ij.})^2}{b} - G \tag{15.18}$$

$$= \frac{(18.5)^2 + (16.9)^2 + \cdots + (16.8)^2}{2} - 1376.38$$

$$= 2.94.$$

The sum of squares for the whole plot error component can be found by

$$SS_{\varepsilon 1} = SS_{RA} - SS_A - SS_R \tag{15.19}$$

$$= 2.94 - 2.80 - 0.11$$

$$= 0.03.$$

The sum of squares for the split plot portion of the experiment is shown below. The sums of squares for factor B (time) is given by

$$SS_B = ra \sum_k \bar{Y}_{..k}^2 - N \bar{Y}_{...}^2 \tag{15.20}$$

$$= \frac{\sum_k (T_{..k})^2}{ra} - G \tag{15.21}$$

$$= \frac{(76.3)^2 + (81.1)^2}{(3)(3)} - 1376.38$$

$$= 1.28$$

and the sum of squares for the subclassing of $A \times B$ can be found by

$$SS_{A\,\mathrm{sub}\,B} = r \sum_j \sum_k \bar{Y}_{.jk}^2 - N \bar{Y}_{...}^2 \tag{15.22}$$

$$= \frac{\sum_j \sum_k (T_{.jk})^2}{r} - G \tag{15.23}$$

$$= \frac{(27.10)^2 + (25.10)^2 + \cdots + (26.00)^2}{3} - 1376.38$$

$$= 4.10.$$

The sums of squares for the interaction of $A \times B$ is given by

$$\begin{aligned} SS_{A\times B} &= SS_{A\,\mathrm{sub}\,B} - SS_A - SS_B \\ &= 4.10 - 2.80 - 1.28 \\ &= 0.02 \end{aligned} \tag{15.24}$$

Finally, the sum of squares for the split plot error component can be found by

$$\begin{aligned} SS_{\varepsilon 2} &= SS_{\mathrm{total}} - SS_R - SS_A - SS_{\varepsilon 1} - SS_B - SS_{A\times B} \\ &= 4.38 - 0.11 - 2.80 - 0.03 - 1.28 - 0.02 \\ &= 0.14. \end{aligned} \tag{15.25}$$

The ANOVA table for split plot design is shown in Table 15.7. The results of the ANOVA test are summarized in Table 15.8.

Table 15.7 ANOVA for a split plot design

Source	*df*	*SS*	*MS*	*EMS*
Replication	$r-1$	SS_R	MS_R	
Factor A	$a-1$	SS_A	MS_A	$\sigma_\varepsilon^2 + b\sigma_\delta^2 + \frac{rb}{a-1}\sum \tau_j^2$
Whole plot error	$(r-1)(a-1)$	$SS_{\varepsilon 1}$	$MS_{\varepsilon 1}$	$\sigma_\varepsilon^2 + b\sigma_\delta^2$
Factor B	$b-1$	SS_B	MS_B	$\sigma_\varepsilon^2 + \frac{ra}{b-1}\sum \beta_k^2$
$A \times B$	$(a-1)(b-1)$	$SS_{A\times B}$	$MS_{A\times B}$	$\sigma_\varepsilon^2 + \frac{r}{(a-1)(b-1)}\sum\sum \tau_j^2\beta_k^2$
Split plot error	$(r-1)a(b-1)$	$SS_{\varepsilon 2}$	$MS_{\varepsilon 2}$	σ_ε^2
Total	rab-1	SS_{total}		

Table 15.8 ANOVA for the wafer manufacturing example

Source	*df*	*SS*	*MS*	*F*	*Significant? ($\alpha = 0.05$)*
Replications	2	0.11	0.055		
$A =$ pressure	2	2.80	1.400	175.00	Yes
Whole plot error	4	0.03	0.008		
$B =$ time	1	1.28	1.280	55.65	Yes
$A \times B$	2	0.02	0.010	0.44	No
Split plot error	6	0.14	0.023		
Total	17	4.38			

15.2.3 Hypothesis testing

To test the significance of whole plot factors (i.e. factor A), the null hypothesis to be tested is H_0: all $\tau_j = 0$ for all j and the alternate hypothesis as H_1: some $\tau_j \neq 0$ for at least one j. Note the expected mean square for factor A, in Table 15.7, contains the term σ_δ^2, therefore the resultant F-test would be based on $F_0 = MS_A(MS_{\varepsilon 1})^{-1}$.

To test the significance of split plot factors and interactions (i.e. factor B and interaction $A \times B$), the null hypothesis to be tested is H_0: all $\beta_k = 0$ (or H_0: all $(\tau\beta)_{jk} = 0$) and the alternate as H_1: some $\beta_k \neq 0$ (or H_1: some $(\tau\beta)_{jk} \neq 0$). The F-tests would be based on $F_0 = MS_B(MS_{\varepsilon 2})^{-1}$ and $F_0 = MS_{A \times B}(MS_{\varepsilon 2})^{-1}$, respectively.

EXERCISES

1. Three automobile manufacturers are being examined for the average miles per gallon (mpg) of the automobiles they produce. Three models of autos were randomly selected from each manufacturer and four autos were randomly selected within each model and tested for one week. The average mpg of each auto is recorded in Table 15.9.
   (a) Describe the appropriate model.
   (b) Conduct an ANOVA and test the hypothesis that the average mpg for all manufacturers are the same.
2. An experiment was conducted to evaluate the quality of high school mathematics education among three different states. Four high schools were randomly selected in each state, and five students were randomly chosen at each school. The college entrance exam scores in mathematics for these students are recorded in Table 15.10. Conduct the ANOVA. Do the average scores of the states differ? Explain.
3. A new account executive for a large company has been told that company sales remain steady throughout any given year, and that annual company sales have not increased in the last three years. To support this claim, he is handed the sales reports for five randomly selected stores within the

Table 15.9

	Manufacturer 1			*Manufacturer 2*			*Manufacturer 3*		
Auto	Model 1	Model 2	Model 3	Model 1	Model 2	Model 3	Model 1	Model 2	Model 3
1	22.5	25.4	26.1	32.3	29.8	25.4	24.3	27.2	18.7
2	24.0	22.3	28.9	30.1	27.5	29.0	22.1	28.3	21.5
3	23.6	26.2	25.3	29.8	28.1	32.0	26.2	26.5	23.2
4	18.9	23.7	24.7	31.3	26.0	31.0	25.4	22.1	20.8

Table 15.10

Student	*High school*			
	1	*2*	*3*	*4*
State: Oklahoma				
1	460	390	430	640
2	520	530	340	560
3	580	300	490	460
4	470	450	710	420
5	420	580	610	380
State: Texas				
1	450	680	480	720
2	530	550	590	420
3	610	430	630	590
4	730	650	640	580
5	580	620	570	630
State: Arkansas				
1	330	420	370	430
2	580	570	560	470
3	450	410	430	360
4	620	490	380	550
5	530	320	550	510

company (Table 15.11). Sales (in thousands of dollars) are broken out by year and quarter.

Do these sales figures support the claims made? Explain.

4. A marketing firm desires to test the effects of packing on a certain product's sales. The experiment will be conducted in four different geographical locations within the United States (NE, SE, Midwest and SW). The product can be packaged in three different size boxes (S, M or L). Small boxes are colored green, white or yellow. Medium boxes are colored blue, red or orange. And large boxes are colored brown, violet or maroon.

Table 15.11

	1988				*1989*				*1990*			
Store	*Qtr 1*	*Qtr 2*	*Qtr 3*	*Qtr 4*	*Qtr 1*	*Qtr 2*	*Qtr 3*	*Qtr 4*	*Qtr 1*	*Qtr 2*	*Qtr 3*	*Qtr 4*
1	80	56	60	91	80	41	62	82	22	47	85	95
2	52	70	32	54	56	24	99	38	74	81	51	22
3	43	28	72	30	89	96	59	43	76	19	62	41
4	29	41	52	78	21	95	84	85	51	87	90	52
5	41	86	71	28	79	43	64	37	83	51	58	77

(a) Set up the appropriate layout for conducting this experiment. Identify the factors.
(b) State the model for this experiment and list the assumptions necessary for conducting ANOVA.
(c) Identify the appropriate hypothesis, expected mean squares and test statistic for testing the effects of box size.

5. A farmer wishes to test the effects of four types of plant food (A, B, C and D) and three types of insecticide (1, 2 and 3) on crop growth. However, the farmer only has 8 acres (3.2 ha) of land in which to conduct the experiment.
   (a) Set up the split plot experimental design to analyze the main and interaction effects, given that the farmer wishes to conduct two replications ($r = 2$).
   (b) List the model and assumptions necessary to conduct the ANOVA.
   (c) How would the design change if the farmer had only 4 acres (1.6 ha) of land in which to conduct this experiment?
6. Describe the difference between nested designs and split plot designs. Set up an example of each.
7. Theoretically, if an experiment were to be conducted over several days, collecting data on each of those days, can the experiment be conducted as a split plot design? Explain.

REFERENCES

1. Milliken, G. A. and Johnson, D. E. (1984) *Analysis of Messy Data*; Vol. I: *Designed Experiments*, Lifetime Learning Publications, Belmont, California.
2. Anderson, V. L. and McLean, R. A. (1974) *Design of Experiments: A Realistic Approach*, Marcel Dekker, New York.
3. Lentner, M. and Bishop, T. (1986) *Experimental Design and Analysis*, Valley Book Company, Blacksburg, Virginia.

FURTHER READING

Chakrabati, M. C. (1962) *Mathematics of Design and Analysis of Experiments*, Asia Publishing House, New York.

Johnson, N. L. and Leone, F. L. (1964) *Statistics and Experimental Design: In Engineering and the Physical Sciences*, Vol. II, John Wiley & Sons, New York.

16 Procedures for selecting fractional factorial designs

Most of the procedures already studied in this book have been developed for the purpose of evaluating main factors and their interactions, on the basis of available experimental data for a selected set of experimental conditions. In particular, Chapter 11 considered a general method for examining the effects of two-level factors and their interactions when a 2^{k-p} design is used. Since in this case there are 2^p possible fractions from which to choose one, it would be very useful to have a logical and systematic approach to decide which fractional design is the most appropriate for a given situation. This chapter is unique in the sense that it focuses exclusively on methods for selecting appropriate experimental designs for specific situations.

Due to the wide availability of powerful microcomputers, the approach taken to formulate and solve complex industrial problems has undergone a substantial transformation during the past few years. In a microcomputer environment engineers can carry out such diverse operations as automatically keeping track of manufacturing processes for statistical control, conducting cost-effective process simulations, performing modern statistical analyses and validating engineering process models. Practicing statisticians can provide insight into the fundamental structure of a range of processes due to the recent development of powerful, robust and simple computerized data analysis techniques.

The subject of experimental design and analysis, particularly in the field of engineering, has been further enhanced over the last 5–10 years by increased research on how the computer can play a significant role not only in the post-data analysis stage, but especially in the pre-data experimental design stage. Specifically, at the present time there exists a need for developing automated procedures that can generate experimental designs/plans in the presence of conflicting requirements related to the number of experimental factors, the number of constraints and the total number of allowable observations. Although the general topic just described goes beyond the scope of this book, this chapter will motivate, summarize and illustrate the use of three procedures for selecting two-level fractional factorial designs.

The procedures studied in this chapter have been developed by Knight [1], Knight and Neuhardt [2], Mount-Campbell and Neuhardt [3] and

Pignatiello [4]. The first procedure is based on a search algorithm that selects an orthogonal fractional factorial design given a list of feasible observations. The second procedure indicates how to partition a set of experimental data in such a way that each partition will correspond to a fractional factorial design having resolution III. The third procedure can be used for selecting fractional factorial designs given the availability and cost of each potential experimental condition.

16.1 SELECTION OF FRACTIONAL FACTORIAL DESIGNS GIVEN A LIST OF FEASIBLE OBSERVATIONS

Before discussing this procedure, examples of infeasible and feasible factorial designs will be given. In these illustrative examples, it is assumed that a database exists with incomplete experimental data from a 2^3 factorial design. The availability or nonavailability of the corresponding experimental conditions is indicated in Table 16.1. As shown in this table, the response value is not available for the first experimental condition in standard order. For this reason, a full 2^3 design is not possible. However, using the generator $X_3 = X_1X_2$ it is possible to select a feasible 2^{3-1} experiment with resolution III. As Table 16.1 indicates, the four experimental results (runs 2, 3, 5 and 8 in standard order) associated with this experiment are available in the database.

The procedure which we will use to choose a fractional factorial design is based on a matrix approach which can be utilized to denote the feasibility of any potential two-level fractional design. The rows of the matrix represent the number of factors k, and the matrix columns represent the number of observations. The maximum number of observations required are given by 2^{k-p} where p is the number of generators. It is assumed that for any value of k the factors have been ranked according to their relative importance. A feasibility matrix for designs having resolution III or more is shown in Figure

Table 16.1 Database for a 2^3 factorial design

Run	X_1	X_2	X_3	*Response value*
1	−	−	−	Unavailable
2	+	−	−	Available
3	−	+	−	Available
4	+	+	−	Available
5	−	−	+	Available
6	+	−	+	Available
7	−	+	+	Available
8	+	+	+	Available

16.1. The matrix shown in this figure has 10 rows and 10 columns. Row 10 corresponds to 1 factor and row 9 corresponds to 2 factors. In general, row i corresponds to $11-i$ factors. Additionally, column 1 corresponds to 2^1 observations and column 2 corresponds to 2^2 observations. In general, column i corresponds to 2^i observations. Each column is also associated with a set of values $p=0,1,2,\ldots$, with the entry belonging to the main diagonal corresponding to $p=0$.

As can be seen in Figure 16.1, the feasibility matrix is divided into the following four mutually exclusive different regions:

Region A: for nonconstructible designs
Region B: for fractional factorial designs

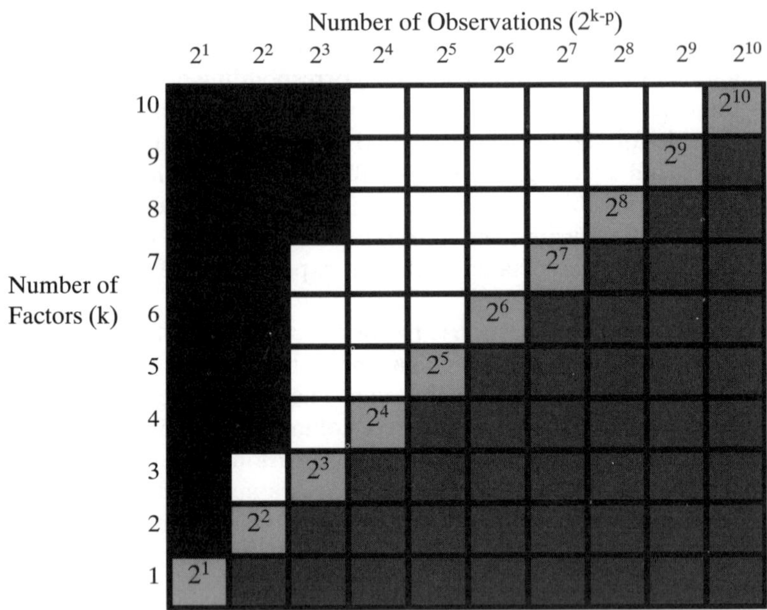

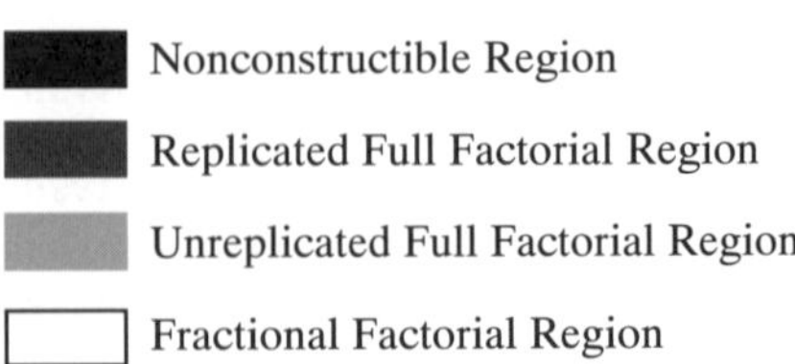

Figure 16.1 Feasibility matrix.

Region C: for unreplicated full factorial designs
Region D: for replicated full factorial designs.

Region A represents all nonconstructible designs since every design from this region has a resolution less than III. If, for example, $k=4$ and $p=2$, this region consists of all 2^{4-2} fractional factorial designs having generators given by

$$I_1 = X_1 X_3^a X_4^b \quad \text{and} \quad I_2 = X_2 X_3^a X_4^b$$

where each of the exponents a and b is equal to either 0 or 1. As can be verified in this case, $I_1 I_2 = X_1 X_2$. This, of course, means that the resolution would be less than III. Region B includes only unreplicated fractional factorial designs; region C corresponds to unreplicated full factorial designs; and region D includes replicated full factorial designs. For example, the cell associated with $k=4$ and $2^5=32$ observations allows 2 replications of a 2^4 design.

A cell of the matrix is considered to be infeasible if the database does not have a response value for at least one of the experimental conditions. For any value of k and any value of p, it is possible to formulate the following rules on the basis of the quantity $F_{kp}=1$ if the 2^{k-p} fractional factorial design is feasible, and $F_{kp}=0$ otherwise:

Rule 1. If $F_{kp}=0$ all cells above and to the right are also infeasible.
Rule 2. If $F_{kp}=1$ all cells to the left are also feasible, with the possible exception of the leftmost cell.

The first rule is based on the observation that if a design is not feasible with fewer runs, it will also be infeasible for a larger number of runs. Therefore, the cells to the right of any infeasible cell will be associated with an infeasible design. The second rule is based on the fact that the cells to the left of any feasible cell (design) will also be feasible as the cells to the left require fewer runs.

Figure 16.2 shows a path as an example of the search algorithm which starts at a full design 2^1 and ends at a 2^{9-5} design. The procedure to generate this path can be decomposed into the following steps.

Step 1

Start at the left-bottom corner of the matrix and verify if the design is feasible. In Figure 16.2 we start at a 2^1 full factorial design corresponding to $k=1$ and $p=0$. If it is feasible, then we proceed to check the next cell, which corresponds to $k=2$ and $p=0$. Successive cells are considered in the region corresponding to full factorial designs by moving along the diagonal until the first infeasible cell is identified. Each entry is marked with a label equal to F_{kp} (1 for feasible designs and 0 for infeasible designs). According to Figure 16.2, the first infeasible cell corresponds to $k=8$ and $p=0$. Therefore, as indicated by rule 1, all cells to the right and above this cell are infeasible.

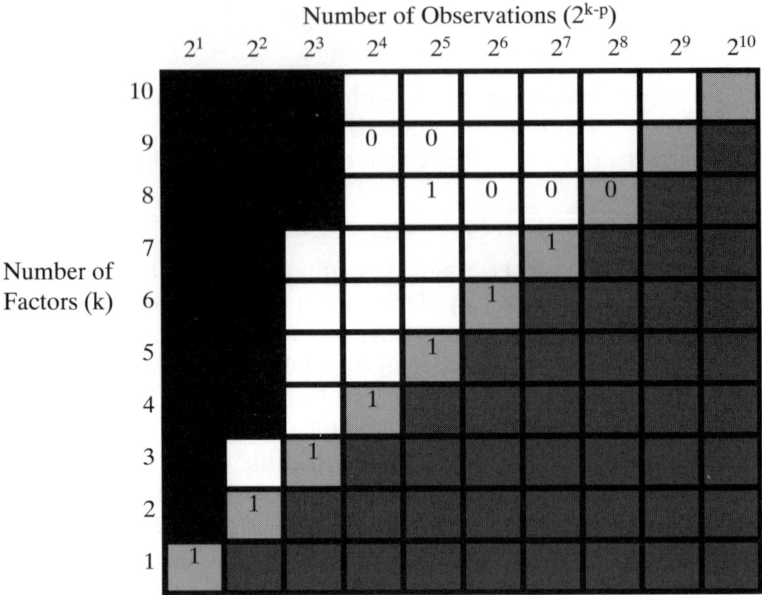

Figure 16.2 Path from search algorithm.

Step 2

Search from right to left starting at the infeasible cell identified in step 1 until the first feasible cell is reached. For this cell set $k = k'$ and $p = p'$. As previously mentioned, the first infeasible cell in step 1 was reached when $k = 8$ and $p = 0$. As shown in Figure 16.2, the cell associated with $k = 8$ and $p = 1$ was first examined and then two more cells associated with $p = 1$ and $p = 2$ were inspected. The first feasible cell in this figure is the cell corresponding to $k = k' = 8$ and $p = p' = 3$.

Step 3

Inspect all cells corresponding to the pairs $(k' + 1, p' + 1)$, $(k' + 2, p' + 2), \ldots,$ etc., until the first infeasible cell is identified. As shown in Figure 16.2, the next cell to be inspected was that associated with $k = 9$ and $p = 4$. This cell is infeasible. For this reason the cell associated with $k = 10$ and $p = 5$ was not inspected.

Step 4

Go back to step 2. The procedure is restarted at the cell associated with $k = 9$ and $p = 4$. If it is infeasible, then we check the cell associated with $k = 9$

and $p = 5$. Proceeding in this fashion, we inspect cells until the nonconstructible region is reached. According to the path shown in Figure 16.2, we stop at the cell associated with $k = 9$ and $p = 5$.

16.2 TOTAL NUMBER OF POSSIBLE 2^{k-p} FRACTIONAL FACTORIAL DESIGNS

The general procedure to determine the factor levels in each run of a 2^{k-p} fractional factorial design consists of solving a system of p linear equations using modulo 2 arithmetic. To illustrate, let vector $\mathbf{X}_i = (x_{i1}, x_{i2}, \ldots, x_{ik})^T$ indicate the levels of factors $X_1, X_2, \ldots, X_k$ in experimental condition i, where $x_{ij} = 0$ when factor X_j is set at its 'low' level in experimental condition i, or $x_{ij} = 1$ when the factor is set at its 'high' level. Here $i = 1, 2, \ldots, 2^{k-p}$. Now let us consider a 0–1 matrix of coefficients $\boldsymbol{\alpha}$ and a 0–1 vector of right-hand sides $\boldsymbol{\pi}$, defined as follows:

$$\boldsymbol{\alpha} = \begin{bmatrix} \alpha_{11} & \alpha_{12} & \cdots & \alpha_{1k} \\ \alpha_{21} & \alpha_{22} & \cdots & \alpha_{2k} \\ \vdots & \vdots & & \vdots \\ \alpha_{p1} & \alpha_{p2} & \cdots & \alpha_{pk} \end{bmatrix} \qquad \boldsymbol{\pi} = \begin{bmatrix} \pi_1 \\ \pi_2 \\ \vdots \\ \pi_p \end{bmatrix}.$$

The linear system under consideration can be formulated as

$$\boldsymbol{\alpha}\mathbf{X} = (\boldsymbol{\pi}, \boldsymbol{\pi}, \ldots, \boldsymbol{\pi})_{pn} \quad (\text{mod } 2) \tag{16.1}$$

where $\mathbf{X} = (\mathbf{X}_1, \mathbf{X}_2, \ldots, \mathbf{X}_n)$ and $n = 2^{k-p}$. Each choice for the matrix of coefficients $\boldsymbol{\alpha}$ actually partitions the full 2^k design into 2^p mutually exclusive orthogonal 2^{k-p} fractional factorials. Since each element of $\boldsymbol{\pi}$ is either 0 or 1, there are 2^p choices for the vector $\boldsymbol{\pi}$. For a specified matrix of coefficients a unique vector of right-hand sides corresponds to one of the 2^p fractions. The solution $\mathbf{X} = (\mathbf{X}_1, \mathbf{X}_2, \ldots, \mathbf{X}_n)$ consists of k-tuples of zeros and ones, indicating the factor levels for each experimental condition or run of the fractional design.

Mount-Campbell and Neuhardt [3] have developed an algorithm to calculate the total number of ways a full 2^k factorial design can be partitioned into 2^p mutually exclusive 2^{k-p} orthogonal fractional factorial designs, each having resolution III. A systematic procedure is provided for the identification of the treatment levels for any possible fractional factorial design with resolution III.

To simplify our presentation, we will consider a 2^{5-2} fractional factorial design. The solution vectors containing the binary indicators of the factors in each run of the fractional design are the 5-tuples $\mathbf{X}_1, \mathbf{X}_2, \ldots, \mathbf{X}_8$. The matrix of coefficients and the vector of right-hand sides are shown below:

$$\boldsymbol{\alpha} = \begin{bmatrix} \alpha_{11} & \alpha_{12} & \alpha_{13} & \alpha_{14} & \alpha_{15} \\ \alpha_{21} & \alpha_{22} & \alpha_{23} & \alpha_{24} & \alpha_{25} \end{bmatrix} \qquad \boldsymbol{\pi} = \begin{bmatrix} \pi_1 \\ \pi_2 \end{bmatrix}$$

Since $p = 2$, there are four possible choices (fractions) for any partition corresponding to a specific choice of the matrix of coefficients. These four possibilities are $(\pi_1, \pi_2)^T = (0, 0)^T$, $(\pi_1, \pi_2)^T = (1, 0)^T$, $(\pi_1, \pi_2)^T = (0, 1)^T$ and $(\pi_1, \pi_2)^T = (1, 1)^T$. For any one of these choices, the treatment levels in each run of the corresponding fraction of a partitioning associated with matrix $\boldsymbol{\alpha}$ are found as the solution to the following system of simultaneous linear equations, using modulo 2 arithmetic:

$$\begin{aligned} \boldsymbol{\alpha}\mathbf{X}_1 &= \boldsymbol{\pi} \pmod 2 \\ \boldsymbol{\alpha}\mathbf{X}_2 &= \boldsymbol{\pi} \pmod 2 \\ &\vdots \\ \boldsymbol{\alpha}\mathbf{X}_8 &= \boldsymbol{\pi} \pmod 2 \end{aligned} \tag{16.2}$$

Any orthogonal fractional factorial design has resolution III if each row and each modulo 2 sum of any number of rows of the matrix $\boldsymbol{\alpha}$ contain at least three entries equal to 1. To test for resolution III, it is convenient to rearrange the matrix of coefficients into a row–echelon form [1, 3]. This form, here referred to as REF, rearranges the entries of the matrix of coefficients in such a way that the following structure is observed:

Rule 1. The p columns of an identity matrix, here referred to as identity columns, appear in the same order they are in the identity matrix, as we scan the matrix of coefficients from left to right. The identity columns do not need to be adjacent.

Rule 2. In the row associated with the unit element of each identity column, there must be at least two entries equal to 1 to the right of the unit element. This means that at most $k - 2$ columns are eligible to become identity columns. As a result, in rule 1 the total number of arrangements of the identity columns, or equivalently the total number of REFs, is equal to $\binom{k-2}{p}$.

Rule 3. In each row all entries to the left of the unit element belonging to the corresponding identity column must be zero.

Rule 4. Once zeros and ones are placed in each row, according to the previous conditions, the remaining elements are completed in such a way that no two rows will have identical completions. Here a completion is a $(k - p)$-tuple with entries equal to zero or one.

Continuing with the analysis of the previously considered 2^{5-2} fractional factorial design, it is noted that the three columns of an identity matrix can be arranged in $\binom{3}{2} = 3$ different ways according to rule 1. These three different arrangements are shown below, where the symbol ● is used to indicate those entries that need to be set equal to zero, as a result of rule 3. Additionally, the symbol ⊕ is used to indicate those entries that still need to be set equal to either zero or one, while satisfying rules 2 and 4. Note that each row completion actually consists of a 3-tuple of zeros and ones (represented by the symbols ● and ⊕). If rows are completed from row p to row 1 (that is, beginning the completion process with the bottom row) the number of possible completions for a row will be reduced by the number of rows completed before it, due to rule 3.

In each of the three REF arrangements shown below, the completion of the second row will be considered first and then that of the first row. In this way the number of possible completions for the first row will be reduced by one, as indicated by rule 4.

Arrangement 1 Candidate 3-tuples for row completion

$$\begin{bmatrix} 1 & 0 & \oplus & \oplus & \oplus \\ 0 & 1 & \oplus & \oplus & \oplus \end{bmatrix} \quad \begin{matrix} [1\ 1\ 1] & [1\ 1\ 0] & [1\ 0\ 1] & [0\ 1\ 1] \\ [1\ 1\ 1] & [1\ 1\ 0] & [1\ 0\ 1] & [0\ 1\ 1] \end{matrix}$$

Arrangement 2 Candidate 3-tuples for row completion

$$\begin{bmatrix} \bullet & 1 & 0 & \oplus & \oplus \\ \bullet & 0 & 1 & \oplus & \oplus \end{bmatrix} \quad \begin{matrix} [0\ 1\ 1] \\ [0\ 1\ 1] \end{matrix}$$

Arrangement 3 Candidate 3-tuples for row completion

$$\begin{bmatrix} 1 & \oplus & 0 & \oplus & \oplus \\ 0 & \bullet & 1 & \oplus & \oplus \end{bmatrix} \quad \begin{matrix} [1\ 1\ 1] & [0\ 1\ 1] & [1\ 0\ 1] & [1\ 1\ 0] \\ [0\ 1\ 1] & & & \end{matrix}$$

In arrangement 1, we see that there are 4 possible completions for the second row, and 3 (4 − 1) possible completions for the first row, hence the number of partitions associated with this arrangement is equal to 12. There is no partition associated with arrangement 2, because the two row completions shown are identical. Similarly, in arrangement 3 the total number of partitions is found to be equal to 3. Thus, the total number of partitions for the 2^5 full design is equal to $T = 12 + 0 + 3 = 15$, and the total number of fractions would be $2^pT = (2^2)(15) = 60$.

As an illustration, let us consider the fraction corresponding to $\pi_1 = 0$ and $\pi_2 = 1$ in the partition associated with the following matrix (corresponding to arrangement 3).

$$\boldsymbol{\alpha} = \begin{bmatrix} 1 & 1 & 0 & 0 & 1 \\ 0 & 0 & 1 & 1 & 1 \end{bmatrix}$$

In this case, the treatment levels in the ith experimental condition are selected from the solution of the system of equations given below:

$$x_{1i} + x_{2i} + x_{5i} = 0 \pmod 2 \quad i = 1, 2, \ldots, 8$$
$$x_{3i} + x_{4i} + x_{5i} = 1 \pmod 2 \quad i = 1, 2, \ldots, 8.$$

Note that for any experimental condition there are two equations in five variables. If we consider x_{1i}, x_{4i} and x_{5i} as independent variables then we obtain the solutions given in Table 16.2.

Table 16.2 Factor levels from solution

X_1	X_4	X_5	X_2	X_3
1	1	1	0	1
0	1	1	1	1
1	0	1	0	0
0	0	1	1	0
1	1	0	1	0
0	1	0	0	0
1	0	0	1	1
0	0	0	0	1

Using the standard notation given in Chapter 11, we can verify that the levels given in Table 16.2 correspond to the generators $X_2 = -X_1X_5$ and $X_3 = +X_4X_5$. The defining relation, therefore, is given by $I = -X_1X_2X_5 = X_3X_4X_5 = -X_1X_2X_3X_4$ which, as expected, indicates that the design has resolution III.

16.3 COST-OPTIMAL FRACTIONAL FACTORIAL DESIGNS

This section describes a procedure developed by Pignatiello [4] to find a cost-optimal fractional factorial design using the feasibility matrix defined in section 16.1 and a cost associated with each experimental condition of a full design. Before describing the procedure, a cost model first proposed by Neuhardt and Mount-Campbell [5] will be used to compute the cost of a fractional factorial design. To simplify our review, we will consider the 2^{5-2} fractional factorial design with the defining relation given below:

$$I = +X_1X_2X_3 = -X_3X_4X_5 = -X_1X_2X_4X_5. \tag{16.3}$$

The cost of the design having the defining relation shown in equation (16.3) can be calculated as

$$T = 2^{5-2}[C(I) + C(X_1X_2X_3) - C(X_3X_4X_5) - C(X_1X_2X_4X_5)]. \tag{16.4}$$

The terms $C(I)$, $C(X_1X_2X_3)$, $C(X_3X_4X_5)$ and $C(X_1X_2X_4X_5)$ are referred to as **cost parameters**. The term $C(I)$ represents the average experimental condition cost. If c_i is the cost of running the ith experimental condition, then $C(I) = \sum_i c_i/(32)$, where $i = 1, 2, \ldots, 32$. Additionally, the term $C(X_1X_2X_3)$ is defined as the difference between the average of the 16 experimental conditions of the full factorial for which $I = +X_1X_2X_3$ and $C(I)$. That is,

$$C(X_1X_2X_3) = \left(\sum_i \frac{c_i}{16}\right) - C(I)$$

where $i = 1, 2, \ldots, 16$. Similarly, the cost parameters associated with $X_3X_4X_5$ and $X_1X_2X_4X_5$ in equation (16.4) are defined as

$$C(X_3X_4X_5) = \left(\sum_i \frac{c_i}{16}\right) - C(I) \quad \text{and} \quad C(X_1X_2X_4X_5) = \left(\sum_i \frac{c_i}{16}\right) - C(I)$$

where the two summations are over those 16 experimental conditions from the full factorial for which $I = +X_3X_4X_5$ and $I = +X_1X_2X_4X_5$, respectively.

The procedure developed by Pignatiello [4] assumes that the specified main and interaction effects are listed according to their relative importance, starting with the most important effect. This procedure begins by defining the following sets:

$\boldsymbol{R}$ = set of effects with the ith element defined as the ith ranked effect.
$\boldsymbol{R}_i$ = subset of **R** containing the top i ($i = 1, 2, \ldots, n_r$) ranked effects in decreasing order of rank.
$\boldsymbol{Q}_i$ = set of factors associated with the main effects or interaction effects in $\boldsymbol{R}_i$.
$\boldsymbol{E}_i$ = set of eligible words with minimal length equal to three. Each word containing effects in $\boldsymbol{R}_i$ is such that no q-effect in $\boldsymbol{R}_i$ will be in a word with less than $3 + q$ other factor effects in $\boldsymbol{R}_i$.

The general procedure for finding a cost-optimal fractional factorial design can be now described as follows [4]:

Step 1. Set $i = 1$, where i refers to the ith row of the feasibility matrix.
Step 2. Determine the cost associated with each experimental condition for the full factorial design among the factors in set $\boldsymbol{Q}_i$. Calculate the cost parameter corresponding to each factorial effect.
Step 3. Construct $\boldsymbol{E}_i$ given the requirement set $\boldsymbol{R}_i$.
Step 4. Use the procedure developed in section 16.1 to find a cost-optimal fractional factorial associated with set $\boldsymbol{E}_i$. Note that the search is restricted to be over $\boldsymbol{E}_i$.
Step 5. Reset $i = i + 1$. If $i > n_r$, stop the procedure; otherwise, go to step 2.

The above procedure will be illustrated by considering a numerical example. This example has been previously considered by Knight and Neuhardt [2] and Pignatiello [4]. Table 16.3 shows a cost associated with each experimental condition of a full 2^5 factorial design. In this table, a cost of 10 000 is associated with each unavailable experimental condition. It is desired to find a cost-optimal fractional factorial design associated with the ordered set

$$\boldsymbol{R} = \{X_1, X_2, X_3, X_4, X_5, X_1X_3, X_1X_4, X_1X_5, X_2X_3, X_3X_4, X_3X_5, X_4X_5\}.$$

If we assume that all main effects must be estimated, then the procedure can be started with $i = 5$ instead of $i = 1$. For each $i = 5, 6, \ldots, 12$, we can find a cost-optimal fractional factorial design using each cost model over set $\boldsymbol{E}_i$. Sets $\boldsymbol{E}_5$ through $\boldsymbol{E}_{12}$ are given in Table 16.4. Treating each cost as the response

Table 16.3 Costs associated with a 2^5 full design

Run	Factors X_1	X_2	X_3	X_4	X_5	Cost
1	−1	−1	−1	−1	−1	10 000
2	1	−1	−1	−1	−1	18
3	−1	1	−1	−1	−1	10 000
4	1	1	−1	−1	−1	20
5	−1	−1	1	−1	−1	6
6	1	−1	1	−1	−1	26
7	−1	1	1	−1	−1	16
8	1	1	1	−1	−1	16
9	−1	−1	−1	1	−1	26
10	1	−1	−1	1	−1	6
11	−1	1	−1	1	−1	16
12	1	1	−1	1	−1	16
13	−1	−1	1	1	−1	18
14	1	−1	1	1	−1	10 000
15	−1	1	1	1	−1	20
16	1	1	1	1	−1	10 000
17	−1	−1	−1	−1	1	26
18	1	−1	−1	−1	1	6
19	−1	1	−1	−1	1	16
20	1	1	−1	−1	1	16
21	−1	−1	1	−1	1	18
22	1	−1	1	−1	1	10 000
23	−1	1	1	−1	1	20
24	1	1	1	−1	1	10 000
25	−1	−1	−1	1	1	10 000
26	1	−1	−1	1	1	18
27	−1	1	−1	1	1	10 000
28	1	1	−1	1	1	20
29	−1	−1	1	1	1	6
30	1	−1	1	1	1	26
31	−1	1	1	1	1	16
32	1	1	1	1	1	16

value in a 2^n factorial design, the cost parameter corresponding to each factorial effect can easily be calculated by using Yates algorithm.

By inspecting sets $\boldsymbol{R}_5 = \{X_1, X_2, X_3, X_4, X_5\}$ and $\boldsymbol{E}_5$, the cost-optimal 2^{5-1} fractional factorial design over $\boldsymbol{E}_5$ can be found by identifying the cost parameter with the largest absolute value and forming the defining relation in such a way that the sign on the cost parameter is negative. In Table 16.5, $C(X_3X_4X_5)$ has the largest absolute value over $\boldsymbol{E}_5$, and, therefore, the cost-optimal 2^{5-1} fractional factorial design over $\boldsymbol{E}_5$ uses the generator $X_5 = +X_3X_4$.

Table 16.4 Sets $\boldsymbol{E}_5$ through $\boldsymbol{E}_{12}$. $\boldsymbol{R} = \{X_1, X_2, X_3, X_4, X_5, X_1X_3, X_1X_4, X_1X_5, X_2X_3, X_3X_4, X_3X_5, X_4X_5\}$

Set	*Eligible words*
$\boldsymbol{E}_5$	$X_1X_2X_3$ $X_1X_2X_4$ $X_1X_3X_4$ $X_2X_3X_4$ $X_1X_2X_3X_4$ $X_1X_2X_5$ $X_1X_3X_5$ $X_2X_3X_5$ $X_1X_2X_3X_5$ $X_1X_4X_5$ $X_2X_4X_5$ $X_1X_2X_4X_5$ $X_3X_4X_5$ $X_1X_3X_4X_5$ $X_2X_3X_4X_5$ $X_1X_2X_3X_4X_5$
$\boldsymbol{E}_6$	$X_1X_2X_4$ $X_2X_3X_5$ $X_1X_2X_3X_4$ $X_1X_2X_5$ $X_2X_3X_5$ $X_1X_2X_3X_5$ $X_1X_4X_5$ $X_2X_4X_5$ $X_1X_2X_4X_5$ $X_3X_4X_5$ $X_1X_3X_4X_5$ $X_2X_3X_4X_5$ $X_1X_2X_3X_4X_5$
$\boldsymbol{E}_7$	$X_2X_3X_4$ $X_1X_2X_3X_4$ $X_1X_2X_5$ $X_2X_3X_5$ $X_1X_2X_3X_5$ $X_2X_4X_5$ $X_1X_2X_4X_5$ $X_3X_4X_5$ $X_1X_3X_4X_5$ $X_2X_3X_4X_5$ $X_1X_2X_3X_4X_5$
$\boldsymbol{E}_8$	$X_2X_3X_4$ $X_1X_2X_3X_4$ $X_2X_3X_5$ $X_1X_2X_3X_5$ $X_2X_4X_5$ $X_1X_2X_4X_5$ $X_3X_4X_5$ $X_1X_3X_4X_5$ $X_2X_3X_4X_5$ $X_1X_2X_3X_4X_5$
$\boldsymbol{E}_9$	$X_1X_2X_3X_4$ $X_2X_4X_5$ $X_1X_2X_4X_5$ $X_3X_4X_5$ $X_1X_3X_4X_5$ $X_2X_3X_4X_5$ $X_1X_2X_3X_4X_5$
$\boldsymbol{E}_{10}$	$X_1X_2X_3X_4$ $X_2X_4X_5$ $X_1X_2X_4X_5$ $X_2X_3X_4X_5$ $X_1X_2X_3X_4X_5$
$\boldsymbol{E}_{11}$	$X_1X_2X_3X_4$ $X_2X_4X_5$ $X_1X_2X_4X_5$ $X_2X_3X_4X_5$ $X_1X_2X_3X_4X_5$
$\boldsymbol{E}_{12}$	$X_1X_2X_3X_4$ $X_1X_2X_4X_5$ $X_1X_2X_3X_4X_5$

If we are interested in a cost-optimal 2^{5-2} fractional factorial over $\boldsymbol{E}_5$, the procedure is rather simple but lengthy. First any two words w_1 and w_2 in $\boldsymbol{E}_5$ are selected and used to define the generators $I_1 = \pm w_1$ and $I_2 = \pm w_2$. For each of the four possible choices, a defining relation $I = I_1 = I_2 = I_1I_2$ can be used to compute the cost of the corresponding fractional factorial design. This procedure is actually repeated for each pair of words and then the design associated with the lowest cost is selected.

It can be verified that in the example under consideration, the cost-optimal 2^{5-2} fractional factorial design over $\boldsymbol{E}_5$ corresponds to the defining relation

$$I = +X_1X_2X_3 = +X_1X_2X_4X_5 = +X_3X_4X_5.$$

After computing each cost parameter, the cost of the fraction is found to be equal to 88.

A summary of the results now follows [4].

(a) For $i = 6$ and $i = 7$ the cost-optimal 2^{5-1} design is found to be the same as the cost-optimal 2^{5-1} design identified when $i = 5$. In each case, two cost-optimal 2^{5-2} fractional factorial designs exist with a cost of 128. The defining relations for these designs are

$$I = +X_3X_4X_5 = +X_1X_2X_5 = +X_1X_2X_3X_4$$

Table 16.5 Cost parameters

Word	C(word)
I	2512.74
X_1	0
X_2	0.25
X_1X_2	0
X_3	0
X_1X_3	2497.75
X_2X_3	0
$X_1X_2X_3$	−2.75
X_4	0
X_1X_4	0
X_2X_4	0
$X_1X_2X_4$	0
X_3X_4	0
$X_1X_3X_4$	0
$X_2X_3X_4$	0
$X_1X_2X_3X_4$	0
X_5	0
X_1X_5	0
X_2X_5	0
$X_1X_2X_5$	0
X_3X_5	0
$X_1X_3X_5$	0
$X_2X_3X_5$	0
$X_1X_2X_3X_5$	0
X_4X_5	0
$X_1X_4X_5$	+2492.75
$X_2X_4X_5$	0
$X_1X_2X_4X_5$	−2.25
$X_3X_4X_5$	−2496.75
$X_1X_3X_4X_5$	0
$X_2X_3X_4X_5$	−0.25
$X_1X_2X_3X_4X_5$	0

and

$$I = +X_3X_4X_5 = -X_1X_2X_5 = -X_1X_2X_3X_4.$$

(b) For $i = 8$, no cost-optimal 2^{5-2} fractional factorial design is feasible, since this design has only 8 degrees of freedom and 9 are required. This is also true for $i = 9$, 10, 11, 12.

(c) The cost-optimal 2^{5-1} fractional factorial design for $i = 8 = 9$ has a generator $X_5 = X_3X_4$ with cost of 256. However, when $i = 10 = 11 = 12$, the cost-optimal 2^{5-1} fractional factorial design is infeasible, because the fraction includes unavailable experimental conditions.

EXERCISES

1. Write the system of equations to determine which experimental conditions are chosen in a 2^{6-2} orthogonal fractional factorial.
2. In Exercise 1, determine the number of 2^{6-2} partitions having resolution III. Calculate the number of fractions. Choose any fraction and verify that it has resolution III.
3. For a 2^{7-3} fractional design calculate the total number of REF arrangements. For the arrangement that starts (from left to right) with an identity matrix, calculate the possible number of completions from each row.
4. Repeat Exercise 3 considering the first, second and third columns of an identity matrix as the first, second and fourth columns of the REF arrangement respectively.
5. Given the set $\boldsymbol{R} = \{X_1, X_2, X_3X_4, X_3, X_4X_5X_6\}$, find sets $\boldsymbol{R}_i$, $\boldsymbol{Q}_i$ and $\boldsymbol{E}_i$ for $i = 1, 2, 3, 4, 5$, such that the cost-optimal fractional factorial designs will satisfy a resolution III requirement.
6. Verify that the optimal 2^{5-2} design over $\boldsymbol{E}_5$ has a cost equal to 88 in the example considered in section 16.3.

REFERENCES

1. Knight, J. W. (1977) On the selection of fractional factorials given a list of feasible observations. Ph.D. Dissertation, The Ohio State University, Department of Industrial and System Engineering.
2. Knight, J. W. and Neuhardt, J. B. (1983) Computer-aided design of fractional factorial experiments given a list of feasible observations. *IIE Transactions*, **15**(2), June, 142–9.
3. Mount-Campbell, C. A. and Neuhardt, J. B. (1981) On the number of 2^{n-p} fractional of resolution III. *Communications in Statistics – Theory and Methods*, **A10**(20), 2101–11.
4. Pignatiello, J. J. (1985) A minimum cost approach for finding fractional factorials. *IIE Transactions*, 212–18.
5. Neuhardt, J. B. and Mount-Campbell, C. A. (1978) Selection of cost optimal 2^{k-p} fractional factorials. *Communications in Statistics – Computations*, **B7**, 369–83.

FURTHER READING

Mount-Campbell, C. A. and Neuhardt, J. B. (1980) Selecting cost optimal main-effect fractions of 3^n factorials. *AIIE Transactions*, **12**, 80–6.

Appendix A
An interactive microcomputer methodology

A.1 INTRODUCTION

This is the manual for the Interactive Microcomputer Methodology for the Statistical Analysis of Industrial Experiments (ID). This manual shows procedures and details of how to use this package. This microcomputer program can be used to significantly enhance the learning environment in a course on design and analysis of industrial experiments.

A.2 DESCRIPTION OF FILES

The software package is available from the authors. For more information see the advertisement. The ID code consists of 14 program files containing the data analysis procedures studied in this book, there is also one data file for each procedure. Figure A.1 shows the program file names of the ID code.

The files on the user disk are compiled versions of the software. The software has one executable file and the chain files are compiled versions of the BASIC program files. The program can be invoked by the ID.EXE file.

ID.EXE: Introduction and identification program of the interactive microcomputer program.

MENU.TBC: Main menu of the interactive microcomputer methodology.

ANOFVA.TBC: Contains the procedure for the one-way analysis of variance (ANOVA).

AFATST.TBC: Contains the after-ANOVA tests for identifying treatments that cause rejection of the null hypothesis: orthogonal contrasts, Scheffe's method and studentized range test.

COMBLO.TBC: Contains the procedure for the two-way ANOVA for a complete block design (each block contains all treatments.)

INCBLO.TBC: Contains the procedure for the two-way ANOVA for a balanced incomplete block design (each block contains missing treatments).

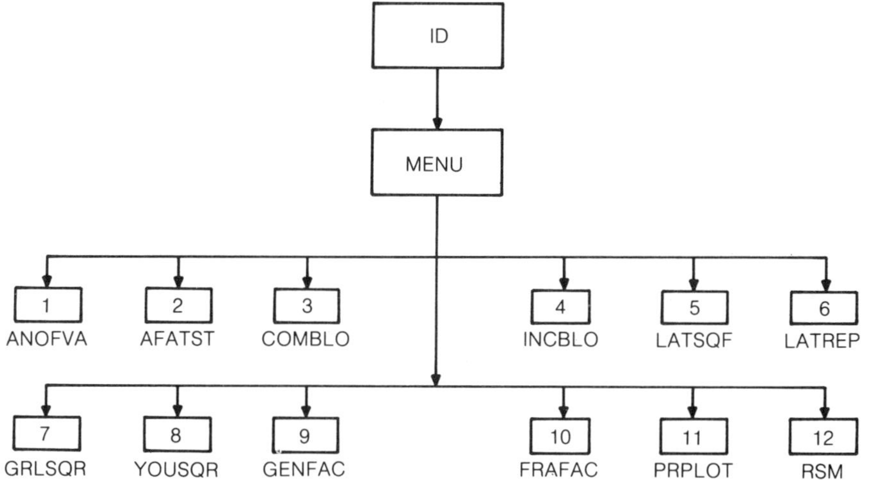

Figure A.1 Program files in the ID code.

LATSQR.TBC: Contains the procedure for a three-factor ANOVA: treatment, block and position. This design is referred to as a Latin square design.

LATREP.TBC: Contains the procedure for performing the ANOVA for designs using replicated Latin Squares. The program allows for different types of replications.

GRLSQR.TBC: Contains the procedure for a four-factor ANOVA: treatment, block, position and Greek letter. This design is referred to as a Graeco-Latin square design.

YOUSQR.TBC: Contains the procedure for a three-factor ANOVA: treatment, block and position with incomplete block design. This design is referred to as a Youden square design.

GENFAC.TBC: Contains a procedure for performing significance tests for main and interactions effects using a general factorial design.

FRAFAC.TBC: Contains a procedure for analyzing fractional experiment designs of size 2^{n-p}. In can be executed with or without mirror image design.

PRPLOT.TBC: Provides normal probability plots to test the significance of main effects and factor interactions using an unreplicated two-level design.

RSM.TBC: Contains a response surface methodology to carry out first-order, second-order and canonical analysis.

HELP.TBC: Contains a description of each module of the program.

The above-mentioned files and chain files are compiled from the source code with the '.bas' extension. The following is the list of source code files:

MENU.BAS
ANOFVA.BAS
AFATST.BARS

COMBLO.BAS
INCOMBLO.BAS
LATSQR.BAS
LATREP.BAS
GRLSQR.BAS
YOUSQR.BAS
GENFAC.BAS
FRAFAC.BAS
RSM.BAS
SUB_1.BAS
HELP.BAS
SUB_2.BAS
SUB_3.BAS

The subroutines SUB_1.BAS, SUB_2.BAS and SUB_3.BAS are linked with the main code during compiling.

A.3 BASIC PURPOSE OF EACH PROGRAM

ANOFVA.TBC

The ANOFVA.TBC (one-way ANOVA) program is used to find which treatments in a one-factor experiment are significantly different. The user should input level of significance (alpha), number of treatments, and a sample of random observations for each treatment.

AFATST.TBC

The AFATST.BC (after-ANOVA tests) program is used to determine which treatments cause the null hypothesis to be rejected. Three different methods are used in the program:

(a) Orthogonal contrasts
(b) Scheffe's method
(c) Range test.

In addition to the ANOFVA.TBC program data requirements, the AFATST.TBC program needs the coefficients of the orthogonal contrasts, the critical value of the F-statistic and the critical values of the studentized range test.

COMBLO.TBC

The COMBLO.TBC (complete block design) program is used to test the significance of treatments and blocks. The following input data are required by the program: level of significance, number of treatments, number of blocks and a sample of random observations for each block.

INCBLO.TBC

The INCBLO.TBC (balanced incomplete block design) program is used to conduct a special ANOVA (adjusted) for treatments and blocks. The following input data are required by the program: level of significance, number of treatments, number of blocks and a sample of random observations for each block (with missing observations in each block).

LATSQR.TBC

The LATSQR.TBC program is used to make the ANOVA more sensitive when there are three sources of systematic variation in addition to random variation. This program requires the following data input: level of significance, number of Latin letters (treatments), number of blocks and the position of letters in each block.

LATREP.TBC

The LATREP.TBC program is used when a Latin square design is replicated. This program requires the following input data: level of significance, dimension of the replicated Latin squares, specification of the blocking variables (rows and columns) which are shared by the replicated squares and Latin letter arrangement for each square.

GRLSQR.TBC

The GRLSQR.TBC program is used to test the significance of treatments (Latin letters) in the presence of three restrictions on randomization: rows, columns and Greek letters. Data required herein are the same as in the Latin square design plus the position of Greek letters. Greek letters should be entered first, then Latin letters and last numerical input data.

YOUSQR.TBC

The YOUSQR.TBC program is used to test the significance of treatments in a balanced incomplete block design where the effect due to the positions in a block can be separated from all other sources of variation. This program requires the following data input: level of significance, number of Latin letters (treatments), number of blocks and the position of letters in each block (with missing observations in each block).

GENFAC.TBC

The GENFAC.TBC program is used to perform a significance test on main effects and interaction effects in a factorial experiment. The technique used is a multi-way analysis of variance. The input data include number of factors, number of replications and a random sample of observations for each level combination.

FRAFAC.TBC

The FRAFAC.TBC program is used to run a 2^{n-p} fractional factorial experiment with or without a mirror image design. The program requires a random sample for each experimental condition, and the generators of the fractional factorial design.

PRPLOT.TBC

The PRPLOT.TBC program is used to determine the extent of the effect of a factor or interaction between factors on the response of a full two-level factorial experiment by using normal probability plots. This subroutine requires one random observation for each experimental condition.

RSM.TBC

The RSM.TBC program is used to carry out the first-order, second-order and canonical analyses of the response surface methodology. A two-level factorial experiment is used for the first-order analysis and a three-level factorial experiment is used for the second-order analysis.

A.4 USER MANUAL

A.4.1 How to run the code

The programs have been constructed for the specific purpose of aiding students in a number of aspects related to the analysis of experimental data. Input errors are checked and guidelines are given. The use of the computerized procedure involves six steps that can be summarized as follows:

Step 1. Insert the program disk in the active drive.

Step 2. Enter ID and hit RETURN to run the code.

Step 3. First screen is the identification of the code. After this hit RETURN to continue. Refer to Figure A.2.

Step 4. The MENU will be displayed on the screen (refer to Figure A.3.)

```
*** WELCOME TO THE DESIGN OF EXPERIMENTS MICROCOMPUTER PACKAGE ***

                    DEVELOPED BY
                 ALBERTO GARCIA-DIAZ
                  DON T. PHILLIPS

                    REVISED 1994

        DEPARTMENT OF INDUSTRIAL ENGINEERING
               TEXAS A&M UNIVERSITY

          PLEASE PRESS RETURN TO CONTINUE
```

Figure A.2 Identification for the DE code.

```
            ****SELECT ANY ONE OF THE FOLLOWING****

 1. ONE WAY ANALYSIS OF VARIANCE ................................ 1
 2. AFTER ANOVA TESTS ........................................... 2
 3. COMPLETE BLOCK DESIGN ....................................... 3
 4. BALANCED INCOMPLETE BLOCK DESIGN ............................ 4
 5. LATIN SQUARE DESIGN ......................................... 5
 6. LATIN SQUARE WITH REPLICATIONS .............................. 6
 7. GRAECO LATIN SQUARE DESIGN .................................. 7
 8. YOUDEN SQUARE DESIGN ........................................ 8
 9. GENERAL FACTORIAL DESIGN .................................... 9
10. TWO LEVEL FACTORIAL DESIGN ..................................10
11. NORMAL PROBABILITY PLOTS ....................................11
12. RESPONSE SURFACE METHODOLOGY ................................12
13. HELP ........................................................13
14. QUIT ........................................................14

              PLEASE ENTER OPTION NUMBER:_
```

Figure A.3 Display of menu.

Step 5. Select a specific program by typing its corresponding number and input data according to the instructions given in each program.

Step 6. The user can return to the DOS system by typing the number corresponding to the QUIT option.

SAMPLE OPTION

A.4.2 One-way analysis of variance

General instructions:

Limitation
Maximum number of treatments = 10
Maximum number of observations per treatment = 30

Step 1. Select any one of the following options:

1. RUN WITH NEW DATA
2. RUN WITH SAMPLE DATA
3. RUN DATA FROM FILE
4. DIRECTORY OF EXISTING FILES
5. MENU
6. QUIT

If option 1 is selected, steps 2–13 will be performed sequentially.

If option 2 is selected, a sample data will be read by the program and steps 5–13 will be performed sequentially.

If option 3 is selected, the user will be asked to input the INPUT FILE NAME and steps 5–13 will be performed sequentially.

If option 4 is selected, the user will be asked to specify the data drive. Then the data file contained in that specified drive will be displayed on the screen.

If option 5 is selected, the MAIN MENU will be displayed on the screen.

If option 6 is selected, the program will be terminated and bring the system back to DOS.

Step 2. Input the following data:

ALPHA=_
NO. OF TREATMENTS=_
IS NUMBER OF OBSERVATIONS EQUAL IN ALL TREATMENTS? (Y/N)_
NO. OF OBSERVATIONS FOR EACH TREATMENT=_

Step 3. DO YOU WANT TO CHANGE ANY OF THESE DATA? (Y/N)
If the answer is YES, steps 2–3 will be performed again.

Step 4. Input the data following the cursor which appears on the screen until all the data are completely input.

Step 5. The input data will be displayed on the screen.

Step 6. DO YOU WANT TO CHANGE THE INPUT DATA? (Y/N)
If the answer is YES, the following information must be provided:

PLEASE ENTER TREATMENT NUMBER=_
PLEASE ENTER OBSERVATION NUMBER=_
OLD VALUE=(GIVEN BY COMPUTER)
NEW VALUE=_

After the new value is given by the user, steps 5–6 will be performed again for additional changes.

Step 7. DO YOU WANT TO HAVE A HARD COPY OF THE INPUT DATA? (Y/N)
If the answer is YES, the input data will be printed by printer connected.

Step 8. DO YOU WANT TO SAVE THE INPUT DATA? (Y/N)
If the answer is YES, the user will be asked to input the OUTPUT FILE NAME.

Step 9. The ANOVA table will be displayed on the screen.

Step 10. WOULD YOU LIKE TO HAVE A HARD COPY OF THE RESULTS? (Y/N)
If the answer is YES, the results will be printed by the printer connected.

Step 11. DO YOU WANT TO CALCULATE THE COMPONENTS OF VARIANCE? (Y/N)
If the answer is YES, the results of COMPONENTS OF VARIANCE will be displayed on the screen.

Step 12. WOULD YOU LIKE TO HAVE A HARD COPY OF THE RESULTS? (Y/N)
If the answer is YES, the results will be printed by the printer conected.

Step 13. SELECT ANY ONE OF THE FOLLOWING OPTIONS:

1. RERUN WITH SAME DATA
2. RERUN WITH SAME DATA BUT NEW ALPHA
3. RERUN DATA FROM FILE
4. RERUN WITH NEW DATA
5. MENU
6. QUIT

If option 1 is selected, the available data will be used again and steps 5–13 will be performed sequentially.

If option 2 is selected, the user will be asked to give the new ALPHA and steps 5–13 will be performed sequentially.

If option 3 is selected, the user will be asked to input the INPUT FILE NAME and steps 5–13 will be performed sequentially.

If option 4 is selected, the system will erase the previous data and steps 2–13 will be performed sequentially.

If option 5 is selected, the MAIN MENU will be displayed on the screen.

If option 6 is selected, the program will be terminated and bring the system back to DOS.

SAMPLE RUN

Sample problem: Example 3.3

The National Business Machine company is putting together a computer system for sale to business. While they manufacture most of their own equipment, they do not wish to take the time at present to develop a letter quality printer (LQP) to go with their system. For this reason, they wish to find the best reasonably priced LQP available to subcontract as a system option. One consideration in the selection is the hours of operation until failure. In order to determine whether there might be some difference between machines, NBM approached three randomly chosen manufacturers, which supplied them with six of the requested models. NBM then tested the three printers from each manufacturer, arriving at the results in hours to first failure given in Table A.1.

Microcomputer solution

The process for solving this sample problem consists of the following steps:

Step 1. SELECT ANY ONE OF THE FOLLOWING OPTIONS: (Refer to Figure A.4.)

Step 2. INPUT THE FOLLOWING DATA: (Refer to Figure A.5.)

Table A.1

Model I	*Model II*	*Model III*
60	102	121
45	96	132
72	105	118
68	99	128
71	103	131
52	95	126

ONE WAY ANALYSIS OF VARIANCE

* * * * SELECT ANY ONE OF THE FOLLOWING * * * *

RUN WITH NEW DATA 1
RUN WITH SAMPLE DATA 2
RUN DATA FROM FILE 3
DIRECTORY OF EXISTING FILES 4
MENU 5
QUIT 6

PLEASE ENTER OPTION NUMBER : 1

Figure A.4 Display of first menu.

* * * *INPUT THE FOLLOWING DATA * * * *

ALPHA = .05
NO. OF TREATMENTS = 3
IS NUMBER OF OBSERVATIONS EQUAL IN ALL TREATMENTS ?(Y/N)Y
NO. OF OBSERVATIONS FOR EACH TREATMENT = 6

DO YOU WANT TO CHANGE ANY OF THESE DATA ? (Y/N)

Figure A.5 Procedure for entering the original data.

Step 3. DO YOU WANT TO CHANGE ANY OF THESE DATA? (Y/N) N
Step 4. Input the data following the cursor which appears on the screen until all the data are completely input. (Refer to Figure A.6.)
Step 5. The input data will be displayed on the screen. Refer to Figure A.7.

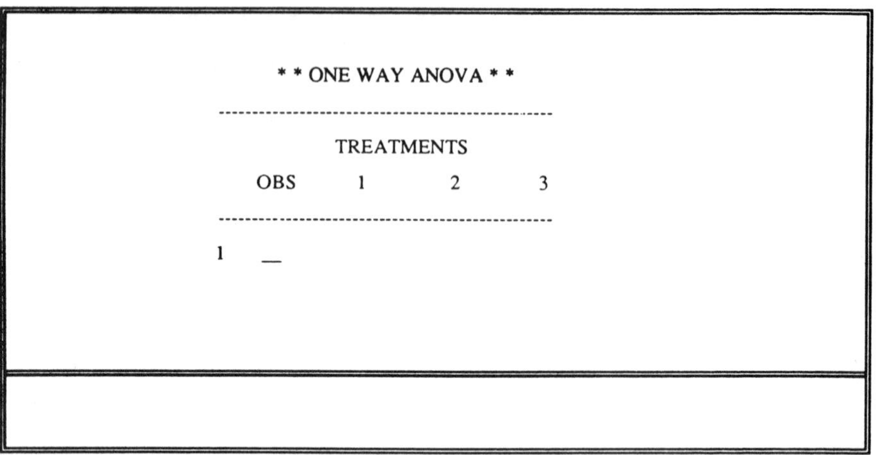

Figure A.6 Procedure for entering the input data.

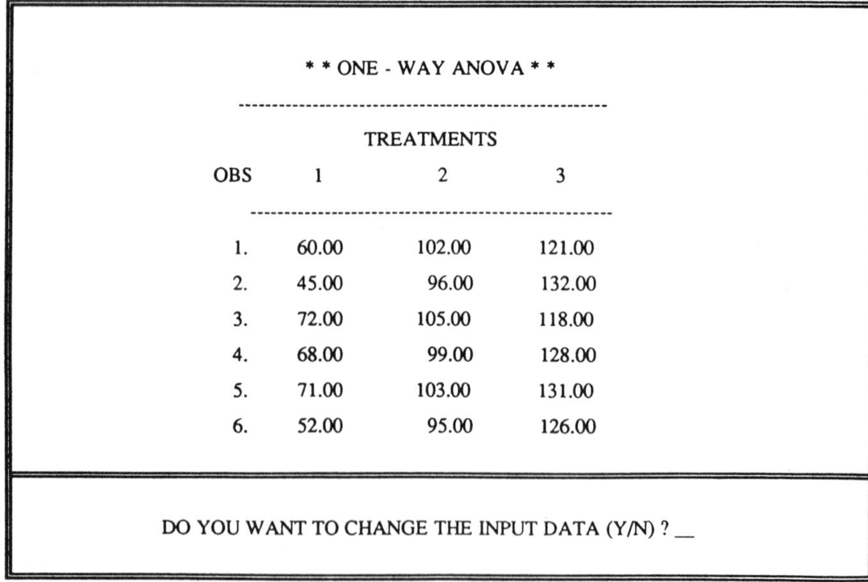

Figure A.7 Display of the input data.

Step 6. DO YOU WANT TO CHANGE THE INPUT DATA? (Y/N) N
Step 7. DO YOU WANT TO HAVE A HARD COPY OF THE INPUT DATA? (Y/N) N
Step 8. DO YOU WANT TO SAVE THE INPUT DATA? (Y/N) N

Step 9. The ANOVA table will be displayed on the screen. Refer to Figure A.8.

```
*******************************************************************************
                        AVOVA TABLE ONE FACTOR
*******************************************************************************
  SOURCE       SS        DF       MS          F       ALP    OBSALP    SIG?
  TREAT     12705.77      2     6352.88     113.26    0.05     0.00     YES
  ERROR       841.34     15       56.09
-------------------------------------------------------------------------------
  TOTAL     13547.54     17
*******************************************************************************
```

Figure A.8. Display of the ANOVA table

Step10. WOULD YOU LIKE TO HAVE A HARD COPY OF THE RESULTS? (Y/N) N

Step 11. DO YOU WANT TO CALCULATE THE COMPONENTS OF VARIANCE? (Y/N) N

Step 12. WOULD YOU LIKE TO HAVE A HARD COPY OF THE RESULTS? (Y/N) N

Step 13. SELECT ANY ONE OF THE FOLLOWING OPTIONS: Refer to Figure A.9.

```
ONE WAY ANALYSIS OF VARIANCE

****SELECT ANY ONE OF THE FOLLOWING****

RERUN WITH SAME DATA ..........................................1
RERUN WITH SAME DATA BUT NEW ALPHA ............................2
RERUN DATA FROM FILE ..........................................3
RERUN WITH NEW DATA ...........................................4
MENU ..........................................................5
QUIT ..........................................................6

PLEASE ENTER OPTION NUMBER:
```

Figure A.9. Display of second menu

Appendix B
Statistical tables

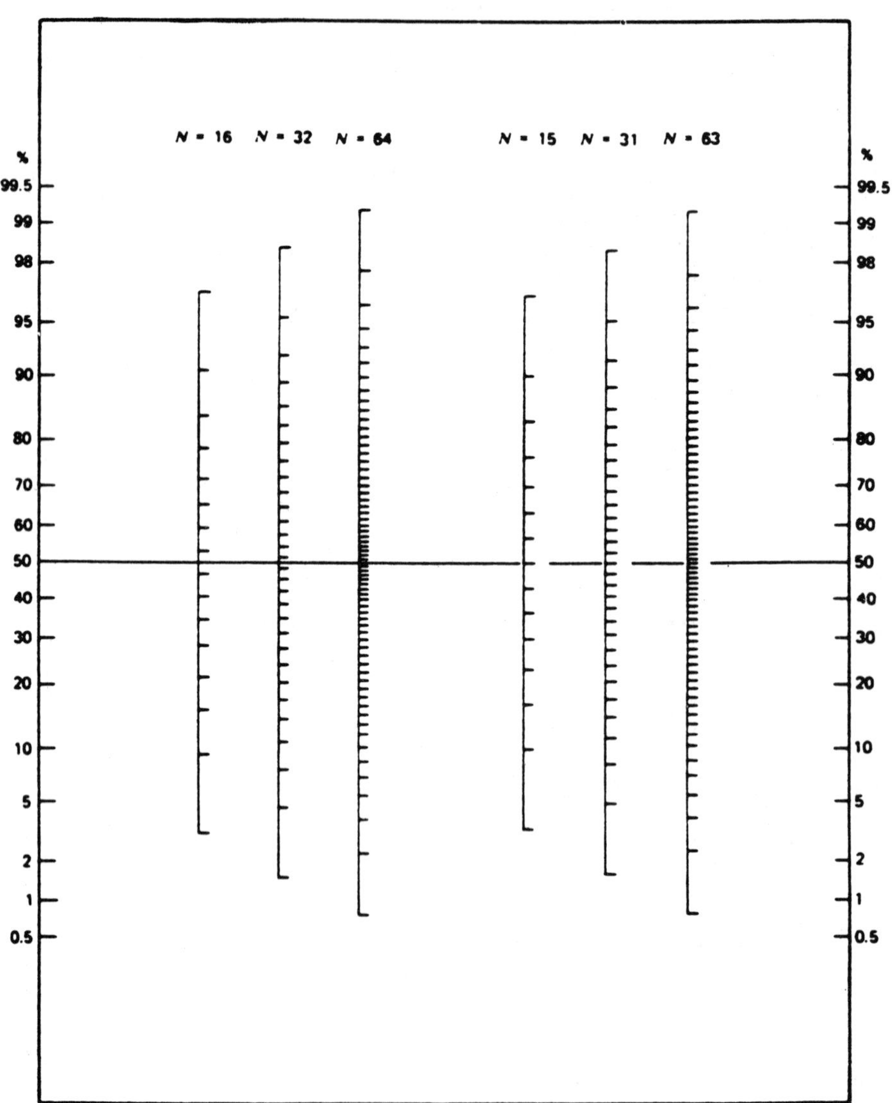

Figure B.1 Scales for normal plots.

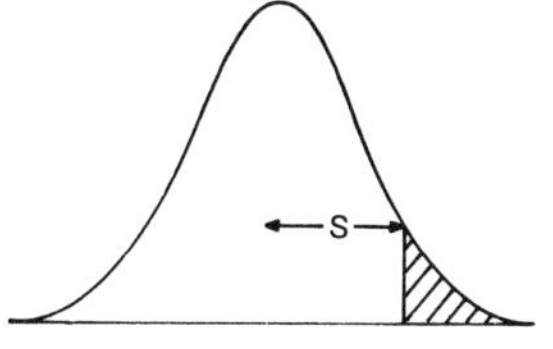

Table B.1 Tail area of unit normal distribution

z	*0.00*	*0.01*	*0.02*	*0.03*	*0.04*	*0.05*	*0.06*	*0.07*	*0.08*	*0.09*
0.0	0.5000	0.4960	0.4920	0.4880	0.4840	0.4801	0.4761	0.4721	0.4681	0.4641
0.1	0.4602	0.4562	0.4522	0.4483	0.4443	0.4404	0.4364	0.4325	0.4286	0.4247
0.2	0.4207	0.4168	0.4129	0.4090	0.4052	0.4013	0.3974	0.3936	0.3897	0.3859
0.3	0.3821	0.3783	0.3745	0.3707	0.3669	0.3632	0.3594	0.3557	0.3520	0.3483
0.4	0.3446	0.3409	0.3372	0.3336	0.3300	0.3264	0.3228	0.3192	0.3156	0.3121
0.5	0.3085	0.3050	0.3015	0.2981	0.2946	0.2912	0.2877	0.2843	0.2810	0.2776
0.6	0.2743	0.2709	0.2676	0.2643	0.2611	0.2578	0.2546	0.2514	0.2483	0.2451
0.7	0.2420	0.2389	0.2358	0.2327	0.2296	0.2266	0.2236	0.2206	0.2177	0.2148
0.8	0.2119	0.2090	0.2061	0.2033	0.2005	0.1977	0.1949	0.1922	0.1894	0.1867
0.9	0.1841	0.1814	0.1788	0.1762	0.1736	0.1711	0.1685	0.1660	0.1635	0.1611
1.0	0.1587	0.1562	0.1539	0.1515	0.1492	0.1469	0.1446	0.1423	0.1401	0.1379
1.1	0.1357	0.1335	0.1314	0.1292	0.1271	0.1251	0.1230	0.1210	0.1190	0.1170
1.2	0.1151	0.1131	0.1112	0.1093	0.1075	0.1056	0.1038	0.1020	0.1003	0.0985
1.3	0.0968	0.0951	0.0934	0.0918	0.0901	0.0885	0.0869	0.0853	0.0838	0.0823
1.4	0.0808	0.0793	0.0778	0.0764	0.0749	0.0735	0.0721	0.0708	0.0694	0.0681
1.5	0.0668	0.0655	0.0643	0.0630	0.0618	0.0606	0.0594	0.0582	0.0571	0.0559
1.6	0.0548	0.0537	0.0526	0.0516	0.0505	0.0495	0.0485	0.0475	0.0465	0.0455
1.7	0.0446	0.0436	0.0427	0.0418	0.0409	0.0401	0.0392	0.0384	0.0375	0.0367
1.8	0.0359	0.0351	0.0344	0.0336	0.0329	0.0322	0.0314	0.0307	0.0301	0.0294
1.9	0.0287	0.0281	0.0274	0.0268	0.0262	0.0256	0.0250	0.0244	0.0239	0.0233
2.0	0.0228	0.0222	0.0217	0.0212	0.0207	0.0202	0.0197	0.0192	0.0188	0.0183
2.1	0.0179	0.0174	0.0170	0.0166	0.0162	0.0158	0.0154	0.0150	0.0146	0.0143
2.2	0.0139	0.0136	0.0132	0.0129	0.0125	0.0122	0.0119	0.0116	0.0113	0.0110
2.3	0.0107	0.0104	0.0102	0.0099	0.0096	0.0094	0.0091	0.0089	0.0087	0.0084
2.4	0.0082	0.0080	0.0078	0.0075	0.0073	0.0071	0.0069	0.0068	0.0066	0.0064
2.5	0.0062	0.0060	0.0059	0.0057	0.0055	0.0054	0.0052	0.0051	0.0049	0.0048
2.6	0.0047	0.0045	0.0044	0.0043	0.0041	0.0040	0.0039	0.0038	0.0037	0.0036
2.7	0.0035	0.0034	0.0033	0.0032	0.0031	0.0030	0.0029	0.0028	0.0027	0.0026
2.8	0.0026	0.0025	0.0024	0.0023	0.0023	0.0022	0.0021	0.0021	0.0020	0.0019
2.9	0.0019	0.0018	0.0018	0.0017	0.0016	0.0016	0.0015	0.0015	0.0014	0.0014
3.0	0.0013	0.0013	0.0013	0.0012	0.0012	0.0011	0.0011	0.0011	0.0010	0.0010
3.1	0.0010	0.0009	0.0009	0.0009	0.0008	0.0008	0.0008	0.0008	0.0007	0.0007
3.2	0.0007	0.0007	0.0006	0.0006	0.0006	0.0006	0.0006	0.0005	0.0005	0.0005
3.3	0.0005	0.0005	0.0005	0.0004	0.0004	0.0004	0.0004	0.0004	0.0004	0.0003
3.4	0.0003	0.0003	0.0003	0.0003	0.0003	0.0003	0.0003	0.0003	0.0003	0.0002
3.5	0.0002	0.0002	0.0002	0.0002	0.0002	0.0002	0.0002	0.0002	0.0002	0.0002
3.6	0.0002	0.0002	0.0001	0.0001	0.0001	0.0001	0.0001	0.0001	0.0001	0.0001
3.7	0.0001	0.0001	0.0001	0.0001	0.0001	0.0001	0.0001	0.0001	0.0001	0.0001
3.8	0.0001	0.0001	0.0001	0.0001	0.0001	0.0001	0.0001	0.0001	0.0001	0.0001
3.9	0.0000	0.0000	0.0000	0.0000	0.0000	0.0000	0.0000	0.0000	0.0000	0.0000

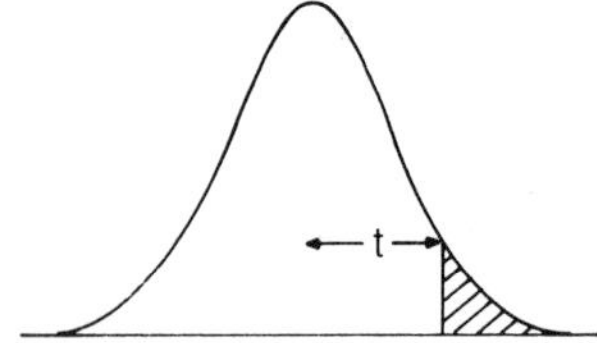

Table B2 Probability points of the t distribution with v degrees of freedom

	Tail area probability									
v	*0.4*	*0.25*	*0.1*	*0.05*	*0.025*	*0.01*	*0.005*	*0.0025*	*0.001*	*0.0005*
1	0.325	1.000	3.078	6.314	12.706	31.821	63.657	127.32	318.31	636.620
2	0.289	0.816	1.886	2.920	4.303	6.965	9.925	14.089	22.326	31.598
3	0.277	0.765	1,638	2.353	3.182	4.541	5.841	7.453	10.213	12.924
4	0.271	0.741	1.533	2.132	2.776	3.747	4.604	5.598	7.173	8.610
5	0.267	0.727	1.476	2.015	2.571	3.365	4.032	4.773	5.893	6.869
6	0.265	0.718	1.440	1.943	2.447	3.143	3.707	4.317	5.208	5.959
7	0.263	0.711	1.415	1.895	2.365	2.998	3.499	4.029	4.785	5.408
8	0.262	0.706	1.397	1.860	2.306	2.896	3.355	3.833	4.501	5.041
9	0.261	0.703	1.383	1.833	2.262	2.821	3.250	3.690	4.297	4.781
10	0.260	0.700	1.372	1.812	2.228	2.764	3.169	3.581	4.144	4.587
11	0.260	0.697	1.363	1.796	2.201	2.718	3.106	3.497	4.025	4.437
12	0.259	0.695	1.356	1.782	2.179	2.681	3.055	3.428	3.930	4.318
13	0.259	0.694	1.350	1.771	2.160	2.650	3.012	3.372	3.852	4.221
14	0.258	0.692	1.345	1.761	2.145	2.624	2.977	3.326	3.787	4.140
15	0.258	0.691	1.341	1.753	2.131	2.602	2.947	3.286	3.733	4.073
16	0.258	0.690	1.337	1.746	2.120	2.583	2.921	3.252	3.686	4.015
17	0.257	0.689	1.333	1.740	2.110	2.567	2.898	3.222	3.646	3.965
18	0.257	0.688	1.330	1.734	2.101	2.552	2.878	3.197	3.610	3.922
19	0.257	0.688	1.328	1.729	2.093	2.539	2.861	3.174	3.579	3.883
20	0.257	0.687	1.325	1.725	2.086	2.528	2.845	3.153	3.552	3.850
21	0.257	0.686	1.323	1.721	2.080	2.518	2.831	3.135	3.527	3.819
22	0.256	0.686	1.321	1.717	2.074	2.508	2.819	3.119	3.505	3.792
23	0.256	0.685	1.319	1.714	2.069	2.500	2.807	3.104	3.485	3.767
24	0.256	0.685	1.318	1.711	2.064	2.492	2.797	3.091	3.467	3.745
25	0.256	0.684	1.316	1.708	2.060	2.485	2.787	3.078	3.450	3.725
26	0.256	0.684	1.315	1.706	2.056	2.479	2.779	3.067	3.435	3.707
27	0.256	0.684	1.314	1.703	2.052	2.473	2.771	3.057	3.421	3.690
28	0.256	0.683	1.313	1.701	2.048	2.467	2.763	3.047	3.408	3.674
29	0.256	0.683	1.311	1.699	2.045	2.462	2.756	3.038	3.396	3.659
30	0.256	0.683	1.310	1.697	2.042	2.457	2.750	3.030	3.385	3.646
40	0.255	0.681	1.303	1.684	2.021	2.423	2.704	2.971	3.307	3.551
60	0.254	0.679	1.296	1.671	2.200	2.390	2.660	2.915	3.232	3.460
120	0.254	0.677	1.289	1.685	1.980	2.358	2.617	2.860	3.160	3.373
∞	0.253	0.674	1.282	1.645	1.960	2.326	2.576	2.807	3.090	3.291

Source: Taken with permission from E.S. Pearson and H.O. Hartley (eds) (1958), *Biometrika Tables for Statisticians*, Vol. 1, Cambridge University Press.

Parts of the table are also taken from Table III of Fisher and Yates: *Statistical Tables for Biological, Agricultural and Medical Research*, published by Longman Group Ltd., London (previously published by Oliver and Boyd, Edinburgh), by permission of the authors and publishers.

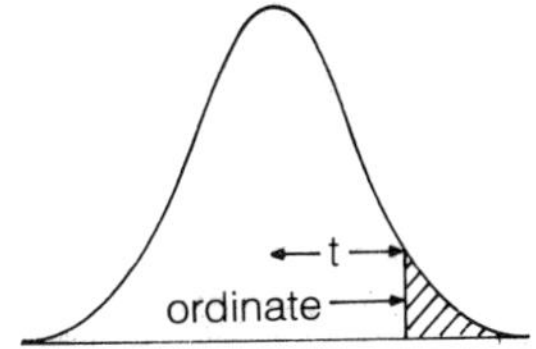

Table B3 Ordinates of t-distribution with v degrees of freedom

v	*0.00*	*0.25*	*0.50*	*0.75*	*1.00*	*1.25*	*1.50*	*1.75*	*2.00*	*2.25*	*2.50*	*2.75*	*3.00*
1	0.318	0.300	0.255	0.204	0.159	0.124	0.098	0.078	0.064	0.053	0.044	0.037	0.032
2	0.354	0.338	0.296	0.244	0.193	0.149	0.114	0.088	0.068	0.053	0.042	0.034	0.027
3	0.368	0.353	0.313	0.261	0.207	0.159	0.120	0.090	0.068	0.051	0.039	0.030	0.023
4	0.375	0.361	0.322	0.270	0.215	0.164	0.123	0.091	0.066	0.049	0.036	0.026	0.020
5	0.380	0.366	0.328	0.276	0.220	0.168	0.125	0.091	0.065	0.047	0.033	0.024	0.017
6	0.383	0.369	0.332	0.280	0.223	0.170	0.126	0.090	0.064	0.045	0.032	0.022	0.016
7	0.385	0.372	0.335	0.283	0.226	0.172	0.126	0.090	0.063	0.044	0.030	0.021	0.014
8	0.387	0.373	0.337	0.285	0.228	0.173	0.127	0.090	0.062	0.043	0.029	0.019	0.013
9	0.388	0.375	0.338	0.287	0.229	0.174	0.127	0.090	0.062	0.042	0.028	0.018	0.012
10	0.389	0.376	0.340	0.288	0.230	0.175	0.127	0.090	0.061	0.041	0.027	0.018	0.011
11	0.390	0.377	0.341	0.289	0.231	0.176	0.128	0.089	0.061	0.040	0.026	0.017	0.011
12	0.391	0.378	0.342	0.290	0.232	0.176	0.128	0.089	0.060	0.040	0.026	0.016	0.010
13	0.391	0.378	0.343	0.291	0.233	0.177	0.128	0.089	0.060	0.039	0.025	0.016	0.010
14	0.392	0.379	0.343	0.292	0.234	0.177	0.128	0.089	0.060	0.039	0.025	0.015	0.010
15	0.392	0.380	0.344	0.292	0.234	0.177	0.128	0.089	0.059	0.038	0.024	0.015	0.009
16	0.393	0.380	0.344	0.293	0.235	0.178	0.128	0.089	0.059	0.038	0.024	0.015	0.009
17	0.393	0.380	0.345	0.293	0.235	0.178	0.128	0.089	0.059	0.038	0.024	0.014	0.009
18	0.393	0.381	0.345	0.294	0.235	0.178	0.129	0.088	0.059	0.037	0.023	0.014	0.008
19	0.394	0.381	0.346	0.294	0.236	0.179	0.129	0.088	0.058	0.037	0.023	0.014	0.008
20	0.394	0.381	0.346	0.294	0.236	0.179	0.129	0.088	0.058	0.037	0.023	0.014	0.008
22	0.394	0.382	0.346	0.295	0.237	0.179	0.129	0.088	0.058	0.036	0.022	0.013	0.008
24	0.395	0.382	0.347	0.296	0.237	0.179	0.129	0.088	0.057	0.036	0.022	0.013	0.007
26	0.395	0.383	0.347	0.296	0.237	0.180	0.129	0.088	0.057	0.036	0.022	0.013	0.007
28	0.395	0.383	0.348	0.296	0.238	0.180	0.129	0.088	0.057	0.036	0.021	0.012	0.007
30	0.396	0.383	0.348	0.297	0.238	0.180	0.129	0.088	0.057	0.035	0.021	0.012	0.007
35	0.396	0.384	0.348	0.297	0.239	0.180	0.129	0.088	0.056	0.035	0.021	0.012	0.006
40	0.396	0.384	0.349	0.298	0.239	0.181	0.129	0.087	0.056	0.035	0.020	0.011	0.006
45	0.397	0.384	0.349	0.298	0.239	0.181	0.129	0.087	0.056	0.034	0.020	0.011	0.006
50	0.397	0.385	0.350	0.298	0.240	0.181	0.129	0.087	0.056	0.034	0.020	0.011	0.006
∞	0.399	0.387	0.352	0.301	0.242	0.183	0.130	0.086	0.054	0.032	0.018	0.009	0.004

(Column headings: *Value of t*)

Table B.4 Probability points of the χ^2-distribution with v degrees of freedom

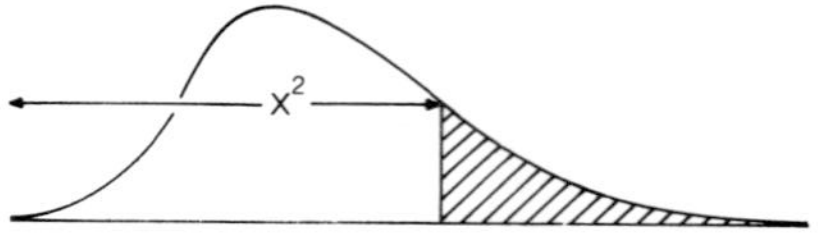

v	*Tail area probability*													
	0.995	*0.99*	*0.975*	*0.95*	*0.9*	*0.75*	*0.5*	*0.25*	*0.1*	*0.05*	*0.025*	*0.01*	*0.005*	*0.001*
1	—	—	—	—	0.016	0.102	0.455	1.32	2.71	3.84	5.02	6.63	7.88	10.8
2	0.010	0.020	0.051	0.103	0.211	0.575	1.39	2.77	4.61	5.99	7.38	9.21	10.6	13.8
3	0.072	0.115	0.216	0.352	0.584	1.21	2.37	4.11	6.25	7.81	9.35	11.3	12.8	16.3
4	0.207	0.297	0.484	0.711	1.06	1.92	3.36	5.39	7.78	9.49	11.1	13.3	14.9	18.5
5	0.412	0.554	0.831	1.15	1.61	2.67	4.35	6.63	9.24	11.1	12.8	15.1	16.7	20.5
6	0.676	0.872	1.24	1.64	2.20	3.45	5.35	7.84	10.6	12.6	14.4	16.8	18.5	22.5
7	0.989	1.24	1.69	2.17	2.83	4.25	6.35	9.04	12.0	14.1	16.0	18.5	20.3	24.3
8	1.34	1.65	2.18	2.73	3.49	5.07	7.34	10.2	13.4	15.5	17.5	20.1	22.0	26.1
9	1.73	2.09	2.70	3.33	4.17	5.90	8.34	11.4	14.7	16.9	19.0	21.7	23.6	27.9
10	2.16	2.56	3.25	3.94	4.87	6.74	9.34	12.5	16.0	18.3	20.5	23.2	25.2	29.6
11	2.60	3.05	3.82	4.57	5.58	7.58	10.3	13.7	17.3	19.7	21.9	24.7	26.8	31.3
12	3.07	3.57	4.40	5.23	6.30	8.44	11.3	14.8	18.5	21.0	23.3	26.2	28.3	32.9
13	3.57	4.11	5.01	5.89	7.04	9.30	12.3	16.0	19.8	22.4	24.7	27.7	29.8	34.5
14	4.07	4.66	5.63	6.57	7.79	10.2	13.3	17.1	21.1	23.7	26.1	29.1	31.3	36.1
15	4.60	5.23	6.26	7.26	8.55	11.0	14.3	18.2	22.3	25.0	27.5	30.6	32.8	37.7
16	5.14	5.81	6.91	7.96	9.31	11.9	15.3	19.4	23.5	26.3	28.8	32.0	34.3	39.3
17	5.70	6.41	7.56	8.67	10.1	12.8	16.3	20.5	24.8	27.6	30.2	33.4	35.7	40.8
18	6.26	7.01	8.23	9.39	10.9	13.7	17.3	21.6	26.0	28.9	31.5	34.8	37.2	42.3
19	6.84	7.63	8.91	10.1	11.7	14.6	18.3	22.7	27.2	30.1	32.9	36.2	38.6	43.8
20	7.43	8.26	9.59	10.9	12.4	15.5	19.3	23.8	28.4	31.4	34.2	37.6	40.0	45.3
21	8.03	8.90	10.3	11.6	13.2	16.3	20.3	24.9	29.6	32.7	35.5	38.9	41.4	46.8
22	8.64	9.54	11.0	12.3	14.0	17.2	21.3	26.0	30.8	33.9	36.8	40.3	42.8	48.3
23	9.26	10.2	11.7	13.1	14.8	18.1	22.3	27.1	32.0	35.2	38.1	41.6	44.2	49.7
24	9.89	10.9	12.4	13.8	15.7	19.0	23.3	28.2	33.2	36.4	39.4	43.0	45.6	51.2
25	10.5	11.5	13.1	14.6	16.5	19.9	24.3	29.3	34.4	37.7	40.6	44.3	46.9	52.6
26	11.2	12.2	13.8	15.4	17.3	20.8	25.3	30.4	35.6	38.9	41.9	45.6	48.3	54.1
27	11.8	12.9	14.6	16.2	18.1	21.7	26.3	31.5	36.7	40.1	43.2	47.0	49.6	55.5
28	12.5	13.6	15.3	16.9	18.9	22.7	27.3	32.6	37.9	41.3	44.5	48.3	51.0	56.9
29	13.1	14.3	16.0	17.7	19.8	23.6	28.3	33.7	39.1	42.6	45.7	49.6	52.3	58.3
30	13.8	15.0	16.8	18.5	20.6	24.5	29.3	34.8	40.3	43.8	47.0	50.9	53.7	59.7

Source: Adapted from Table 8, Percentage points of the χ^2 distribution, in E. S. Pearson and H. O. Hartley (ed) (1966), *Biometrika Tables for Statisticians*, Vol. 1, 3rd edn, Cambridge University Press. Used by permission.

Table B.5 Percentage points of the F-distribution: upper 25% points

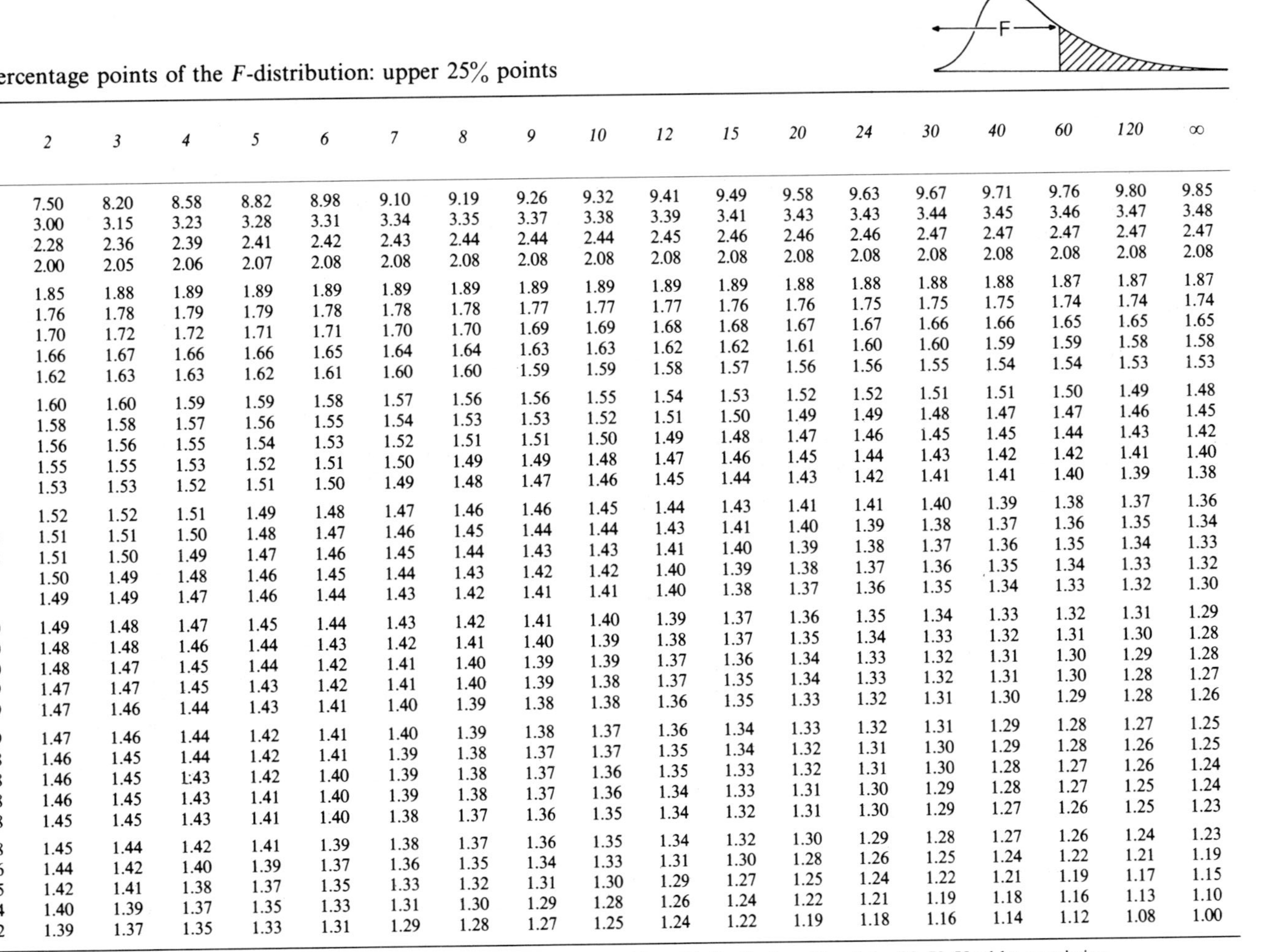

ν_2 \ ν_1	1	2	3	4	5	6	7	8	9	10	12	15	20	24	30	40	60	120	∞
1	5.83	7.50	8.20	8.58	8.82	8.98	9.10	9.19	9.26	9.32	9.41	9.49	9.58	9.63	9.67	9.71	9.76	9.80	9.85
2	2.57	3.00	3.15	3.23	3.28	3.31	3.34	3.35	3.37	3.38	3.39	3.41	3.43	3.43	3.44	3.45	3.46	3.47	3.48
3	2.02	2.28	2.36	2.39	2.41	2.42	2.43	2.44	2.44	2.44	2.45	2.46	2.46	2.46	2.47	2.47	2.47	2.47	2.47
4	1.81	2.00	2.05	2.06	2.07	2.08	2.08	2.08	2.08	2.08	2.08	2.08	2.08	2.08	2.08	2.08	2.08	2.08	2.08
5	1.69	1.85	1.88	1.89	1.89	1.89	1.89	1.89	1.89	1.89	1.89	1.89	1.88	1.88	1.88	1.88	1.87	1.87	1.87
6	1.62	1.76	1.78	1.79	1.79	1.78	1.78	1.78	1.77	1.77	1.77	1.76	1.76	1.75	1.75	1.75	1.74	1.74	1.74
7	1.57	1.70	1.72	1.72	1.71	1.71	1.70	1.70	1.69	1.69	1.68	1.68	1.67	1.67	1.66	1.66	1.65	1.65	1.65
8	1.54	1.66	1.67	1.66	1.66	1.65	1.64	1.64	1.63	1.63	1.62	1.62	1.61	1.60	1.60	1.59	1.59	1.58	1.58
9	1.51	1.62	1.63	1.63	1.62	1.61	1.60	1.60	1.59	1.59	1.58	1.57	1.56	1.56	1.55	1.54	1.54	1.53	1.53
10	1.49	1.60	1.60	1.59	1.59	1.58	1.57	1.56	1.56	1.55	1.54	1.53	1.52	1.52	1.51	1.51	1.50	1.49	1.48
11	1.47	1.58	1.58	1.57	1.56	1.55	1.54	1.53	1.53	1.52	1.51	1.50	1.49	1.49	1.48	1.47	1.47	1.46	1.45
12	1.46	1.56	1.56	1.55	1.54	1.53	1.52	1.51	1.51	1.50	1.49	1.48	1.47	1.46	1.45	1.45	1.44	1.43	1.42
13	1.45	1.55	1.55	1.53	1.52	1.51	1.50	1.49	1.49	1.48	1.47	1.46	1.45	1.44	1.43	1.42	1.42	1.41	1.40
14	1.44	1.53	1.53	1.52	1.51	1.50	1.49	1.48	1.47	1.46	1.45	1.44	1.43	1.42	1.41	1.41	1.40	1.39	1.38
15	1.43	1.52	1.52	1.51	1.49	1.48	1.47	1.46	1.46	1.45	1.44	1.43	1.41	1.41	1.40	1.39	1.38	1.37	1.36
16	1.42	1.51	1.51	1.50	1.48	1.47	1.46	1.45	1.44	1.44	1.43	1.41	1.40	1.39	1.38	1.37	1.36	1.35	1.34
17	1.42	1.51	1.50	1.49	1.47	1.46	1.45	1.44	1.43	1.43	1.41	1.40	1.39	1.38	1.37	1.36	1.35	1.34	1.33
18	1.41	1.50	1.49	1.48	1.46	1.45	1.44	1.43	1.42	1.42	1.40	1.39	1.38	1.37	1.36	1.35	1.34	1.33	1.32
19	1.41	1.49	1.49	1.47	1.46	1.44	1.43	1.42	1.41	1.41	1.40	1.38	1.37	1.36	1.35	1.34	1.33	1.32	1.30
20	1.40	1.49	1.48	1.47	1.45	1.44	1.43	1.42	1.41	1.40	1.39	1.37	1.36	1.35	1.34	1.33	1.32	1.31	1.29
21	1.40	1.48	1.48	1.46	1.44	1.43	1.42	1.41	1.40	1.39	1.38	1.37	1.35	1.34	1.33	1.32	1.31	1.30	1.28
22	1.40	1.48	1.47	1.45	1.44	1.42	1.41	1.40	1.39	1.39	1.37	1.36	1.34	1.33	1.32	1.31	1.30	1.29	1.28
23	1.39	1.47	1.47	1.45	1.43	1.42	1.41	1.40	1.39	1.38	1.37	1.35	1.34	1.33	1.32	1.31	1.30	1.28	1.27
24	1.39	1.47	1.46	1.44	1.43	1.41	1.40	1.39	1.38	1.38	1.36	1.35	1.33	1.32	1.31	1.30	1.29	1.28	1.26
25	1.39	1.47	1.46	1.44	1.42	1.41	1.40	1.39	1.38	1.37	1.36	1.34	1.33	1.32	1.31	1.29	1.28	1.27	1.25
26	1.38	1.46	1.45	1.44	1.42	1.41	1.39	1.38	1.37	1.37	1.35	1.34	1.32	1.31	1.30	1.29	1.28	1.26	1.25
27	1.38	1.46	1.45	1.43	1.42	1.40	1.39	1.38	1.37	1.36	1.35	1.33	1.32	1.31	1.30	1.28	1.27	1.26	1.24
28	1.38	1.46	1.45	1.43	1.41	1.40	1.39	1.38	1.37	1.36	1.34	1.33	1.31	1.30	1.29	1.28	1.27	1.25	1.24
29	1.38	1.45	1.45	1.43	1.41	1.40	1.38	1.37	1.36	1.35	1.34	1.32	1.31	1.30	1.29	1.27	1.26	1.25	1.23
30	1.38	1.45	1.44	1.42	1.41	1.39	1.38	1.37	1.36	1.35	1.34	1.32	1.30	1.29	1.28	1.27	1.26	1.24	1.23
40	1.36	1.44	1.42	1.40	1.39	1.37	1.36	1.35	1.34	1.33	1.31	1.30	1.28	1.26	1.25	1.24	1.22	1.21	1.19
60	1.35	1.42	1.41	1.38	1.37	1.35	1.33	1.32	1.31	1.30	1.29	1.27	1.25	1.24	1.22	1.21	1.19	1.17	1.15
120	1.34	1.40	1.39	1.37	1.35	1.33	1.31	1.30	1.29	1.28	1.26	1.24	1.22	1.21	1.19	1.18	1.16	1.13	1.10
∞	1.32	1.39	1.37	1.35	1.33	1.31	1.29	1.28	1.27	1.25	1.24	1.22	1.19	1.18	1.16	1.14	1.12	1.08	1.00

Source: M. Merrington and C. M. Thompson (1943), Tables of percentage points of the inverted beta (F) distribution, *Biometrika*, **33**, 73. Used by permission.

Table B.5 (*continued*), Percentage points of the *F*-distribution: upper 10% points

ν_2 \ ν_1	1	2	3	4	5	6	7	8	9	10	12	15	20	24	30	40	60	120	∞
1	39.86	49.50	53.59	55.83	57.24	58.20	58.91	59.44	59.86	60.19	60.71	61.22	61.74	62.00	62.26	62.53	62.79	63.06	63.33
2	8.53	9.00	9.16	9.24	9.29	9.33	9.35	9.37	9.38	9.39	9.41	9.42	9.44	9.45	9.46	9.47	9.47	9.48	9.49
3	5.54	5.46	5.39	5.34	5.31	5.28	5.27	5.25	5.24	5.23	5.22	5.20	5.18	5.18	5.17	5.16	5.15	5.14	5.13
4	4.54	4.32	4.19	4.11	4.05	4.01	3.98	3.95	3.94	3.92	3.90	3.87	3.84	3.83	3.82	3.80	3.79	3.78	3.76
5	4.06	3.78	3.62	3.52	3.45	3.40	3.37	3.34	3.32	3.30	3.27	3.24	3.21	3.19	3.17	3.16	3.14	3.12	3.10
6	3.78	3.46	3.29	3.18	3.11	3.05	3.01	2.98	2.96	2.94	2.90	2.87	2.84	2.82	2.80	2.78	2.76	2.74	2.72
7	3.59	3.26	3.07	2.96	2.88	2.83	2.78	2.75	2.72	2.70	2.67	2.63	2.59	2.58	2.56	2.54	2.51	2.49	2.47
8	3.46	3.11	2.92	2.81	2.73	2.67	2.62	2.59	2.56	2.54	2.50	2.46	2.42	2.40	2.38	2.36	2.34	2.32	2.29
9	3.36	3.01	2.81	2.69	2.61	2.55	2.51	2.47	2.44	2.42	2.38	2.34	2.30	2.28	2.25	2.23	2.21	2.18	2.16
10	3.29	2.92	2.73	2.61	2.52	2.46	2.41	2.38	2.35	2.32	2.28	2.24	2.20	2.18	2.16	2.13	2.11	2.08	2.06
11	3.23	2.86	2.66	2.54	2.45	2.39	2.34	2.30	2.27	2.25	2.21	2.17	2.12	2.10	2.08	2.05	2.03	2.00	1.97
12	3.18	2.81	2.61	2.48	2.39	2.33	2.28	2.24	2.21	2.19	2.15	2.10	2.06	2.04	2.01	1.99	1.96	1.93	1.90
13	3.14	2.76	2.56	2.43	2.35	2.28	2.23	2.20	2.16	2.14	2.10	2.05	2.01	1.98	1.96	1.93	1.90	1.88	1.85
14	3.10	2.73	2.52	2.39	2.31	2.24	2.19	2.15	2.12	2.10	2.05	2.01	1.96	1.94	1.91	1.89	1.86	1.83	1.80
15	3.07	2.70	2.49	2.36	2.27	2.21	2.16	2.12	2.09	2.06	2.02	1.97	1.92	1.90	1.87	1.85	1.82	1.79	1.76
16	3.05	2.67	2.46	2.33	2.24	2.18	2.13	2.09	2.06	2.03	1.99	1.94	1.89	1.87	1.84	1.81	1.78	1.75	1.72
17	3.03	2.64	2.44	2.31	2.22	2.15	2.10	2.06	2.03	2.00	1.96	1.91	1.86	1.84	1.81	1.78	1.75	1.72	1.69
18	3.01	2.62	2.42	2.29	2.20	2.13	2.08	2.04	2.00	1.98	1.93	1.89	1.84	1.81	1.78	1.75	1.72	1.69	1.66
19	2.99	2.61	2.40	2.27	2.18	2.11	2.06	2.02	1.98	1.96	1.91	1.86	1.81	1.79	1.76	1.73	1.70	1.67	1.63
20	2.97	2.59	2.38	2.25	2.16	2.09	2.04	2.00	1.96	1.94	1.89	1.84	1.79	1.77	1.74	1.71	1.68	1.64	1.61
21	2.96	2.57	2.36	2.23	2.14	2.08	2.02	1.98	1.95	1.92	1.87	1.83	1.78	1.75	1.72	1.69	1.66	1.62	1.59
22	2.95	2.56	2.35	2.22	2.13	2.06	2.01	1.97	1.93	1.90	1.86	1.81	1.76	1.73	1.70	1.67	1.64	1.60	1.57
23	2.94	2.55	2.34	2.21	2.11	2.05	1.99	1.95	1.92	1.89	1.84	1.80	1.74	1.72	1.69	1.66	1.62	1.59	1.55
24	2.93	2.54	2.33	2.19	2.10	2.04	1.98	1.94	1.91	1.88	1.83	1.78	1.73	1.70	1.67	1.64	1.61	1.57	1.53
25	2.92	2.53	2.32	2.18	2.09	2.02	1.97	1.93	1.89	1.87	1.82	1.77	1.72	1.69	1.66	1.63	1.59	1.56	1.52
26	2.91	2.52	2.31	2.17	2.08	2.01	1.96	1.92	1.88	1.86	1.81	1.76	1.71	1.68	1.65	1.61	1.58	1.54	1.50
27	2.90	2.51	2.30	2.17	2.07	2.00	1.95	1.91	1.87	1.85	1.80	1.75	1.70	1.67	1.64	1.60	1.57	1.53	1.49
28	2.89	2.50	2.29	2.16	2.06	2.00	1.94	1.90	1.87	1.84	1.79	1.74	1.69	1.66	1.63	1.59	1.56	1.52	1.48
29	2.89	2.50	2.28	2.15	2.06	1.99	1.93	1.89	1.86	1.83	1.78	1.73	1.68	1.65	1.62	1.58	1.55	1.51	1.47
30	2.88	2.49	2.28	2.14	2.05	1.98	1.93	1.88	1.85	1.82	1.77	1.72	1.67	1.64	1.61	1.57	1.54	1.50	1.46
40	2.84	2.44	2.23	2.09	2.00	1.93	1.87	1.83	1.79	1.76	1.71	1.66	1.61	1.57	1.54	1.51	1.47	1.42	1.38
60	2.79	2.39	2.18	2.04	1.95	1.87	1.82	1.77	1.74	1.71	1.66	1.60	1.54	1.51	1.48	1.44	1.40	1.35	1.29
120	2.75	2.35	2.13	1.99	1.90	1.82	1.77	1.72	1.68	1.65	1.60	1.55	1.48	1.45	1.41	1.37	1.32	1.26	1.19
∞	2.71	2.30	2.08	1.94	1.85	1.77	1.72	1.67	1.63	1.60	1.55	1.49	1.42	1.38	1.34	1.30	1.24	1.17	1.00

Source: M. Merrington and C. M. Thompson (1943), Tables of percentage points of the inverted beta (*F*) distribution, *Biometrika*, **33**, 73. Used by permission.

Table B.5 (*continued*), Percentage points of the F-distribution: upper 5% points

ν_1 / ν_1	*1*	*2*	*3*	*4*	*5*	*6*	*7*	*8*	*9*	*10*	*12*	*15*	*20*	*24*	*30*	*40*	*60*	*120*	∞
1	161.4	199.5	215.7	224.6	230.2	234.0	236.8	238.9	240.5	241.9	243.9	245.9	248.0	249.1	250.1	251.1	252.2	253.3	254.3
2	18.51	19.00	19.16	19.25	19.30	19.33	19.35	19.37	19.38	19.40	19.41	19.43	19.45	19.45	19.46	19.47	19.48	19.49	19.50
3	10.13	9.55	9.28	9.12	9.01	8.94	8.89	8.85	8.81	8.79	8.74	8.70	8.66	8.64	8.62	8.59	8.57	8.55	8.53
4	7.71	6.94	6.59	6.39	6.26	6.16	6.09	6.04	6.00	5.96	5.91	5.86	5.80	5.77	5.75	5.72	5.69	5.66	5.63
5	6.61	5.79	5.41	5.19	5.05	4.95	4.88	4.82	4.77	4.74	4.68	4.62	4.56	4.53	4.50	4.46	4.43	4.40	4.36
6	5.99	5.14	4.76	4.53	4.39	4.28	4.21	4.15	4.10	4.06	4.00	3.94	3.87	3.84	3.81	3.77	3.74	3.70	3.67
7	5.59	4.74	4.35	4.12	3.97	3.87	3.79	3.73	3.68	3.64	3.57	3.51	3.44	3.41	3.38	3.34	3.30	3.27	3.23
8	5.32	4.46	4.07	3.84	3.69	3.58	3.50	3.44	3.39	3.35	3.28	3.22	3.15	3.12	3.08	3.04	3.01	2.97	2.93
9	5.12	4.26	3.86	3.63	3.48	3.37	3.29	3.23	3.18	3.14	3.07	3.01	2.94	2.90	2.86	2.83	2.79	2.75	2.71
10	4.96	4.10	3.71	3.48	3.33	3.22	3.14	3.07	3.02	2.98	2.91	2.85	2.77	2.74	2.70	2.66	2.62	2.58	2.54
11	4.84	3.98	3.59	3.36	3.20	3.09	3.01	2.95	2.90	2.85	2.79	2.72	2.65	2.61	2.57	2.53	2.49	2.45	2.40
12	4.75	3.89	3.49	3.26	3.11	3.00	2.91	2.85	2.80	2.75	2.69	2.62	2.54	2.51	2.47	2.43	2.38	2.34	2.30
13	4.67	3.81	3.41	3.18	3.03	2.92	2.83	2.77	2.71	2.67	2.60	2.53	2.46	2.42	2.38	2.34	2.30	2.25	2.21
14	4.60	3.74	3.34	3.11	2.96	2.85	2.76	2.70	2.65	2.60	2.53	2.46	2.39	2.35	2.31	2.27	2.22	2.18	2.13
15	4.54	3.68	3.29	3.06	2.90	2.79	2.71	2.64	2.59	2.54	2.48	2.40	2.33	2.29	2.25	2.20	2.16	2.11	2.07
16	4.49	3.63	3.24	3.01	2.85	2.74	2.66	2.59	2.54	2.49	2.42	2.35	2.28	2.24	2.19	2.15	2.11	2.06	2.01
17	4.45	3.59	3.20	2.96	2.81	2.70	2.61	2.55	2.49	2.45	2.38	2.31	2.23	2.19	2.15	2.10	2.06	2.01	1.96
18	4.41	3.55	3.16	2.93	2.77	2.66	2.58	2.51	2.46	2.41	2.34	2.27	2.19	2.15	2.11	2.06	2.02	1.97	1.92
19	4.38	3.52	3.13	2.90	2.74	2.63	2.54	2.48	2.42	2.38	2.31	2.23	2.16	2.11	2.07	2.03	1.98	1.93	1.88
20	4.35	3.49	3.10	2.87	2.71	2.60	2.51	2.45	2.39	2.35	2.28	2.20	2.12	2.08	2.04	1.99	1.95	1.90	1.84
21	4.32	3.47	3.07	2.84	2.68	2.57	2.49	2.42	2.37	2.32	2.25	2.18	2.10	2.05	2.01	1.96	1.92	1.87	1.81
22	4.30	3.44	3.05	2.82	2.66	2.55	2.46	2.40	2.34	2.30	2.23	2.15	2.07	2.03	1.98	1.94	1.89	1.84	1.78
23	4.28	3.42	3.03	2.80	2.64	2.53	2.44	2.37	2.32	2.27	2.20	2.13	2.05	2.01	1.96	1.91	1.86	1.81	1.76
24	4.26	3.40	3.01	2.78	2.62	2.51	2.42	2.36	2.30	2.25	2.18	2.11	2.03	1.98	1.94	1.89	1.84	1.79	1.73
25	4.24	3.39	2.99	2.76	2.60	2.49	2.40	2.34	2.28	2.24	2.16	2.09	2.01	1.96	1.92	1.87	1.82	1.77	1.71
26	4.23	3.37	2.98	2.74	2.59	2.47	2.39	2.32	2.27	2.22	2.15	2.07	1.99	1.95	1.90	1.85	1.80	1.75	1.69
27	4.21	3.35	2.96	2.73	2.57	2.46	2.37	2.31	2.25	2.20	2.13	2.06	1.97	1.93	1.88	1.84	1.79	1.73	1.67
28	4.20	3.34	2.95	2.71	2.56	2.45	2.36	2.29	2.24	2.19	2.12	2.04	1.96	1.91	1.87	1.82	1.77	1.71	1.65
29	4.18	3.33	2.93	2.70	2.55	2.43	2.35	2.28	2.22	2.18	2.10	2.03	1.94	1.90	1.85	1.81	1.75	1.70	1.64
30	4.17	3.32	2.92	2.69	2.53	2.42	2.33	2.27	2.21	2.16	2.09	2.01	1.93	1.89	1.84	1.79	1.74	1.68	1.62
40	4.08	3.23	2.84	2.61	2.45	2.34	2.25	2.18	2.12	2.08	2.00	1.92	1.84	1.79	1.74	1.69	1.64	1.58	1.51
60	4.00	3.15	2.76	2.53	2.37	2.25	2.17	2.10	2.04	1.99	1.92	1.84	1.75	1.70	1.65	1.59	1.53	1.47	1.39
120	3.92	3.07	2.68	2.45	2.29	2.17	2.09	2.02	1.96	1.91	1.83	1.75	1.66	1.61	1.55	1.50	1.43	1.35	1.25
∞	3.84	3.00	2.60	2.37	2.21	2.10	2.01	1.94	1.88	1.83	1.75	1.67	1.57	1.52	1.46	1.39	1.32	1.22	1.00

Table B.5 (*continued*), Percentage points of the F-distribution: upper 1% points

ν_2 \ ν_1	1	2	3	4	5	6	7	8	9	10	12	15	20	24	30	40	60	120	∞
1	4052	4999.50	5403	5625	5764	5859	5928	5982	6022	6056	6106	6157	6209	6235	6261	6287	6313	6339	6366
2	98.50	99.00	99.17	99.25	99.30	99.33	99.36	99.37	99.39	99.40	99.42	99.43	99.45	99.46	99.47	99.47	99.48	99.49	99.50
3	34.12	30.82	29.46	28.71	28.24	27.91	27.67	27.49	27.35	27.23	27.05	26.87	26.69	26.60	26.50	26.41	26.32	26.22	26.13
4	21.20	18.00	16.69	15.98	15.52	15.21	14.98	14.80	14.66	14.55	14.37	14.20	14.02	13.93	13.84	13.75	13.65	13.56	13.46
5	16.26	13.27	12.06	11.39	10.97	10.67	10.46	10.29	10.16	10.05	9.89	9.72	9.55	9.47	9.38	9.29	9.20	9.11	9.02
6	13.75	10.92	9.78	9.15	8.75	8.47	8.26	8.10	7.98	7.87	7.72	7.56	7.40	7.31	7.23	7.14	7.06	6.97	6.88
7	12.25	9.55	8.45	7.85	7.46	7.19	6.99	6.84	6.72	6.62	6.47	6.31	6.16	6.07	5.99	5.91	5.82	5.74	5.65
8	11.26	8.65	7.59	7.01	6.63	6.37	6.18	6.03	5.91	5.81	5.67	5.52	5.36	5.28	5.20	5.12	5.03	4.95	4.86
9	10.56	8.02	6.99	6.42	6.06	5.80	5.61	5.47	5.35	5.26	5.11	4.96	4.81	4.73	4.65	4.57	4.48	4.40	4.31
10	10.04	7.56	6.55	5.99	5.64	5.39	5.20	5.06	4.94	4.85	4.71	4.56	4.41	4.33	4.25	4.17	4.08	4.00	3.91
11	9.65	7.21	6.22	5.67	5.32	5.07	4.89	4.74	4.63	4.54	4.40	4.25	4.10	4.02	3.94	3.86	3.78	3.69	3.60
12	9.33	6.93	5.95	5.41	5.06	4.82	4.64	4.50	4.39	4.30	4.16	4.01	3.86	3.78	3.70	3.62	3.54	3.45	3.36
13	9.07	6.70	5.74	5.21	4.86	4.62	4.44	4.30	4.19	4.10	3.96	3.82	3.66	3.59	3.51	3.43	3.34	3.25	3.17
14	8.86	6.51	5.56	5.04	4.69	4.46	4.28	4.14	4.03	3.94	3.80	3.66	3.51	3.43	3.35	3.27	3.18	3.09	3.00
15	8.68	6.36	5.42	4.89	4.56	4.32	4.14	4.00	3.89	3.80	3.67	3.52	3.37	3.29	3.21	3.13	3.05	2.96	2.87
16	8.53	6.23	5.29	4.77	4.44	4.20	4.03	3.89	3.78	3.69	3.55	3.41	3.26	3.18	3.10	3.02	2.93	2.84	2.75
17	8.40	6.11	5.18	4.67	4.34	4.10	3.93	3.79	3.68	3.59	3.46	3.31	3.16	3.08	3.00	2.92	2.83	2.75	2.65
18	8.29	6.01	5.09	4.58	4.25	4.01	3.84	3.71	3.60	3.51	3.37	3.23	3.08	3.00	2.92	2.84	2.75	2.66	2.57
19	8.18	5.93	5.01	4.50	4.17	3.94	3.77	3.63	3.52	3.43	3.30	3.15	3.00	2.92	2.84	2.76	2.67	2.58	2.49
20	8.10	5.85	4.94	4.43	4.10	3.87	3.70	3.56	3.46	3.37	3.23	3.09	2.94	2.86	2.78	2.69	2.61	2.52	2.42
21	8.02	5.78	4.87	4.37	4.04	3.81	3.64	3.51	3.40	3.31	3.17	3.03	2.88	2.80	2.72	2.64	2.55	2.46	2.36
22	7.95	5.72	4.82	4.31	3.99	3.76	3.59	3.45	3.35	3.26	3.12	2.98	2.83	2.75	2.67	2.58	2.50	2.40	2.31
23	7.88	5.66	4.76	4.26	3.94	3.71	3.54	3.41	3.30	3.21	3.07	2.93	2.78	2.70	2.62	2.54	2.45	2.35	2.26
24	7.82	5.61	4.72	4.22	3.90	3.67	3.50	3.36	3.26	3.17	3.03	2.89	2.74	2.66	2.58	2.49	2.40	2.31	2.21
25	7.77	5.57	4.68	4.18	3.85	3.63	3.46	3.32	3.22	3.13	2.99	2.85	2.70	2.62	2.54	2.45	2.36	2.27	2.17
26	7.72	5.53	4.64	4.14	3.82	3.59	3.42	3.29	3.18	3.09	2.96	2.81	2.66	2.58	2.50	2.42	2.33	2.23	2.13
27	7.68	5.49	4.60	4.11	3.78	3.56	3.39	3.26	3.15	3.06	2.93	2.78	2.63	2.55	2.47	2.38	2.29	2.20	2.10
28	7.64	5.45	4.57	4.07	3.75	3.53	3.36	3.23	3.12	3.03	2.90	2.75	2.60	2.52	2.44	2.35	2.26	2.17	2.06
29	7.60	5.42	4.54	4.04	3.73	3.50	3.33	3.20	3.09	3.00	2.87	2.73	2.57	2.49	2.41	2.33	2.23	2.14	2.03
30	7.56	5.39	4.51	4.02	3.70	3.47	3.30	3.17	3.07	2.98	2.84	2.70	2.55	2.47	2.39	2.30	2.21	2.11	2.01
40	7.31	5.18	4.31	3.83	3.51	3.29	3.12	2.99	2.89	2.80	2.66	2.52	2.37	2.29	2.20	2.11	2.02	1.92	1.80
60	7.08	4.98	4.13	3.65	3.34	3.12	2.95	2.82	2.72	2.63	2.50	2.35	2.20	2.12	2.03	1.94	1.84	1.73	1.60
120	6.85	4.79	3.95	3.48	3.17	2.96	2.79	2.66	2.56	2.47	2.34	2.19	2.03	1.95	1.86	1.76	1.66	1.53	1.38
∞	6.63	4.61	3.78	3.32	3.02	2.80	2.64	2.51	2.41	2.32	2.18	2.04	1.88	1.79	1.70	1.59	1.47	1.32	1.00

Table B.5 (*continued*). Percentage points of the F-distribution: upper 0.1% points

ν_2 \ ν_1	1	2	3	4	5	6	7	8	9	10	12	15	20	24	30	40	60	120	∞
1	4053*	5000*	5404*	5625*	5764*	5859*	5929*	5981*	6023*	6056*	6107*	6158*	6209*	6235*	6261*	6287*	6313*	6340*	6366*
2	998.5	999.0	999.2	999.2	999.3	999.3	999.4	999.4	999.4	999.4	999.4	999.4	999.4	999.5	999.5	999.5	999.5	999.5	999.5
3	167.0	148.5	141.1	137.1	134.6	132.8	131.6	130.6	129.9	129.2	128.3	127.4	126.4	125.9	125.4	125.0	124.5	124.0	123.5
4	74.14	61.25	56.18	53.44	51.71	50.53	49.66	49.00	48.47	48.05	47.41	46.76	46.10	45.77	45.43	45.09	44.75	44.40	44.05
5	47.18	37.12	33.20	31.09	29.75	28.84	28.16	27.64	27.24	26.92	26.42	25.91	25.39	25.14	24.87	24.60	24.33	24.06	23.79
6	35.51	27.00	23.70	21.92	20.81	20.03	19.46	19.03	18.69	18.41	17.99	17.56	17.12	16.89	16.67	16.44	16.21	15.99	15.75
7	29.25	21.69	18.77	17.19	16.21	15.52	15.02	14.63	14.33	14.08	13.71	13.32	12.93	12.73	12.53	12.33	12.12	11.91	11.70
8	25.42	18.49	15.83	14.39	13.49	12.86	12.40	12.04	11.77	11.54	11.19	10.84	10.48	10.30	10.11	9.92	9.73	9.53	9.33
9	22.86	16.39	13.90	12.56	11.71	11.13	10.70	10.37	10.11	9.89	9.57	9.24	8.90	8.72	8.55	8.37	8.19	8.00	7.81
10	21.04	14.91	12.55	11.28	10.48	9.92	9.52	9.20	8.96	8.75	8.45	8.13	7.80	7.64	7.47	7.30	7.12	6.94	6.76
11	19.69	13.81	11.56	10.35	9.58	9.05	8.66	8.35	8.12	7.92	7.63	7.32	7.01	6.85	6.68	6.52	6.35	6.17	6.00
12	18.64	12.97	10.80	9.63	8.89	8.38	8.00	7.71	7.48	7.29	7.00	6.71	6.40	6.25	6.09	5.93	5.76	5.59	5.42
13	17.81	12.31	10.21	9.07	8.35	7.86	7.49	7.21	6.98	6.80	6.52	6.23	5.93	5.78	5.63	5.47	5.30	5.14	4.97
14	17.14	11.78	9.73	8.62	7.92	7.43	7.08	6.80	6.58	6.40	6.13	5.85	5.56	5.41	5.25	5.10	4.94	4.77	4.60
15	16.59	11.34	9.34	8.25	7.57	7.09	6.74	6.47	6.26	6.08	5.81	5.54	5.25	5.10	4.95	4.80	4.64	4.47	4.31
16	16.12	10.97	9.00	7.94	7.27	6.81	6.46	6.19	5.98	5.81	5.55	5.27	4.99	4.85	4.70	4.54	4.39	4.23	4.06
17	15.72	10.66	8.73	7.68	7.02	6.56	6.22	5.96	5.75	5.58	5.32	5.05	4.78	4.63	4.48	4.33	4.18	4.02	3.85
18	15.38	10.39	8.49	7.46	6.81	6.35	6.02	5.76	5.56	5.39	5.13	4.87	4.59	4.45	4.30	4.15	4.00	3.84	3.67
19	15.08	10.16	8.28	7.26	6.62	6.18	5.85	5.59	5.39	5.22	4.97	4.70	4.43	4.29	4.14	3.99	3.84	3.68	3.51
20	14.82	9.95	8.10	7.10	6.46	6.02	5.69	5.44	5.24	5.08	4.82	4.56	4.29	4.15	4.00	3.86	3.70	3.54	3.38
21	14.59	9.77	7.94	6.95	6.32	5.88	5.56	5.31	5.11	4.95	4.70	4.44	4.17	4.03	3.88	3.74	3.58	3.42	3.26
22	14.38	9.61	7.80	6.81	6.19	5.76	5.44	5.19	4.99	4.83	4.58	4.33	4.06	3.92	3.78	3.63	3.48	3.32	3.15
23	14.19	9.47	7.67	6.69	6.08	5.65	5.33	5.09	4.89	4.73	4.48	4.23	3.96	3.82	3.68	3.53	3.38	3.22	3.05
24	14.03	9.34	7.55	6.59	5.98	5.55	5.23	4.99	4.80	4.64	4.39	4.14	3.87	3.74	3.59	3.45	3.29	3.14	2.97
25	13.88	9.22	7.45	6.49	5.88	5.46	5.15	4.91	4.71	4.56	4.31	4.06	3.79	3.66	3.52	3.37	3.22	3.06	2.89
26	13.74	9.12	7.36	6.41	5.80	5.38	5.07	4.83	4.64	4.48	4.24	3.99	3.72	3.59	3.44	3.30	3.15	2.99	2.82
27	13.61	9.02	7.27	6.33	5.73	5.31	5.00	4.76	4.57	4.41	4.17	3.92	3.66	3.52	3.38	3.23	3.08	2.92	2.75
28	13.50	8.93	7.19	6.25	5.66	5.24	4.93	4.69	4.50	4.35	4.11	3.86	3.60	3.46	3.32	3.18	3.02	2.86	2.69
29	13.39	8.85	7.12	6.19	5.59	5.18	4.87	4.64	4.45	4.29	4.05	3.80	3.54	3.41	3.27	3.12	2.97	2.81	2.64
30	13.29	8.77	7.05	6.12	5.53	5.12	4.82	4.58	4.39	4.24	4.00	3.75	3.49	3.36	3.22	3.07	2.92	2.76	2.59
40	12.61	8.25	6.60	5.70	5.13	4.73	4.44	4.21	4.02	3.87	3.64	3.40	3.15	3.01	2.87	2.73	2.57	2.41	2.23
60	11.97	7.76	6.17	5.31	4.76	4.37	4.09	3.87	3.69	3.54	3.31	3.08	2.83	2.69	2.55	2.41	2.25	2.08	1.89
120	11.38	7.32	5.79	4.95	4.42	4.04	3.77	3.55	3.38	3.24	3.02	2.78	2.53	2.40	2.26	2.11	1.95	1.76	1.54
∞	10.83	6.91	5.42	4.62	4.10	3.74	3.47	3.27	3.10	2.96	2.74	2.51	2.27	2.13	1.99	1.84	1.66	1.45	1.00

*Multiply these entries by 100.

Table B.6 Upper 5-percent points of studentized range q*

n_2	$p^\dagger$: 2	3	4	5	6	7	8	9	10	11	12	13	14	15	16	17	18	19	20
1	18.0	26.7	32.8	37.2	40.5	43.1	45.4	47.3	49.1	50.6	51.9	53.2	54.3	55.4	56.3	57.2	58.0	58.8	59.6
2	6.09	8.28	9.80	10.89	11.73	12.43	13.03	13.54	13.99	14.39	14.75	15.08	15.38	15.65	15.91	16.14	16.36	16.57	16.77
3	4.50	5.88	6.83	7.51	8.04	8.47	8.85	9.18	9.46	9.72	9.95	10.16	10.35	10.52	10.69	10.84	10.98	11.12	11.24
4	3.93	5.00	5.76	6.31	6.73	7.06	7.35	7.60	7.83	8.03	8.21	8.37	8.52	8.67	8.80	8.92	9.03	9.14	9.24
5	3.61	4.54	5.18	5.64	5.99	6.28	6.52	6.74	6.93	7.10	7.25	7.39	7.52	7.64	7.75	7.86	7.95	8.04	8.13
6	3.46	4.34	4.90	5.31	5.63	5.89	6.12	6.32	6.49	6.65	6.79	6.92	7.04	7.14	7.24	7.34	7.43	7.51	7.59
7	3.34	4.16	4.68	5.06	5.35	5.59	5.80	5.99	6.15	6.29	6.42	6.54	6.65	6.75	6.84	6.93	7.01	7.08	7.16
8	3.26	4.04	4.53	4.89	5.17	5.40	5.60	5.77	5.92	6.05	6.18	6.29	6.39	6.48	6.57	6.65	6.73	6.80	6.87
9	3.20	3.95	4.42	4.76	5.02	5.24	5.43	5.60	5.74	5.87	5.98	6.09	6.19	6.28	6.36	6.44	6.51	6.58	6.65
10	3.15	3.88	4.33	4.66	4.91	5.12	5.30	5.46	5.60	5.72	5.83	5.93	6.03	6.12	6.20	6.27	6.34	6.41	6.47
11	3.11	3.82	4.26	4.58	4.82	5.03	5.20	5.35	5.49	5.61	5.71	5.81	5.90	5.98	6.06	6.14	6.20	6.27	6.33
12	3.08	3.77	4.20	4.51	4.75	4.95	5.12	5.27	5.40	5.51	5.61	5.71	5.80	5.88	5.95	6.02	6.09	6.15	6.21
13	3.06	3.73	4.15	4.46	4.69	4.88	5.05	5.19	5.32	5.43	5.53	5.63	5.71	5.79	5.86	5.93	6.00	6.06	6.11
14	3.03	3.70	4.11	4.41	4.64	4.83	4.99	5.13	5.25	5.36	5.46	5.56	5.64	5.72	5.79	5.86	5.92	5.98	6.03
15	3.01	3.67	4.08	4.37	4.59	4.78	4.94	5.08	5.20	5.31	5.40	5.49	5.57	5.65	5.72	5.79	5.85	5.91	5.96
16	3.00	3.65	4.05	4.34	4.56	4.74	4.90	5.03	5.15	5.26	5.35	5.44	5.52	5.59	5.66	5.73	5.79	5.84	5.90
17	2.98	3.62	4.02	4.31	4.52	4.70	4.86	4.99	5.11	5.21	5.31	5.39	5.47	5.55	5.61	5.68	5.74	5.79	5.84
18	2.97	3.61	4.00	4.28	4.49	4.67	4.83	4.96	5.07	5.17	5.27	5.35	5.43	5.50	5.57	5.63	5.69	5.74	5.79
19	2.96	3.59	3.98	4.26	4.47	4.64	4.79	4.92	5.04	5.14	5.23	5.32	5.39	5.46	5.53	5.59	5.65	5.70	5.75
20	2.95	3.58	3.96	4.24	4.45	4.62	4.77	4.90	5.01	5.11	5.20	5.28	5.36	5.43	5.30	5.56	5.61	5.66	5.71
24	2.92	3.53	3.90	4.17	4.37	4.54	4.68	4.81	4.92	5.01	5.10	5.18	5.25	5.32	5.38	5.44	5.50	5.55	5.59
30	2.89	3.48	3.84	4.11	4.30	4.46	4.60	4.72	4.83	4.92	5.00	5.08	5.15	5.21	5.27	5.33	5.38	5.43	5.48
40	2.86	3.44	3.79	4.04	4.23	4.39	4.52	4.63	4.74	4.82	4.90	4.98	5.05	5.11	5.17	5.22	5.27	5.32	5.36
60	2.83	3.40	3.74	3.98	4.16	4.31	4.44	4.55	4.65	4.73	4.81	4.88	4.94	5.00	5.06	5.11	5.15	5.20	5.24
120	2.80	2.36	3.69	3.92	4.10	4.24	4.36	4.47	4.56	4.64	4.71	4.78	4.84	4.90	4.95	5.00	5.04	5.09	5.13
∞	2.77	3.32	3.63	3.86	4.03	4.17	4.29	4.39	4.47	4.55	4.62	4.68	4.74	4.80	4.84	4.89	4.93	4.97	5.01

*from J. M. May (1952) Extended and corrected tables of the upper percentage points of the studentized Range. *Biometrika*, **39**, 192–3. Reproduced by permission of the trustees of *Biometrika*.

† p is the number of quantities (for example, means) whose range is involved. n_2 is the degrees of freedom in the error estimate.

Table B.7 Upper 1-percent points of studentized range q

n_2^*	p^* 2	3	4	5	6	7	8	9	10	11	12	13	14	15	16	17	18	19	20
1	90.0	135	164	186	202	216	227	237	246	253	260	266	272	227	282	286	290	294	298
2	14.0	19.0	22.3	24.7	26.6	28.2	29.5	30.7	31.7	32.6	33.4	34.1	34.8	35.4	36.0	36.5	37.0	37.5	37.9
3	8.26	10.6	12.2	13.3	14.2	15.0	15.6	16.2	16.7	17.1	17.5	17.9	18.2	18.5	18.8	19.1	19.3	19.5	19.8
4	6.51	8.12	9.17	9.96	10.6	11.1	11.5	11.9	12.3	12.6	12.8	13.1	13.3	13.5	13.7	13.9	14.1	14.2	14.4
5	5.70	6.97	7.80	8.42	8.91	9.32	9.67	9.97	10.24	10.48	10.70	10.89	11.08	11.24	11.40	11.55	11.68	11.81	11.93
6	5.24	6.33	7.03	7.56	7.97	8.32	8.61	8.87	9.10	9.30	9.49	9.65	9.81	9.95	10.08	10.21	10.32	10.43	10.54
7	4.95	5.92	6.54	7.01	7.37	7.68	7.94	8.17	8.37	8.55	8.71	8.86	9.00	9.12	9.24	9.35	9.46	9.55	9.65
8	4.74	5.63	6.20	6.63	6.96	7.24	7.47	7.68	7.87	8.03	8.18	8.31	8.44	8.55	8.66	8.76	8.85	8.94	9.03
9	4.60	5.43	5.96	6.35	6.66	6.91	7.13	7.32	7.49	7.65	7.78	7.91	8.03	8.13	8.23	9.32	8.41	8.49	8.57
10	4.48	5.27	5.77	6.14	6.43	6.67	6.87	7.05	7.21	7.36	7.48	7.60	7.71	7.81	7.91	7.99	8.07	8.15	8.22
11	4.39	5.14	5.62	5.97	6.25	6.48	6.67	6.84	6.99	7.13	7.25	7.36	7.46	7.56	7.65	7.73	7.81	7.88	7.95
12	4.32	5.04	5.50	5.84	6.10	6.32	6.51	6.67	6.81	6.94	7.06	7.17	7.26	7.36	7.44	7.52	7.59	7.66	7.73
13	4.26	4.96	5.40	5.73	5.98	6.19	6.37	6.53	6.67	6.79	6.90	7.01	7.10	7.19	7.27	7.34	7.42	7.48	7.55
14	4.21	4.89	5.32	5.63	5.88	6.08	6.26	6.41	6.54	6.66	6.77	6.87	6.96	7.05	7.12	7.20	7.27	7.33	7.39
15	4.17	4.83	5.25	5.56	5.80	5.99	6.16	6.31	6.44	6.55	6.66	6.76	6.84	6.93	7.00	7.07	7.14	7.20	7.26
16	4.13	4.78	5.19	5.49	5.72	5.92	6.08	6.22	6.35	6.46	6.56	6.66	6.74	6.82	6.90	6.97	7.03	7.09	7.15
17	4.10	4.74	5.14	5.43	5.66	5.85	6.01	6.15	6.27	6.38	6.48	6.57	6.66	6.73	6.80	6.87	6.94	7.00	7.05
18	4.07	4.70	5.09	5.28	5.60	5.79	5.94	6.08	6.20	6.31	6.41	6.50	6.58	6.65	6.72	6.79	6.85	6.91	6.96
19	4.05	4.67	5.05	5.33	5.55	5.73	5.89	6.02	6.14	6.25	6.34	6.43	6.51	6.58	6.65	6.72	6.78	6.84	6.89
20	4.02	4.64	5.02	5.29	5.51	5.69	5.84	5.97	6.09	6.19	6.29	6.37	6.45	6.52	6.59	6.65	6.71	6.76	6.82
24	3.96	4.54	4.91	5.17	5.37	5.54	5.69	5.81	5.92	6.02	6.11	6.19	6.26	6.33	6.39	6.45	6.51	6.56	6.61
30	3.89	4.45	4.80	5.05	5.24	5.40	5.54	5.65	5.76	5.85	5.93	6.01	6.08	6.14	6.20	6.26	6.31	6.36	6.41
40	3.82	4.37	4.70	4.93	5.11	5.27	5.39	5.50	5.60	5.69	5.77	5.84	5.90	5.96	6.02	6.07	6.12	6.17	6.21
60	3.76	4.28	4.60	4.82	4.99	5.13	5.25	5.36	5.45	5.53	5.60	5.67	5.73	5.79	5.84	5.89	5.93	5.98	6.02
120	3.70	4.20	4.50	4.71	4.87	5.01	5.12	5.21	5.30	5.38	5.44	5.51	5.56	5.61	5.66	5.71	5.75	5.79	5.83
∞	3.64	4.12	4.40	4.60	4.76	4.88	4.99	5.08	5.16	5.23	5.29	5.35	5.40	5.45	5.49	5.54	5.57	5.61	5.65

* p is the number of quantities (for example, means) whose range is involved. n_2 is the degrees of freedom in the error estimate.

Table B.8 Values for Duncan's multiple range test, $\alpha = 0.01$

v	*p, the rank order of means*																
	2	3	4	5	6	7	8	9	10	11	12	13	14	15	16	18	20	30
2	14.04	14.04	14.04	14.04	14.04	14.04	14.04	14.04	14.04	14.04	14.04	14.04	14.04	14.04	14.04	14.04	14.04	14.04
3	8.261	8.321	8.321	8.321	8.321	8.321	8.321	8.321	8.321	8.321	8.321	8.321	8.321	8.321	8.321	8.321	8.321	8.321
4	6.512	6.677	6.740	6.756	6.758	6.756	6.756	6.756	6.756	6.756	6.756	6.756	6.756	6.756	6.756	6.756	6.756	6.756
5	5.702	5.893	5.989	6.040	6.065	6.074	6.074	6.074	6.074	6.074	6.074	6.074	6.074	6.074	6.074	6.074	6.074	6.074
6	5.243	5.439	5.549	5.614	5.655	5.680	5.694	5.701	5.703	5.703	5.703	5.703	5.703	5.703	5.703	5.703	5.703	5.703
7	4.949	5.145	5.260	5.334	5.383	5.416	5.439	5.454	5.464	5.470	5.472	5.472	5.472	5.472	5.472	5.472	5.472	5.472
8	4.746	4.939	5.057	5.135	5.189	5.227	5.256	5.276	5.291	5.302	5.309	5.314	5.316	5.317	5.317	5.317	5.317	5.317
9	4.596	4.787	4.906	4.986	5.043	5.086	5.118	5.142	5.160	5.174	5.185	5.193	5.199	5.203	5.205	5.206	5.206	5.206
10	4.482	4.671	4.790	4.871	4.931	4.975	5.010	5.037	5.058	5.074	5.088	5.098	5.106	5.112	5.117	5.122	5.124	5.124
11	4.392	4.579	4.697	4.780	4.841	4.887	4.924	4.952	4.975	4.994	5.009	5.021	5.031	5.039	5.045	5.054	5.059	5.061
12	4.320	4.504	4.622	4.706	4.767	4.815	4.852	4.883	4.907	4.927	4.944	4.958	4.969	4.978	4.986	4.998	5.006	5.011
13	4.260	4.442	4.560	4.644	4.706	4.755	4.793	4.824	4.850	4.872	4.889	4.904	4.917	4.928	4.937	4.950	4.960	4.972
14	4.210	4.391	4.508	4.591	4.654	4.704	4.743	4.775	4.802	4.824	4.843	4.859	4.872	4.884	4.894	4.910	4.921	4.940
15	4.168	4.347	4.463	4.547	4.610	4.660	4.700	4.733	4.760	4.783	4.803	4.820	4.834	4.846	4.857	4.874	4.887	4.914
16	4.131	4.309	4.425	4.509	4.572	4.622	4.663	4.696	4.724	4.748	4.768	4.786	4.800	4.813	4.825	4.844	4.858	4.890
17	4.099	4.275	4.391	4.475	4.539	4.589	4.630	4.664	4.693	4.717	4.738	4.756	4.771	4.785	4.797	4.816	4.832	4.869
18	4.071	4.246	4.362	4.445	4.509	4.560	4.601	4.635	4.664	4.689	4.711	4.729	4.745	4.759	4.772	4.792	4.808	4.850
19	4.046	4.220	4.335	4.419	4.483	4.534	4.575	4.610	4.639	4.665	4.686	4.705	4.722	4.736	4.749	4.771	4.788	4.833
20	4.024	4.197	4.312	4.395	4.459	4.510	4.552	4.587	4.617	4.642	4.664	4.684	4.701	4.716	4.729	4.751	4.769	4.818
24	3.956	4.126	4.239	4.322	4.386	4.437	4.480	4.516	4.546	4.573	4.596	4.616	4.634	4.651	4.665	4.690	4.710	4.770
30	3.889	4.056	4.168	4.250	4.314	4.366	4.409	4.445	4.477	4.504	4.528	4.550	4.569	4.586	4.601	4.628	4.650	4.721
40	3.825	3.988	4.098	4.180	4.244	4.296	4.339	4.376	4.408	4.436	4.461	4.483	4.503	4.521	4.537	4.566	4.591	4.671
60	3.762	3.922	4.031	4.111	4.174	4.226	4.270	4.307	4.340	4.368	4.394	4.417	4.438	4.456	4.474	4.504	4.530	4.620
120	3.702	3.858	3.965	4.044	4.107	4.158	4.202	4.239	4.272	4.301	4.327	4.351	4.372	4.392	4.410	4.442	4.469	4.568
∞	3.643	3.796	3.900	3.978	4.040	4.091	4.135	4.172	4.205	4.235	4.261	4.285	4.307	4.327	4.345	4.379	4.408	4.514

Table B.9 Values for Duncan's multiple range test, $\alpha = 0.05$

v	*p, the rank order of means*																	
	2	*3*	*4*	*5*	*6*	*7*	*8*	*9*	*10*	*11*	*12*	*13*	*14*	*15*	*16*	*18*	*20*	*30*
2	6.085	6.085	6.085	6.085	6.085	6.085	6.085	6.085	6.085	6.085	6.085	6.085	6.085	6.085	6.085	6.085	6.085	6.085
3	4.501	4.516	4.516	4.516	4.516	4.516	4.516	4.516	4.516	4.516	4.516	4.516	4.516	4.516	4.516	4.516	4.516	4.516
4	3.927	4.013	4.033	4.033	4.033	4.033	4.033	4.033	4.033	4.033	4.033	4.033	4.033	4.033	4.033	4.033	4.033	4.033
5	3.635	3.749	3.797	3.814	3.814	3.814	3.814	3.814	3.814	3.814	3.814	3.814	3.814	3.814	3.814	3.814	3.814	3.814
6	3.461	3.587	3.649	3.680	3.694	3.697	3.697	3.697	3.697	3.697	3.697	3.697	3.697	3.697	3.697	3.697	3.697	3.697
7	3.344	3.477	3.548	3.588	3.611	3.622	3.626	3.626	3.626	3.626	3.626	3.626	3.626	3.626	3.626	3.626	3.626	3.626
8	3.261	3.399	3.475	3.521	3.549	3.566	3.575	3.579	3.579	3.579	3.579	3.579	3.579	3.579	3.579	3.579	3.579	3.579
9	3.199	3.339	3.420	3.470	3.502	3.523	3.536	3.544	3.547	3.547	3.547	3.547	3.547	3.547	3.547	3.547	3.547	3.547
10	3.151	3.293	3.376	3.430	3.465	3.489	3.505	3.516	3.522	3.525	3.526	3.526	3.526	3.526	3.526	3.526	3.526	3.526
11	3.113	3.256	3.342	3.397	3.435	3.462	3.480	3.493	3.501	3.506	3.509	3.510	3.510	3.510	3.510	3.510	3.510	3.510
12	3.082	3.225	3.313	3.370	3.410	3.439	3.459	3.474	3.484	3.491	3.496	3.498	3.499	3.499	3.499	3.499	3.499	3.499
13	3.055	3.200	3.289	3.348	3.389	3.419	3.442	3.458	3.470	3.478	3.484	3.488	3.490	3.490	3.490	3.490	3.490	3.490
14	3.033	3.178	3.268	3.329	3.372	3.403	3.426	3.444	3.457	3.467	3.474	3.479	3.482	3.484	3.484	3.485	3.485	3.485
15	3.014	3.160	3.250	3.312	3.356	3.389	3.413	3.432	3.446	3.457	3.465	3.471	3.476	3.478	3.480	3.481	3.481	3.481
16	2.998	3.144	3.235	3.298	3.343	3.376	3.402	3.422	3.437	3.449	3.458	3.465	3.470	3.473	3.477	3.478	3.478	3.478
17	2.984	3.130	3.222	3.285	3.331	3.366	3.392	3.412	3.429	3.441	3.451	3.459	3.465	3.469	3.473	3.476	3.476	3.476
18	2.971	3.118	3.210	3.274	3.321	3.356	3.383	3.405	3.421	3.435	3.445	3.454	3.460	3.465	3.470	3.474	3.474	3.474
19	2.960	3.107	3.199	3.264	3.311	3.347	3.375	3.397	3.415	3.429	3.440	3.449	3.456	3.462	3.467	3.472	3.474	3.474
20	2.950	3.097	3.190	3.255	3.303	3.339	3.368	3.391	3.409	3.424	3.436	3.445	3.453	3.459	3.464	3.470	3.473	3.474
24	2.919	3.066	3.160	3.226	3.276	3.315	3.345	3.370	3.390	3.406	3.420	3.432	3.441	3.449	3.456	3.465	3.471	3.477
30	2.888	3.035	3.131	3.199	3.250	3.290	3.322	3.349	3.371	3.389	3.405	3.418	3.430	3.439	3.447	3.460	3.470	3.486
40	2.858	3.006	3.102	3.171	3.224	3.266	3.300	3.328	3.352	3.373	3.390	3.405	3.418	3.429	3.439	3.456	3.469	3.500
60	2.829	2.976	3.073	3.143	3.198	3.241	3.277	3.307	3.333	3.355	3.374	3.391	3.406	3.419	3.431	3.451	3.467	3.515
120	2.800	2.947	3.045	3.116	3.172	3.217	3.254	3.287	3.314	3.337	3.359	3.377	3.394	3.409	3.423	3.446	3.466	3.532
∞	2.772	2.918	3.017	3.089	3.146	3.193	3.232	3.265	3.294	3.320	3.343	3.363	3.382	3.399	3.414	3.442	3.466	3.550

Table B.10 Values for Dunnett's two-sided comparison test

(a) For two-sided comparisons made with $\alpha = 0.01$

k, the number of means, excluding the control

ν	1	2	3	4	5	6	7	8	9
5	4.03	4.63	4.98	5.22	5.41	5.56	5.69	5.80	5.89
6	3.71	4.21	4.51	4.71	4.87	5.00	5.10	5.20	5.28
7	3.50	3.95	4.21	4.39	4.53	4.64	4.74	4.82	4.89
8	3.36	3.77	4.00	4.17	4.29	4.40	4.48	4.56	4.62
9	3.25	3.63	3.85	4.01	4.12	4.22	4.30	4.37	4.43
10	3.17	3.53	3.74	3.88	3.99	4.08	4.16	4.22	4.28
11	3.11	3.45	3.65	3.79	3.89	3.98	4.05	4.11	4.16
12	3.05	3.39	3.58	3.71	3.81	3.89	3.96	4.02	4.07
13	3.01	3.33	3.52	3.65	3.74	3.82	3.89	3.94	3.99
14	2.98	3.29	3.47	3.59	3.69	3.76	3.83	3.88	3.93
15	2.95	3.25	3.43	3.55	3.64	3.71	3.78	3.83	3.88
16	2.92	3.22	3.39	3.51	3.60	3.67	3.73	3.78	3.83
17	2.90	3.19	3.36	3.47	3.56	3.63	3.69	3.74	3.79
18	2.88	3.17	3.33	3.44	3.53	3.60	3.66	3.71	3.75
19	2.86	3.15	3.31	3.42	3.50	3.57	3.63	3.68	3.72
20	2.85	3.13	3.29	3.40	3.48	3.55	3.60	3.65	3.69
24	2.80	3.07	3.22	3.32	3.40	3.47	3.52	3.57	3.61
30	2.75	3.01	3.15	3.25	3.33	3.39	3.44	3.49	3.52
40	2.70	2.95	3.09	3.19	3.26	3.32	3.37	3.41	3.44
60	2.66	2.90	3.03	3.12	3.19	3.25	3.29	3.33	3.37
120	2.62	2.85	2.97	3.06	3.12	3.18	3.22	3.26	3.29
∞	2.58	2.79	2.92	3.00	3.06	3.11	3.15	3.19	3.22

(b) for two-sided comparisons made with $\alpha = 0.05$

k, the number of means, excluding the cofntrol

ν	1	2	3	4	5	6	7	8	9
5	2.57	3.03	3.29	3.48	3.62	3.73	3.82	3.90	3.97
6	2.45	2.86	3.10	3.26	3.39	3.49	3.57	3.64	3.71
7	2.36	2.75	2.97	3.12	3.24	3.33	3.41	3.47	3.53
8	2.31	2.67	2.88	3.02	3.13	3.22	3.29	3.35	3.41
9	2.26	2.61	2.81	2.95	3.05	3.14	3.20	3.26	3.32
10	2.23	2.57	2.76	2.89	2.99	3.07	3.14	3.19	3.24
11	2.20	2.53	2.72	2.84	2.94	3.02	3.08	3.14	3.19
12	2.18	2.50	2.68	2.81	2.90	2.98	3.04	3.09	3.14
13	2.16	2.48	2.65	2.78	2.87	2.94	3.00	3.06	3.10
14	2.14	2.46	2.63	2.75	2.84	2.91	2.97	3.02	3.07
15	2.13	2.44	2.61	2.73	2.82	2.89	2.95	3.00	3.04
16	2.12	2.42	2.59	2.71	2.80	2.87	2.92	2.97	3.02
17	2.11	2.41	2.58	2.69	2.78	2.85	2.90	2.95	3.00
18	2.10	2.40	2.56	2.68	2.76	2.83	2.89	2.94	2.98
19	2.09	2.39	2.55	2.66	2.75	2.81	2.87	2.92	2.96
20	2.09	2.38	2.54	2.65	2.73	2.80	2.86	2.90	2.95
24	2.06	2.35	2.51	2.61	2.70	2.76	2.81	2.86	2.90
30	2.04	2.32	2.47	2.58	2.66	2.72	2.77	2.82	2.86
40	2.02	2.29	2.44	2.54	2.62	2.68	2.73	2.77	2.81
60	2.00	2.27	2.41	2.51	2.58	2.64	2.69	2.73	2.77
120	1.98	2.24	2.38	2.47	2.55	2.60	2.65	2.69	2.73
∞	1.96	2.21	2.35	2.44	2.51	2.57	2.61	2.65	2.69

Table B.11 Values for Dunnett's one-sided comparison test

(a) For one-sided comparisons made with $\alpha = 0.01$
k, the number of means, excluding the control

v	*1*	*2*	*3*	*4*	*5*	*6*	*7*	*8*	*9*
5	3.37	3.90	4.21	4.43	4.60	4.73	4.85	4.94	5.03
6	3.14	3.61	3.88	4.07	4.21	4.33	4.43	4.51	4.59
7	3.00	3.42	3.66	3.83	3.96	4.07	4.15	4.23	4.30
8	2.90	3.29	3.51	3.67	3.79	3.88	3.96	4.03	4.09
9	2.82	3.19	3.40	3.55	3.66	3.75	3.82	3.89	3.94
10	2.76	3.11	3.31	3.45	3.56	3.64	3.71	3.78	3.83
11	2.72	3.06	3.25	3.38	3.48	3.56	3.63	3.69	3.74
12	2.68	3.01	3.19	3.32	3.42	3.50	3.56	3.62	3.67
13	2.65	2.97	3.15	3.27	3.37	3.44	3.51	3.56	3.61
14	2.62	2.94	3.11	3.23	3.32	3.40	3.46	3.51	3.56
15	2.60	2.91	3.08	3.20	3.29	3.36	3.42	3.47	3.52
16	2.58	2.88	3.05	3.17	3.26	3.33	3.39	3.44	3.48
17	2.57	2.86	3.03	3.14	3.23	3.30	3.36	3.41	3.45
18	2.55	2.84	3.01	3.12	3.21	3.27	3.33	3.38	3.42
19	2.54	2.83	2.99	3.10	3.18	3.25	3.31	3.36	3.40
20	2.53	2.81	2.97	3.08	3.17	3.23	3.29	3.34	3.38
24	2.49	2.77	2.92	3.03	3.11	3.17	3.22	3.27	3.31
30	2.46	2.72	2.87	2.97	3.05	3.11	3.16	3.21	3.24
40	2.42	2.68	2.82	2.92	2.99	3.05	3.10	3.14	3.18
60	2.39	2.64	2.78	2.87	2.94	3.00	3.04	3.08	3.12
120	2.36	2.60	2.73	2.82	2.89	2.94	2.99	3.03	3.06
∞	2.33	2.56	2.68	2.77	2.84	2.89	2.93	2.97	3.00

(b) For one-sided comparisons made with $\alpha = 0.05$.
k, the number of means, excluding the control

v	*1*	*2*	*3*	*4*	*5*	*6*	*7*	*8*	*9*
5	2.02	2.44	2.68	2.85	2.98	3.08	3.16	3.24	3.30
6	1.94	2.34	2.56	2.71	2.83	2.92	3.00	3.07	3.12
7	1.89	2.27	2.48	2.62	2.73	2.82	2.89	2.95	3.01
8	1.86	2.22	2.42	2.55	2.66	2.74	2.81	2.87	2.92
9	1.83	2.18	2.37	2.50	2.60	2.68	2.75	2.81	2.86
10	1.81	2.15	2.34	2.47	2.56	2.64	2.70	2.76	2.81
11	1.80	2.13	2.31	2.44	2.53	2.60	2.67	2.72	2.77
12	1.78	2.11	2.29	2.41	2.50	2.58	2.64	2.69	2.74
13	1.77	2.09	2.27	2.39	2.48	2.55	2.61	2.66	2.71
14	1.76	2.08	2.25	2.37	2.46	2.53	2.59	2.64	2.69
15	1.75	2.07	2.24	2.36	2.44	2.51	2.57	2.62	2.67
16	1.75	2.06	2.23	2.34	2.43	2.50	2.56	2.61	2.65
17	1.74	2.05	2.22	2.33	2.42	2.49	2.54	2.59	2.64
18	1.73	2.05	2.21	2.32	2.41	2.48	2.53	2.58	2.62
19	1.73	2.03	2.20	2.31	2.40	2.47	2.52	2.57	2.61
20	1.72	2.03	2.19	2.30	2.39	2.46	2.51	2.56	2.60
24	1.71	2.01	2.17	2.28	2.36	2.43	2.48	2.53	2.57
30	1.70	1.99	2.15	2.25	2.33	2.40	2.45	2.50	2.54
40	1.68	1.97	2.13	2.23	2.31	2.37	2.42	2.47	2.51
60	1.67	1.95	2.10	2.21	2.28	2.35	2.39	2.44	2.48
120	1.66	1.93	2.08	2.18	2.26	2.32	2.37	2.41	2.45
∞	1.64	1.92	2.06	2.16	2.23	2.29	2.34	2.38	2.42

Index